Differential Rotation and Stellar Convection

G. Rüdiger

Differential Rotation and Stellar Convection

Sun and Solar-type Stars

With 84 Figures and 35 Tables

Akademie-Verlag Berlin 1989

Author:
Dr. sc. rer. nat. Günther Rüdiger
Zentralinstitut für Astrophysik der Akademie der Wissenschaften der DDR
Sternwarte Babelsberg

Akademie-Verlag Berlin ISBN 3-05-500422-1
Gordon and Breach, Science Publishers ISBN 2-88124-066-6

Erschienen im Akademie-Verlag Berlin, Leipziger Straße 3—4, Berlin, DDR-1086

Lizenznummer: 202 · 100/440/88
Printed in the German Democratic Republic
Gesamtherstellung: VEB Druckhaus „Maxim Gorki", Altenburg, 7400
Lektor: Dipl.-Met. Heide Deutscher
LSV 1495
Bestellnummer: 7637711 (9087)

06300

Truth is the child of its time
Brecht, 1938
Die Sonne ist das, was keiner begräbt
Fühmann, 1981

To Sten, my son

Contents

Part II. Macroscopy

Part III. Microscopy

Preface

When I was approached in 1982 by Prof. P. H. Roberts and asked to write a book about the differential rotation of the Sun, my immediate response was negative: "No, no... it's true that we know a lot of mechanisms which produce non-uniform stellar rotation, however, which ones are at work in the Sun?" He countered with "True, Describe and explain them all. That in itself will be no small contribution."

After months of ponderous activity I set to work. At first it was only a mountain of work during which the individual contributions of the pioneers and precursors became ever clearer. The outcomes of these events, which assumed the character of an adventure, are described in the Part I of this book. A small field of exploration but yet full of attempted solutions, as well as small and large steps forward!

In the main 'Part II' of the book, on the basis of heuristic turbulence theory, the various sources of differential rotation are put into their physical context and discussed in detail. Naturally this essentially deals with the turbulent transports of energy and rotational momentum. Part III is intended to verify the previously used formulations with a simple linearised turbulence theory, as well as to make statements about the turbulence mechanism active in the solar convection zone. Whether these closing chapters do justice to their assignment is not, with final certainty, clear. Possibly there are advantages here with the pure numerics of the magnificent American computer programmes which seek to convey, by laminar but sufficiently complicated flow patterns, the large-scale hydrodynamics of the solar convection zone. This is equally fascinating and problematic. Not that calculations can be carried out without physical profit. Careful simulations even lead to an understanding of the causality between the small-scale and large-scale circulations. There is, however, also the danger of numerical "overkill" which fails to enlighten. What has one understood when the computer delivers the expected profiles: That the basic physics is correct?

In the present book we have tried, by means long established, to work out a systematic starting point with formulae and manipulations. This frequently implies manifold forms of linearisation, so that this method also has its problems. We have never considered whether the time of "analytical physics" (the work with clumsy formulae) is at the beginning of its end. We have simply attempted by these means to find out why the sun rotates as it does.

During all these efforts my teachers and superiors Profs F. Krause and K.-H. Schmidt have given me support which I couldn't have done without. Hardly a formula or thought has not been discussed in Krause's study at the Astrophysical Observatory Potsdam and later at the Sternwarte Babelsberg.

Enormous assistance has come from my colleagues, K.-O. Eschrich and Mrs. B. Nader. They carried out numerical research (K.-O. E.) and typed the manuscript (B. N.). Colin de la Motte-Sherman (Potsdam) and Mrs. S. Hills (Edinburgh) probably agonized over the language during the course of many weeks. Special commitment was shown by my two friends Rudolph Tschäpe (Potsdam) and Ilkka Tuominen (Helsinki). The latter first made it finally clear that the numbers given in Chapter 11 really correspond to the solution of the equation system put forward in Chapter 10. Moreover, the form and content of Chapters 11 and 12 are based on intensive discussions and close collaboration with nim over a number of years. Since these Chapters were written further collaboration has developed the ideas outlined here. He also gave the final form to Chapter 3 while carrying the basic observations.

The crowning glory, however, belongs to the spiritual father without whom this book would have become merely another publication. Paul Roberts (Los Angeles) thought up the work and set the finishing touches. In the meantime he consoled me in my doubts over the slow progress of the thought and writing processes.

When I say, "Thank you to all of you" this should also convey the deep relations which we, who live in these decades of science and the changeability of life, have experienced over the years, and it is these relations which, I think, we urgently need in whatever form it comes.

Günther Rüdiger
April 1986

Part I
Preliminaries

Chapter 1. The Sunspot Story

But all, except their Sun, is set.
Byron, 1819

1.1. Black Spots on a Shining Disk

Strikingly often amateurs contribute to the foundation of a science. A few years before Helmholtz's ingenious work "Über die Erhaltung der Kraft", the general practitioner Robert Mayer formulated the energy principle. Neither paper was, however, accepted for publication, and both were privately printed.

The beginnings of modern solar physics are notable for events which occurred within a few months of each other. On this occasion it was a medical student, Johann Fabricius, who wrested priority from the authorities of his time and, like the doctor from Heilbronn later on, he did not escape gruesome punishment. Johann Fabricius, the eldest son of the parson and astronomer David Fabricius, was born in Resterhave near Dornum in East Friesland on January 8, 1587. In early childhood he survived small-pox, measles and the plague. No other information about his youth is recorded. On May 1, 1605 Johann travelled from East Friesland to Helmstedt to commence University studies. It is not known why the elder Fabricius sent his son to Helmstedt. Maybe it was because he himself had studied there. Johann, however, did not stay in Helmstedt very long, for in the following year he enrolled at the University at Wittenberg. Three years later, in December 1609, he went to Leyden to study medicine. It was probably here the he acquired the telescope which he later used to make astronomical discoveries at his parents' home near Osteel.

As the Moon and Jupiter seemed to have been fully explored he looked for other objects of study. He chose the Sun which, as his father remarked, showed certain irregularities at its edge. While he was attentively observing these, he noticed a blackish spot, of considerable size, and somewhat faded on one side. Initially he thought that this was a cloud, but when he repeated the observation through telescopes of various sizes he became sure that this was not the case. He reported later, with dramatic vividness, how still fearing self-deception, he called his father, how the latter confirmed this discovery, and how the phenomenon stimulated both to continue their observations. He spent the night full of doubts. In the morning he got up excitedly, and confirmed his discovery with the first observation of the day. Father and son did not tire of watching this spectacle, but new doubts arose when they detected an apparent motion of the spot. Meanwhile their eyes had become dazzled and they had to devise some other way to continue their observations. Eventually they hit on the idea of watching the Sun's reflection on a sheet of paper through a narrow opening in a darkened room. Now systematic observations commenced. The large spot moved from East to West across the Sun, then a new spot appeared on the eastern limb of the Sun; the spots moved across the solar disk, disappearing at the western limb. After ten (!) days the large spot started to re-appear at the eastern limb of the Sun followed by the smaller ones. Johann Fabricius recognized that a certain revolution of the sunspots had taken place. Soon he had a daring thought: the spots were not moving around the Sun but belonged to it. What they had seen was the Sun's rotation on its own axis, "which Giordano Bruno claimed, and recently Kepler vehemently defended, in his commentaries on the movement of Mars". Having carried out a series of observations Fabricius compiled his results in a short paper (Fig. 1.1) dedicated to Enno von Friesland. The date of the dedication coincides with the visit he made to Wittenberg in order to organise its private printing.

The records show that Johann Fabricius became a Master of Philosophy in September 1611. He never acquired any other degree from a University faculty. Mathematics and astronomy were prominent among the subjects he had to study and, stimulated by his father, he dedicated all his energy to astronomy, as witnessed by a strange letter to Kepler "For how long and how devotedly have I dealt with astronomy! How shall I bear this, how shall I commend my thoughts?" He promised information "on a reliable method of weather forecasting which I have sent to my father who acknowledges its correctness. I believe, however, that there has been no comparable method known for centuries... It is quite reliable for global weather conditions as well as for individual weather phenomena, to such a degree that, even if the weather changes four times in a day, nevertheless no wrong forecast ever occurs. A knowledge of the winds and their constant changes of direction is precisely foretold so that we cannot thank God enough. In addition to all this, the movement of the planets is being researched a little more exactly to improve on the understanding provided by the Ptolemaic tables". Of course, Kepler remained reserved and replied that Fabricius should let him have his discovery but he feared that it might contain more eagerness than truth.

Through the booklet published at the time of his graduation, a few copies of which still exist today, Fabricius assured himself of a place in posterity. Once published, "Narratio de Maculis in Sole observatis..." almost disappeared, like the author. Presumably, he died, not yet 30 years old, sometime between March 1616 and May 1617, and it was his father who, in

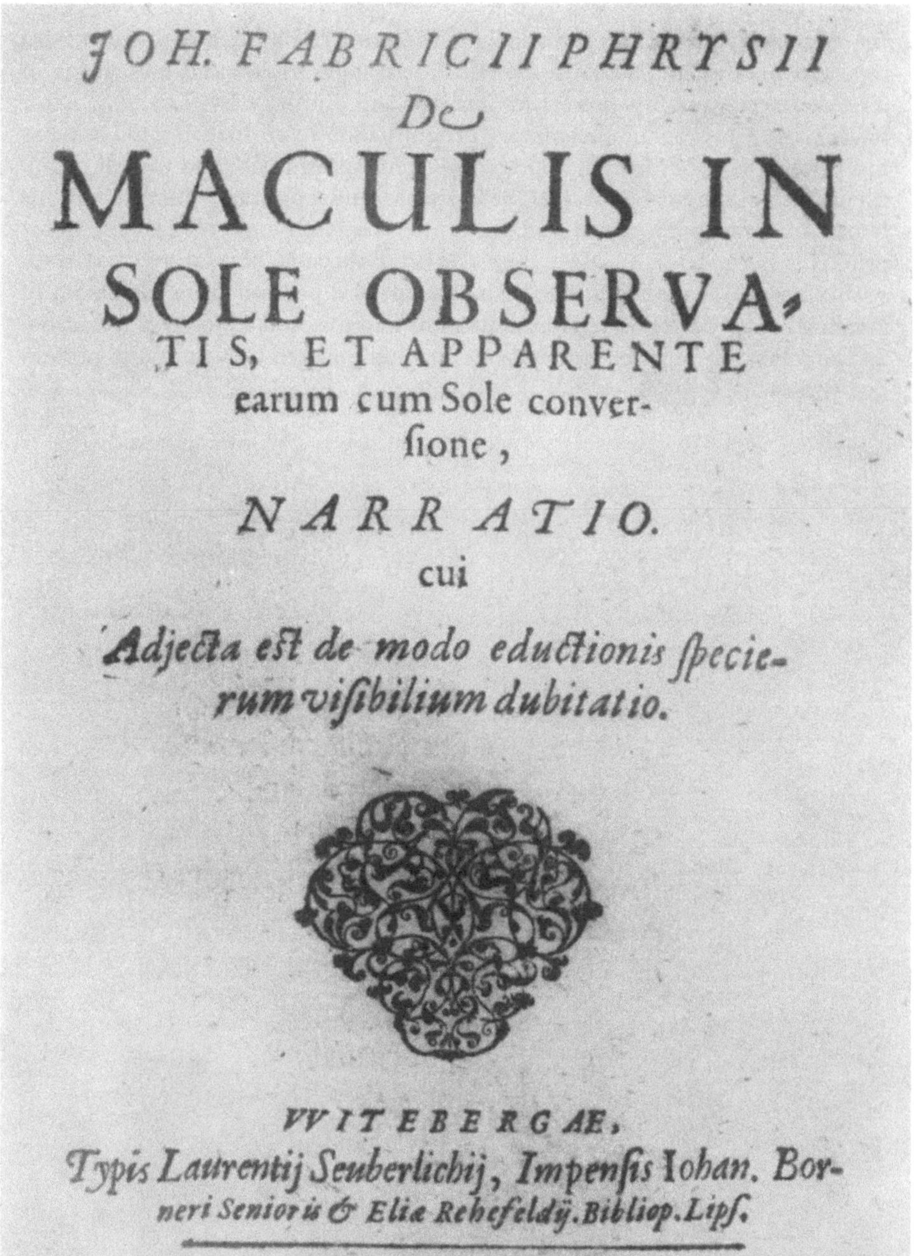
JOH. FABRICII PHRYSII

De

MACULIS IN
SOLE OBSERVA-
TIS, ET APPARENTE
earum cum Sole conver-
ſione,

NARRATIO.

cui

*Adjecta est de modo eductionis ſpecie-
rum viſibilium dubitatio.*

VVITEBERGAE,
Typis Laurentij Seuberlichij, Impenſis Iohan. *Bor-
neri Senioris & Eliæ Rehefeldij. Bibliop. Lipſ.*

ANNO M. DC. XI.

Fig. 1.1. The pioneering paper of sunspot research by J. Fabricius with a dedication, dated June 13 1611, to Enno von Friesland, a friend and sponsor of the family.

the same way as he had recorded for the family the birth of his son, also reported his death. The book, however, had gone astray and knowledge of its existence was only preserved through the obituary by Kepler: "After having read your announcement ('Prognostikon') for 1618 let me express my grief. I see you, friend, deprived of your upright son, and myself of my favorite. Yet, more honorable than praises and epitaphs, there is left his booklet on sunspots, which will guarantee his fame and contains our joint grief...". Soon after, in May 1617, David Fabricius, the father, was slain with a spade. From the pulpit he had accused a peasant from his parish of stealing some of his geese, and the peasant had exacted a bloody retribution. In the following year the counter-reformation plunged Germany into prolonged war.

Fig. 1.2. Christoph Scheiner (1575—1650)

When exactly were sunspots discovered? The "Narratio" does not reveal anything about the first observation; the correspondence with Kepler had ceased; the Prognostikon for 1618, which announced Johann's death, is missing. Abraham Gotthelf Kästner (1796), who re-instated the forgotten Fabricius, said: "their observations were made in the summer months of the year 1610". According to Wolf (1877) the observations took place later "and that it was probably on a December morning in 1610". By re-finding the Prognostikon astrologicum of David Fabricius for the year 1615, Berthold (1894) felt himself in a position to answer the question: "The sunspots were discovered on March 9, 1611". David Fabricius gives the end of 1611 as the publication date of the "Narratio"; Berthold, however, makes it more precise: it appeared at the Frankfurt Autumn Fair in the year 1611. The Fair catalogue was undoubtedly read by all scholars. The prolix title of the "Narratio" that it contained can be thought of as a "Summary" in our present sense. Redundancy was a peculiarity of this time but the essentials, the "maculis in sole" (Fig. 1.1), probably escaped no one.

It was at just this time that the diligent Jesuit, Christoph Scheiner (Fig. 1.2) the inventor of the pantograph, also found spots on the Sun. The news that he had seen something, of which nothing was to be read in Aristotle, was not at all appreciated. Only a few months later he returned to the telescope but, as a precaution, assumed that the black spots on the Sun were small planets passing very close to the Sun. This mathematics professor of the Society of Jesus later meticulously compiled the bulky "Rosa Ursina" (Fig. 1.3) which, it is often said, uncovered the phenomena which were only much later of any interest, namely differential rotation and activity belt migration. For decades he disputed with Galileo in a by no means humble way, and claimed priority for his discoveries. But they, and their influential allies, were in full agreement that the young Fabricius should be completely ignored.

Of course, Galileo (Fig. 1.4a) was, next to Kepler, the most outstanding astronomer of his day. Soon he had the satisfaction of proving that the spots belonged to the Sun's surface. The daily, irregular changes of position fully agreed with the projection of regular segments of a solar parallel of latitude onto the plane perpendicular to the line of vision. His friends, among them the members of the Academy of Lynxes, idolized him. His first published spot observations date from June 2, 1612 (Fig. 1.4b). By this time the mathematician and scholar, Thomas Herriot, had already embarked on a 15 month long observing schedule. His first observations at the beginning of 1611 were unfortunately unsuccessful since occasionally there was not a single spot to be seen. Galileo's paper "Istoria e demonstrazioni alle machi solari ..." (published in a limited edition of 400 copies in Rome in 1613) revealed for the first time that the author was a member of the Academy of Lynxes. It did not, however, contain a single reference to the original discoverer, Johann Fabricius. The great Galileo, who was the first with so many new ideas, also wanted to be the one who made all the original

Fig. 1.3. The cover of "Rosa Ursina" (1630) by Pater Scheiner. Despite its meticulous scholarship, it contains no reference in it to Fabricius. A rose and a bear formed the badge of the Duke of Orsini to whom the book was dedicated.

observations. He was not scrupulous in giving credit to earlier work, and did not even take notice of Kepler's planetary laws.

It was Scheiner who first built the telescopes that had been theoretically designed by Kepler, and just such an instrument was used in Scheiner's Heliotropium Telioscopium to project the black solar spots.

There were, of course, many other notable dates. The final argument about who had priority is not our concern, nor are we much interested in specific dates. Ignorance is the opposite of reality. What is important is how the new scientists treated one another. We can show no picture of the young Fabricius in this book; almost certainly no-one painted his portrait.

Intermezzo

In the 17th century interest in sunspots was widespread. The discovery of a large spot was announced immediately. On the occasion of such an observation the astronomer Cassini wrote in 1661 the strange sentence, which is often quoted today: "It is twenty years since one could last see sunspots although they appeared more often before that time" (see Eddy, 1976). The years 1645 to 1715 (the reign of Louis XIV!), during which there was only "a handful" of large spots, later became known as the Maunder minimum. Descriptions such as those by Athanasius Kirchner in his "Himmlische Entzückungsreise" are to be understood in the same sense. In the middle of the Maunder minimum he recorded the size, duration and apparent movement of the spots, yet abandoned precision in one respect: "The number of spots differs and is uncertain". Although large spots reappeared at the beginning of the 18th century, interest in them seems paradoxically to have rapidly waned. For one hundred years nobody had any ambition either to systematise the individual observations that were made or to observe the Sun over long periods. Many an astronomer thus lost his chance to become the discoverer of the solar cycle. All that would have been required was to use, as the layman Schwabe did later, a (relatively small) telescope not at night but during the day. But the observations would have had to have been carried out practically every day over possibly many years. Among the few records of sunspots in the eighteenth century are those of the "astronomically assiduous" Palitzsch from Prohlis near Dresden. In 1761 he reported, on the occasion of the much observed passage of Venus in front of the Sun, "There were a few sunspots, yet Venus neither approached nor covered one". Palitzsch, a successful farmer, was not much interested in the Sun. He did, however, direct his attention to the atmosphere of Venus.

1.2. The Golden Century

As late as the year 1849, Frederick Wilhelm Bessel stated, "Everything that can still be learned about celestial bodies, for example the appearance and shape of their surfaces, are not unworthy of our attention: they are not

Fig. 1.4.(a) The sunspot paper of Galileo (1613) reveals that the author was a member of the Academy of Lynxes. (b) The earliest observations published are dated 2. 6. 1612, no mention being made of Fabricius.

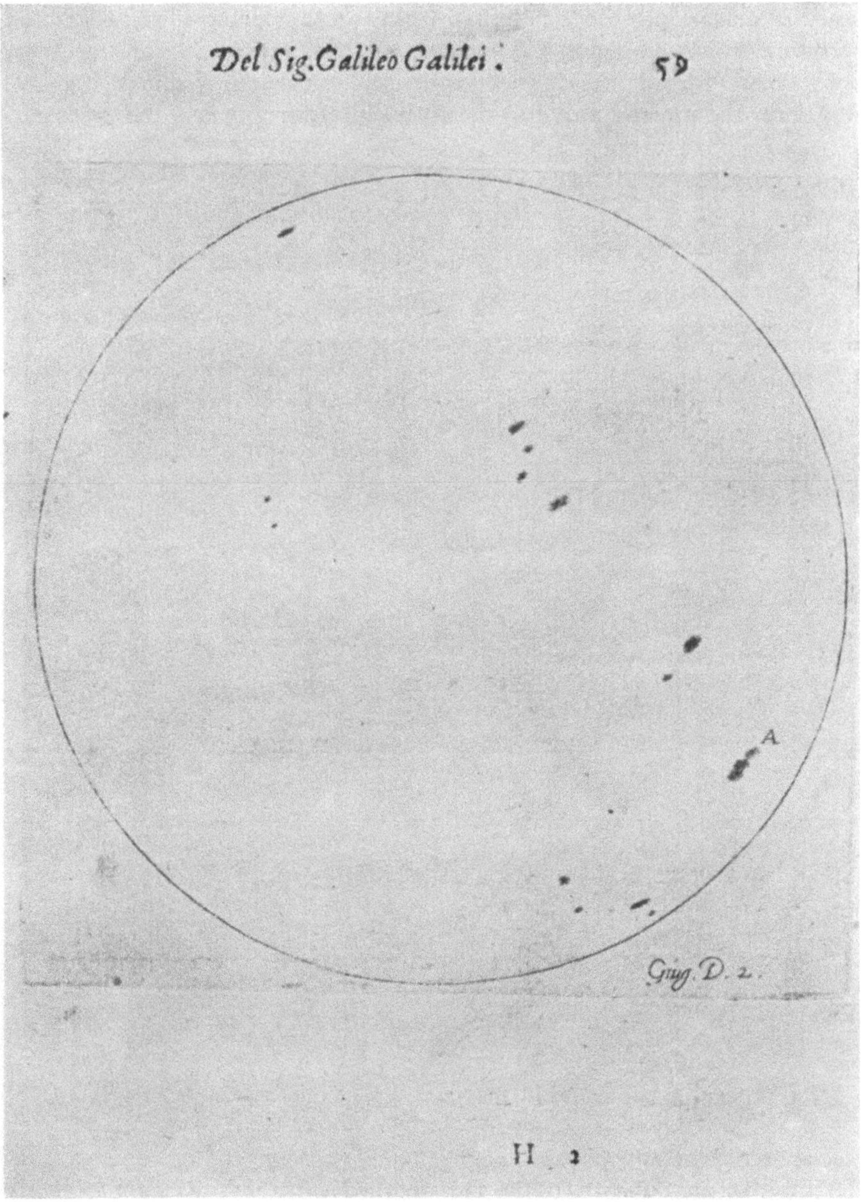

Fig. 1.4.(b)

however our proper concern in astronomy". A phalanx of amateurs and private astronomers ignored this mathematician's prejudice for a century. Some years earlier, the former regimental musician, William Herschel (Fig. 1.5), had already provided most careful observations of the Sun's sur-

Fig. 1.5. William Herschel (1738—1822).

face in an almost astrophysical paper: "Observations tending to investigate the Nature of the Sun ..." (1801). This detailed series of drawings of minutely observed spots ("openings") on the Sun bear witness to the depth of his interest. Indeed, he seems to have been the first person to focus on the annual number of spots on the Sun over a long period, as he was convinced that there was a connection between the number of spots and the weather. Being hampered by a lack of meteorological statistics, he substituted grain prices compiled by the pioneering economist, Adam Smith. Because he supposed that the spots were ray-translucent openings the frequency of whose appearance should coincide with abundances of food, Herschel looked for a correlation between the number of spots and the price of grain (i.e. the weather). He found, as he expected, high grain prices when the Sun was spot-free. And the Sun was far more frequently spot-free than not:

"Our historical account of the disappearance of the spots on the Sun contains five very irregular and very unequal periods. The first takes in a series of 21 years, from 1650 to 1670, both inclusive ... Our next period is very much better ascertained. It begins in December 1676, which year therefore we should not take in, and goes to April, 1684; in all of which time, Flamsteed, who was then observing, saw no spots on the Sun ... A third but very short period, is from the years 1686 to 1688, in which time Cassini could find no spots on the Sun ... The fourth period on record is from the year 1695 to 1700, in which time no spots could be found on the Sun. This makes a period of five years, for in 1700 the spots were seen again ... The fifth period extends from 1710 to 1713; but here there was one spot seen in 1710, none in 1711 and 1712, and again only one spot 1713 ... It will be thought remarkable, that no later periods of the disappearance of the solar spots can be found. The reason however is obvious ..." (Herschel, 1801)

Herschel was very definitely close to discovering the solar cycle — but what he in fact found was the Maunder minimum. The favorites of the Gods seldom leave empty-handed, even when, as in this case, inadequate observations are linked to untenable theories. For Herschel, sunspots were randomly occurring distortions of the Sun's atmosphere, and a regularity in their appearance was bound to seem illogical to him. The later discovery, that about every 11 years a particularly large number of holes are torn open on the Sun, would surely have amazed him.

The breakthrough was the achievement of the son of a physician from Dessau, Heinrich Samuel Schwabe, an unhappy pharmacist (see Fig. 1.6). During his school years he had, in his own words, to "assist my father with surgical operations and make paperbags for my grandfather". Probably on the advice of a friend, he directed a small newly-purchased instrument towards the Sun. At much the same time he won a second telescope in a lottery, and it was not long before he sold his pharmacy to begin his "true life", by devoting himself to astronomy and botany. From the very beginning he drew up record books in readiness for decades of observations.

Originally he expected to find a planet extremely close to the Sun that could be seen only when it passed in front of the solar disk. For more than 40 years he observed the Sun daily, and recorded the temperature of the air. After years rich with spots an uneventful period followed in 1833—34. When, ten years later, there were still hardly any spots to be counted he (courageously!) took up his pen on New Year's Eve, 1843: "Comparing now the number of groups with the number of spot-free days one finds that sunspots had a period of about 10 years and that they appeared so frequently for 5 years that during this time there were but a few or no spot-free days ...

The future will show whether this period shows some consistency, whether the smallest activity of the Sun in producing spots lasts for one or two years, and whether this activity increases more rapidly than it decreases. On

Fig. 1.6. Heinrich Samuel Schwabe (1789—1875).

April 10 and 11, and on May 10, the Sun was so extraordinarily clear, there being only a slightly overcast sky, that the paler light of its limb stood out very clearly." (Schwabe, 1844).

This was rather limited information on which to found an entire branch of science. Also, hardly anything influenced the science of his day: once more a new discovery met ignorance and indifference. But Schwabe, the extrovert socialiser, was stronger than Fabricius. Although confined to his bed every winter since he was forty because of gout, he rises "after each new attack as if newly born ... drinks neither wine nor beer ..." In his words: "I can count back to my father'sgrand father — they all died of gout in their 60th year — and I have now reached my 80th year, though I have always lived with pain." He could not climb up to his observatory any more, and had to stop

Jahr.	Gruppen.	Fleckenfreie Tage.	Beobachtungs-Tage.
1826	118	22	277
1827	161	2	273
1828	225	0	282
1829	199	0	244
1830	190	1	217
1831	149	3	239
1832	84	49	270
1833	33	139	267
1834	51	120	273
1835	173	18	244
1836	272	0	200
1837	333	0	168
1838	282	0	202
1839	162	0	205
1840	152	3	263
1841	102	15	283
1842	68	64	307
1843	34	149	312
1844	52	111	321
1845	114	29	332
1846	157	1	314
1847	257	0	276
1848	330	0	278
1849	238	0	285
1850	186	2	308

Fig. 1.7. Schwabe's success: Humboldt publishes the hitherto ignored sunspot count in Volume 3 of his "Kosmos".

making observations. But fate did not deny him a well-deserved appreciation, and even at an advanced age he could still read "the smallest print without spectacles, both by day-light and by lamp-light". In 1850 Humboldt published Schwabe's statistics in the third volume of his "Kosmos" (Fig. 1.7). The strength of his cause was now understood not only by his colleagues, Schmidt and Wolf, but also by others: Carrington, Secci, Spörer ... The Royal Astronomical Society awarded him, in 1857, its Gold Medal, at the same time accepting Wolf as a member. Carrington took the medal to Dessau. He had already visited Schwabe the previous year, to find out about — and inspect — his observing equipment. "His observatory is, Mr. Carrington tells me, a small room on the roof of his house", the President would report in his extensive praise of the "Discovery of the periodicity of sunspots". Carrington was several times a guest in enlightened and social Dessau, exchanging ideas with Schwabe through friends who could act as interpreters.

Humboldt enjoyed the special esteem of the local ducal family. On a visit in 1833 he became acquainted with Schwabe and introduced him to the Duchess. From that time onwards Schwabe was a regular visitor to the palace and gave the Duchess and her children lessons in astronomy and botany. One of the other teachers became his wife. She also had connections with astronomy through her brother who was a friend of Encke.

Schwabe was an amateur with astonishing determination. When he lent his observing notebooks to the Royal Astronomical Society at their request, there was a total of 39 volumes! He died in his home-town in 1875 after he had bequethed his scientific tools to the local high school, "as I am all on my own, and have nobody to whom I may leave my instruments".

The first to take Schwabe's discovery seriously had been Rudolf Wolf. Born in 1816, he was a mathematics teacher, an amateur historian, and the head of the Observatory of Berne (later Zürich). After he (and others) had discerned, in 1852, parallels between solar activity and the Earth's magnetism he asked himself the important question, "whether sunspot frequency has always varied periodically and, if so, what actually is the mean duration of its period. The latter investigation, which at that time was the more difficult since for most of the ... series there was as yet only partial (or indeed no) knowledge, I undertook myself and, by collecting all the relevant notes contained in journals, academic collections and individual works, I could find initially ... twelve periods, and then show that the mean sunspot period was 11.111 years, so that in one century just nine periods take place, — and I could finally demonstrate that all observations that had become known to me fitted quite well into the 22 periods that have taken place since sunspots were first discovered and I could even predict the occurrence of a new minimum in 1855" (Wolf, 1877).

Schwabe and Wolf were successful by mere counting. After them the measuring began. It was still possible to ignore the physical peculiarities: the positions and the movements of the spots still contained unknown facts. In particular a strange phenomenon in the proper motion of sunspots was soon seen which was destined to change ideas about the Sun radically.

The man who did more than merely count, Richard Christopher Carrington was born in the same year in which Schwabe used a telescope for the first time. The son of a rich brewer it was intended that he should study theology at Trinity College, Cambridge, but he preferred to become an astronomer. Despite his wealth he served a three-year apprenticeship as an observer, but soon had his own observatory built at Redhill in Surrey. During its construction he received the results of Schwabe and Wolf: "I thought I could very well appropriate the Solar Spots to myself at Redhill for the next eleven years' period, estimated to commence in 1855 and end in 1866". The plan was well conceived. The task was not beyond the capabilities of a private astronomer (assistant included) and, despite its topicality, Carrington remained almost the only person working in the field. Each night he measured circumpolar stars, and the following day the spots on the Sun. He seems to have concerned himself little with his father's liquid products. When his father died in 1858, however, the son had at last to take an interest himself in alcoholic matters. Observations of the Sun were halted prematurely in 1861, but the seven years work had yielded abundant fruit. Carrington had determined the motion of spots by latitude and longitude. Both measurements were made for thousands of cases and, after averaging, their dependence on heliographical latitude was analysed. He saw "... that the diurnal motions in longitude are subject to a wellmarked law of variation depending on the latitude, while it is not apparent that the motions tabulated for the latitudes are anything beyond the accidental differences of observations. Trial readily shows that no parabolic curve or expression of the form $\sin l$ or $\sin^2 l$ will satisfy the above values, but that the whole table of results for longitude may very fairly be represented by the expression $14' - 165' \sin^{7/4} l$" (Carrington, 1863; Fig. 1.8).

Although the power 1.75 was not supported by later work, we here meet Carrington as the discoverer of solar differential rotation. He did not detect a meridional motion of spots, and in fact, as we know, it is extremely slow. Carrington's third important result concerns the positions of the spots. They hardly ever occurred at the equator or at the poles. Rather, they appeared in two belts between latitudes 6° and 35° North and South with the zones contracting towards the equator and ultimately vanishing when they reached the equator. His observational program "... was to show a great contraction in the limiting parallels between which spots were formed for two years previously to the minimum of 1856, and soon after this epoch the apparent commencement of two fresh belts of spots in high latitudes North and South, which have in the subsequent years shown a tendency to coalesce and ultimately to contract as before to extinction. Whether this is what occurs at each period of increase and decrease of frequency of the spots must be left to observers who may follow me to show ... the origin of this phenomenon and instruct us on the question "What is a Sun"?"

It may be recognized today that Carrington made both the important discoveries of early solar physics: differential rotation and spot-belt migration. He had undertaken an enormous amount of work and, perhaps because it had been too much for him, he fell chronically ill, had to sell his brewery, and died before his fiftieth birthday.

Lat.	D. Motion		Weight.	Lat.	D. Motion.		Weight.
+36°	−′	−′	−	−36°	−50′	+ 6′	1
35	−	−	−	35	−	−	−
34	−43	+ 4	5	34	−41	− 1	7
33	−33	+ 7	2	33	−36	−10	1
32	−30	− 2	1	32	−52	− 5	1
31	−34	+ 5	6	31	−	−	−
+30	−20	− 1	5	−30	−38	+ 2	4
29	−34	+ 8	2	29	−34	+ 1	14
28	−30	+ 6	10	28	−35	+ 1	8
27	−27	+ 3	5	27	−40	0	4
26	−22	− 0	17	26	−30	+ 1	6
25	−12	+ 3	2	25	−20	+ 3	12
24	−15	+ 3	9	24	−24	+ 5	5
23	−19	+ 2	14	23	−14	− 0	2
22	−12	− 1	13	22	−14	− 0	28
21	−14	− 0	14	21	−17	+ 3	10
+20	−10	+ 1	12	−20	−11	+ 2	16
19	−10	− 1	19	19	−14	+ 1	8
18	− 6	− 1	3	18	− 5	− 1	19
17	−11	− 3	5	17	−10	+ 1	13
16	− 3	+ 1	7	16	−15	− 3	3
15	+ 0	+ 2	18	15	− 9	− 0	11
14	− 3	− 2	12	14	− 4	− 1	11
13	− 2	− 1	10	13	+ 1	0	1
12	+18	− 3	7	12	+ 1	+ 1	38
11	+ 5	− 0	16	11	+ 8	− 1	7
+10	+ 2	+ 0	9	−10	+ 1	0	8
9	+ 5	−10	4	9	+12	+ 1	17
8	+ 9	− 0	27	8	+ 6	+ 3	14
7	+ 8	− 2	21	7	+14	+ 1	5
6	+ 9	− 3	7	6	−	−	−
5	+19	+ 6	2	5	−	−	−
4	+15	+ 2	3	4	+ 6	− 3	1
3	+38	− 2	1	3	0	0	4
2	−	−	−	2	−17	+ 9	1
1	−	−	−	1	− 4	0	1

Fig. 1.8. The discovery of solar differential rotation by Carrington. The first column gives the latitude and the second column shows the diurnal motion which is maximal in the equatorial region. (We had no success locating Carrington's portrait.)

The vacancy at the center of solar studies was filled by a teacher from Anklam. Born in Berlin in 1822 it was only at the age of 36 that Gustav Spörer (Fig. 1.9), began to observe the Sun carefully. Independently of Carrington he looked for a rotation law, and by 1861 had already found its

Fig. 1.9. Gustav Spörer (1822—1896).

characteristic structure. After this great success his name became well-known and his small Pommeranian solar observatory was equipped with a 5-inch refractor by the later Kaiser Frederick III. Spörer's sunspot statistics for the minimum of 1867 also confirmed Carrington's law of zonal migration. Such results spoke for themselves, and for Spörer. It was this, amongst other factors, that led to the foundation of the Astrophysical Observatory in Potsdam to which he was called in 1874. It was there that he recognized possibly the strangest phenomenon of the solar cycle, namely, "that during

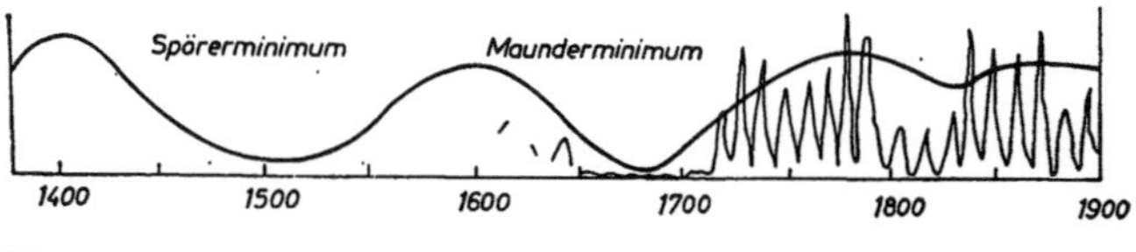

Fig. 1.10. Representation of the long-period variation of the solar cycle. Thin line: sunspot numbers; thick line: ^{14}C abundance.

the mid-17th century over a very long period essentially different conditions prevailed than have in modern times" (Spörer, 1887). Again these are the years of the Maunder minimum (Fig. 1.10) already discussed by Herschel. Spörer worked unceasingly. His lengthy analyses in the Publications of the Astrophysical Observatory of Potsdam contain rich material. In 1882 he became Chief Observer. Unshakeable health permitted him to work relentlessly until old age. At 73, in the first year of his retirement and without any preliminary symptoms, he died suddenly on a journey to visit his children.

Chapter 2. The Flow of Concepts

... a joy of life — provided
man is not master of the world
and does not seek to be so.
Christa Wolf, 1983

2.1. The Foundations: Zöllner's Flow and Lebedinski's Anisotropy

Turbulence in nature is the rule rather than the exception. Nevertheless there has been only a slow laminar flow towards today's understanding of solar rotation. After a promising launching more than a hundred years ago, this branch of scholarship entered the doldrums and it was a long time before a better understanding was reached, and particularly in one important aspect that is still imperfectly understood today: the relationship of the turbulence on the Sun to its non-uniform rotation.

The origins of the theory go back to the great seer and polymath Frederick Zöllner. Reading today his treatise on the Sun, one can scarcely believe one's eyes. With great ease he proceeds from the observed cyclical occurrence of sunspots to the "periodicity of solar magnetism"; he raises the possibility that "the presence of sunspots must modify the rotation law". This is what Spörer had observed between 1861 and 1864. Again, Zöllner writes, "All rotating celestial bodies have magnetic poles that are not coincident with their poles of rotation". He was particularly impressed by the rotation of celestial bodies and saw it "as the single obvious reason for the latitude-dependence of physical quantities. Rotation produces a streaming in the outer layers of a centrally heated liquid sphere, resulting in a negative pole-equator temperature difference." This is exactly what Secci had observed. One thing elevates Zöllner above the rest: he looked for the causes in the celestial bodies themselves. Only rarely (if ever?) did he resort to the hypotheses of extra-solar and fossil fields. This made him the astrophysicist who re-instated the discredited "Physische Himmelskunde". Zöllner's publication "Über das Rotationsgesetz der Sonne und der großen Planeten" explained how a meridional flow could affect rotation: he argued that the flow towards the equator of a thin layer of fluid on the surface of a rigidly rotating sphere would produce a frictional force on the sphere's surface. Under several preconditions Zöllner calculated the increase in the zonal velocity of the liquid. Expressed as an angular velocity, he wrote

$$\Omega = (A - B\cos^2\theta)/\sin\theta \tag{1}$$

with A and B being constants and θ the solar co-latitude. Application of this formula yielded (he claimed) a better account of the observations than the relation given by Carrington himself.

Another peculiarity may be noted: from the existence of differential rotation, Zöllner thoughtfully inferred the "rigid state of sunspots", which otherwise would be sheared into spiral stripes. A large spot, after only one rotation, would have become a stripe covering 120° of longitude. The calculation was impressive but, alas, he who worked so much with Carrington's figures used a value for $\mathrm{d}\Omega/\mathrm{d}\theta$ which was too large by a factor of 60: 1.6 degrees instead of 1.6 minutes. In a reprint of his publications made after the mistake had been noticed, he preferred to give a 15-line commentary rather than to correct it.

One person immediately decided that it was not necessary to seek an origin for (1). If the viscosity, ν, is sufficiently small in the outer parts of the Sun, the viscous decay time corresponding to its large dimension, R, will be so long that the circulation may be considered as a fossil flow. Wilsing (1891) in Potsdam argued against the validity of Zöllner's assumption of a "rigid or glowing-liquid solar surface", without mentioning Zöllner's remark that it was not essential to his theory. He wanted to show "that under some hypotheses the rotation law in the outermost layers of the Sun can be understood as the relic of an original flow. It remains to determine the decay time of the velocity differences which existed at a certain epoch." This time is of the order R^2/ν. Wilsing, working with viscosity coefficients for gases, derived time-scales of 2 million years for a smoothing of the velocity by 7%.

Emden in his famous "Gaskugeln" did not accept the relic theory: "The constancy of the Sun's effective temperature requires convective flows to mix the material. Cooled surface elements are replaced by new warmer elements. Any equatorial acceleration would be smoothed out by the mixing process ... Only a steady source can maintain the equatorial acceleration against the action of eddies." Like Helmholtz he postulated the existence of eddies which were thought to emerge from certain discontinuity surfaces within the Sun. These eddies, which destroyed any regularity in the flow, also prevented solid-body rotation from occurring. "The angular velocity must vary in space and time." Radial mixing leads to $r^2\Omega$ = constant which, however, cannot explain the observations. Only if the thickness of the mixing zone were latitude-dependent, could an accelerated equator be explained. Indeed, the discontinuities considered by Emden, which are said to break the "convective mixing", are highly non-spherical. No wonder he states: "The equatorial region of the Sun's surface layers must rotate faster than the polar ones." The same trick was used by Jeans (1929) who derived the sub-rotation law r^{-2} from the radial momentum transport by radiation. In order to obtain the observed latitudinal dependence of the angular velocity he postulated, in addition, an oblate solar interior, with an axis ratio of 0.92. If the surface of this ellipsoid rotated with a uniform angular velocity, equatorial acceleration would be almost trivially explained.

In 1891, the director of the Gotha Observatory, Harzer, presented in a complex paper appearing in the *Astronomische Nachrichten* an idea that apparently is not unconnected with the "thermal wind" of meteorology. Despite the "von Herrn Auwers gütigst mitgeteilten Tatsache", that the equatorial radius of the Sun does not differ from its polar radius by more than one part in 19,000, he nevertheless proceeded to assume that rotationally-produced flattening generated meridional flows that created the observed equatorial acceleration. According to him, the degree of flattening required was not so large that the Sun would be detectably oblate.

R. A. Sampson (1895) from Newcastle upon Tyne demonstrated in great detail how the Legendre polynomials P_n^1 could be used to satisfy the frictional part of the ϕ-component of the momentum equation. By adjusting his solutions in such a way that they fitted the surface observations he considered the matter solved. In addition, one can even demand that the angular velocity increases with depth, as seemed to be suggested by experiments in a rotating glass ball of water (Belopolsky, 1890). The crux must lie in the boundary conditions. It can be shown that Legendre polynomials provide eigensolutions only if the lateral eddy viscosity is negative. There have been many subsequent rotation laws derived from the isolated viscosity term of the momentum equation, $\mathrm{div}\,(r^2 \sin^2\theta\, \nabla\Omega) = 0$. This is a differential equation of second order, and uniform rotation is therefore not its only solution. Its dependence on boundary conditions provides different ways of introducing a second solution. In this context, the work of Giäo and Wehrlé (1932) may be singled out from amongst many. Their claim "that the famous mystery of the solar equatorial acceleration is now completely cleared up" was vehemently disputed by Emden (1936). He suggested that to make the statement true the word "completely" would have to be replaced by "in no way". He argued that (i) the rotation period fluctuates over timescales of a few years and (ii) $\mathrm{d}\Omega/\mathrm{d}r$ does not vanish, in contradiction to the author's assumption. Emden summarized many solutions of the viscous part of the momentum equation, all of which could be adjusted to reproduce the observed rotational θ-profile. "None, however, reveals anything about the solar rotation law".

Nothing could be said against the hypothesis of a fossil rotation profile without dramatic changes in the effective viscosity. In turbulent media, however, everything is totally different.

In 1895 Reynolds pointed to the existence of additional, turbulently-generated stresses which are basic to many areas of fluid dynamics, and not merely to the subject of this book. Boussinesq (1897), Taylor (1915) and Wilhelm Schmidt (1917) related the Reynolds stresses to the gradients of the mean flow. The coefficient connecting the stress to the rate of strain, the "Austausch" (Schmidt, 1917), has the dimensions [g cm^{-1} s^{-1}] of a dynamic viscosity. Quotients of turbulent fluxes and mean gradients yielded consistent values. In current notation the immediate consequence of this idea is the relation

$$Q_{ij} = -\nu_T(\overline{u}_{i,j} + \overline{u}_{j,i}), \tag{2}$$

where $Q_{ij} = \overline{u_i' u_j'}$ is the correlation tensor and $\varrho \nu_T$ is the coefficient referred to above (ϱ density). If the mean flow $\overline{u}$ consisted of a global rotation with angular velocity Ω, the "old Boussinesq" relation (2) would yield

$$Q_{r\varphi} = -\nu_T r \sin\theta \, \partial\Omega/\partial r, \qquad Q_{\theta\varphi} = -\nu_T \sin\theta \, \partial\Omega/\partial\theta, \tag{3}$$

for the "turbulent angular momentum transport" or the "zonal momentum fluxes". In this form it played a really basic role in the theory of rotation laws for celestial bodies. In (3), spherical polar coordinates have been introduced: r is the radius, θ the polar angle and ϕ the longitude.

It was Prandtl who, in 1925 in Göttingen, introduced the concept of the mixing length with the help of which the Austausch became calculable. The eddy viscosity always exceeded the molecular viscosity by many orders of magnitude. Prandtl argued that the fluctuations u' and v' (say) must be expressible in terms of quantities connected to the mean flow. He suggested $u' \sim L\, \mathrm{d}\overline{u}/\mathrm{d}y$ and $v' \sim L\, \mathrm{d}\overline{u}/\mathrm{d}y$, and hence

$$\overline{u'v'} \simeq -L^2 \, |\mathrm{d}\overline{u}/\mathrm{d}y| \, \mathrm{d}\overline{u}/\mathrm{d}y, \tag{4}$$

where L is the "mixing length". Initially the mixing-length concept was intimately connected with the geometry defining the hydrodynamic problem. Prandtl saw an analogy between turbulence as a consequence of dynamic instability on the one hand and gas-kinetic processes on the other. The mixing length modelled the distance travelled by a turbulent element before it lost its individuality. Later on, however, it proved convenient to use the local scale-height of the density or pressure.

And what about the Sun? Schwarzschild begins his well-known work of 1906 with the words: "The Sun's surface exhibits, through granulation, sunspots and prominences, changing conditions and strong variations." From his famous text on the stability of radiative equilibria (i. e. of energy transport by the diffusion of photons), he derived his well-known criterion, from which he concluded: "that possibly in the solar interior there is a nearly adiabatic zone of rising and falling flows...". Shortly before this, Janssen (1896) had produced first-class photographs of solar granulation. After just a few minutes the granules lost their identity. From this material, Hansky (1905) estimated that the granules lived for only 5 minutes. This strongly suggested turbulence. A model of the solar granulation was provided by the processes of steady laminar convective transport of heat in liquids that had been investigated by Bénard at the beginning of this century. The determination of both the Reynolds and Rayleigh numbers for granulation revealed, however, that these motions had to be both unsteady and turbulent. The careful application of Schwarzschild's criterion to certain zones in a stellar interior led, in the early 1930's, to the belief that convective equilibrium might obtain, rather than the radiative equilibrium previously postulated by Eddington. Unsöld was the first to show in 1930 that the ionisation of hydrogen lowered the adiabatic gradient $g/C_p = (\gamma - 1)g\mu/\gamma\mathfrak{R}$. The value of the ratio of specific heats, $\gamma = C_p/C_v$, for a partially ionised gas is less than 5/3, which is the value for monatomic gases, i. e. neutral or totally ionised hydrogen. In regions of the solar atmosphere where hydrogen is only partially ionised, the radiative temperature gradient need exceed the adiabatic one only very slightly for convection to take place. Unsöld's convection zones were, however, very thin.

In 1932, Ludwig Biermann in Göttingen wrote some thoughtful papers in which he developed the theory of "Konvektionszonen im Inneren der Sterne". These texts, which still read well today, introduced the concept of the mixing length into astrophysics: he was in close contact with Prandtl. The mixing length was estimated to be 10^4 km. Biermann also discussed turbulently-induced enhancements of the viscosity and thermal conductivity by as much as a factor of a million. Here, and also in Prandtl (1932), one already finds the expression $\varrho L C_p u' \Delta \nabla T$ for the heat transported by turbulence. "In contrast to the corresponding expression in the theory of gases, C_p appears here rather than C_v since mixing at the eddy's edge with its surroundings takes place under constant pressure; $\Delta \nabla T$ appears rather than ∇T, since the relevant temperature difference is the one that exists between the eddy and its surroundings." A very detailed investigation of Biermann's findings was also undertaken much later by Spiegel and Veronis (1960).

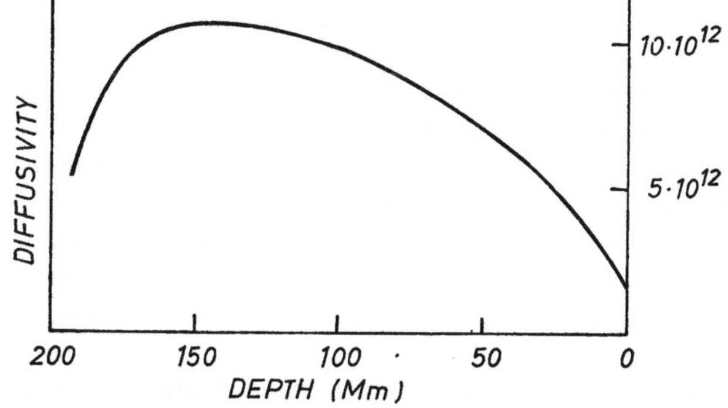

Fig. 2.1. Turbulent diffusivity $u'L/3$ in the SCZ (after Stix, 1976a). The data of Baker and Temesvàry (1966) were used for the computation. In all Figures c. g. s. units are employed and T is in Kelvin, except where noted.

Already Schmidt (1925, p. 17) felt that the diffusion ansatz ∇T was too simple, and he worked with the "potential temperature" rather the actual temperature. For both temperature gradients, Biermann estimated the deviation to be $\Delta \nabla T \simeq 10^{-8}$ K/cm ($\nabla T^{\mathrm{ad}} \simeq 10^{-3}$). The turbulent velocity reached only 10 m/s, and the eddy viscosity was always 10^{12} cm²/s (see Fig. 2.1 for a comparison with modern calculations). In contrast to (4), the Austausch A (i. e. the coefficient relating stress to rate of strain) was already known to be the product of the density, the mixing length and the root mean square (r. m. s.) velocity of the turbulent element, i. e.

$$\nu_T = A/\varrho \simeq Lu' \simeq L^2/\tau_c, \tag{5}$$

where τ_c is the lifetime, or turnover time, of the element.

In several papers Siedentopf (1933) of Jena also investigated "Konvektion in Sternatmosphären". The Solar Convection Zone (SCZ) was still believed to be very thin (1,000 km), but he quite rightly estimated the upward velocities within it as 1—2 km/s, and it seemed reasonable to assume that the convective elements would overshoot the surface, enter the photosphere, and be seen as granulation. He gave 10^2—10^3 s for the lifetime of granules, in full agreement with Hansky's observations. In 1938, at the Babelsberg Observatory, Biermann once again took up the study of the "Theorie der Granulation und der Wasserstoffkonvektionszone der Sonne". In contrast to his previous studies, he could now assume a significantly higher percentage of hydrogen. The increased gas pressure led to a "thickness of the SCZ of the order of 10^5 km. The convective flux transports the whole energy, hence the SCZ has an adiabatic structure. It attains temperatures of 10^6—10^7 K". The velocity at the upper boundary reached 10^5 cm/s, which fitted well with Allen's observations (1.6 km/s). The excess in temperature gradient was $10^{-6.8}$ and the lifetime of granules was 10^3 s.

All this had been calculated a short time before the discovery of the negative hydrogen ion as the main source of opacity. More recent investigations confirm the findings of the pioneers: that stochastic motions take place in the Sun's outer layers, that these are large-scale motions, and that they do not necessarily have the same properties in all directions. Was it possible that differential rotation could now be understood on this basis? It was exactly this question which A. J. Lebedinski asked himself in Leningrad. His publication, in Russian, dates back to June 1940. It is based on two ideas. With acute perception he recognised that it was Faye's rotational profile of 1865, namely $\Omega = \Omega_e\,(1 - 0.216\cos^2\theta)$, which should be explained. This is the only form that highlights spherical functions in the solution of the hydrodynamical equations.

The other idea concerns the mechanism of angular momentum transport. Presumably as a consequence of being associated with the famous school of Soviet theoreticians working on turbulence, Lebedinski treated turbulence and Reynolds stresses as the dominant physical phenomena at and below the solar surface. Because of the enormous scales the "Austausch", so he argued, must necessarily be anisotropic. "Two coefficients should be distinguished, a radial one and a horizontal one (A_V and A_H). The non-isotropic character of turbulent friction is brought about because the mean values of the velocity and the mixing (Mischungsweg) are different in different directions." For the interesting velocity correlations Lebedinski wrote

$$\begin{aligned} \varrho Q_{r\phi} &= -\,[rA_V\,\partial\Omega/\partial r + 2(A_V - A_H)\Omega]\sin\theta\,, \\ \varrho Q_{\theta\phi} &= -\,A_H(\partial\Omega/\partial\theta)\sin\theta\,. \end{aligned} \tag{6}$$

Expressions like these have frequently arisen since his time. Also the conservation of angular momentum, the ϕ-component of the Reynolds equation, was given in the same form as it is usually seen in today. Lebedinski demonstrated that rigid rotation is not the appropriate solution of this equation. The solar surface should be considered to be force-free, and therefore impermeable to angular momentum. Yet, strangely enough, he chose as a second boundary condition, as had Sampson before him, the observed rotational profile. But this is precisely the quantity being sought! One can even show that the equation based on (6) does not have a latitude dependent Ω-solution (if A_H is positive) when the bottom of the SCZ is either force-free or rigidly rotating. Lebedinski's rotation profiles must therefore violate the inner boundary conditions. Nonetheless, he was the first to make the non-vanishing of the angular momentum transport for $\Omega = \text{const.}$ the basis of the theory of differential rotation. In this sense we may regard him as the discoverer of the phenomenon which we today call the Λ-effect, where we use the symbol Λ to remind us of his contribution to Astrophysics.

In 1946 in Oslo, Jeremi Wasiutyński's book "Studies in Hydrodynamics and Structure of Stars and Planets" appeared. Besides being a very extensive survey of the relevant literature, the first chapter dealt with turbulence using tensor calculus. Of particular interest is the third section "Turbulent viscosity by rotation in 3 dimensions", in which he suggested that different viscosities apply to the three principal directions of the spherical coordinates. Wasiutyński calculated the consequences of this assumption on the correlation tensor Q_{ij} and arrived at the expressions

$$\begin{aligned} \varrho Q_{r\phi} &= -\,r^{-1}A_r\partial(r^2\Omega\sin\theta)/\partial r + 2A_\phi\Omega\sin\theta\,, \\ \varrho Q_{\theta\phi} &= -\,(A_\theta/\sin\theta)\,\partial(\Omega\sin^2\theta)/\partial\theta + 2A_\phi\Omega\cos\theta\,, \end{aligned} \tag{7}$$

which, for $A_\phi = A_\theta$, are similar to those given by Lebedinski. For rigid rotations one obtains

$$\varrho Q^{(r)}_{r\phi} = 2(A_\phi - A_r)\,\Omega\sin\theta\,, \qquad \varrho Q^{(r)}_{\theta\phi} = 2(A_\phi - A_\theta)\Omega\cos\theta\,, \tag{8}$$

for the true representation of the Λ-effect. Although the second equation is problematical (due to nonzero correlation at the poles), (8) already contains an essential difference from the Boussinesq relations (3): anisotropic turbulence transports angular momentum even when the mean rotation is rigid,

i. e. Ω = constant no longer satisfies the angular momentum conservation law. Although this was recognized by Wasiutyński, he uncharacteristically did not pursue the matter with an explicit theory. This is somewhat surprising for he was never deterred by lengthy calculations. He may, perhaps, have noticed that expressions (7) do not produce *latitudinal* variations in rotation.

Another important statement by Wasiutyński concerns the turbulent energy flux $F^{\mathrm{conv}} = \varrho C_p \overline{u'T'}$. He generalized the already well-known expressions "to three dimensions":

$$F_\mu{}^{\mathrm{conv}} = C_p A_{\mu\nu}(\partial T^{\mathrm{ad}}/\partial x_\nu - \partial T/\partial x_\nu). \tag{9}$$

It was only much later that this apparently trivial extension proved to be important in the evaluation of the turbulent heat flux in a rotating fluid. Unlike that of Lebedinski, Wasiutynski's work had a considerable impact on future developments. For example, in 1966 Elsässer discussed the conditions under which the viscosity tensor takes the form given by Wasiutynski.

Csada, in a detailed paper in 1949, developed the complete relationship between "the differential rotation and the large-scale meridional motion in stars". Both the generation of a meridional circulation by nonuniform rotation and the reverse process were studied: the turbulence aspects remained, however, unconsidered. He had, therefore, no chance of discovering the true origins of the phenomena observed.

2.2. The Mean-Field Concept: Eddy Viscosity and the Lambda Effect

After the Λ-effect had been discovered for a second time, it was used unfailingly to consider problems concerning rotation. Independently of his forerunners, Biermann (1951) first stressed that gases rotate rigidly because the velocities of the molecules are normally isotropic. Anisotropic velocities are expected for "turbulence created by the instability of a thermally-created stratification. All accelerations due to density differences are directed towards or away from the rotation axis and are roughly parallel to the radial direction. Sideways motion only occurs because of the demands of the divergence condition. In contrast, dynamic turbulence is essentially isotropic". We note that, here, Biermann speaks of anisotropic turbulence, not anisotropic viscosity. He treated turbulent angular momentum transport in a way similar to Wasiutynski with the help of two coefficients A_1 (isotropic) and A_2 (anisotropic), and it was he who was the first to find the rotation law,

$$\partial\Omega/\partial r = -2A_2(A_1 + A_2)^{-1}\Omega r^{-1}, \tag{10}$$

which depended, however, on depth and not latitude. A dilemma now arose: anisotropy in the turbulence led to a differential rotation, but not to one of the desired form. It seemed that one should either return to Zöllner or investigate turbulence more deeply.

Both possibilities required some physical reasoning. After it had been proved that anisotropic viscosity produced only radially varying rotation, Kippenhahn (1963) opted for meridional circulation. For the first time the correct boundary conditions for a force-free surface were included. For a depth-dependent angular velocity, the centrifugal force is not the gradient of a potential, even for constant density: meridional circulations must therefore be present. For the case when $\partial\Omega/\partial r > 0$ (known as "super-rotation"),

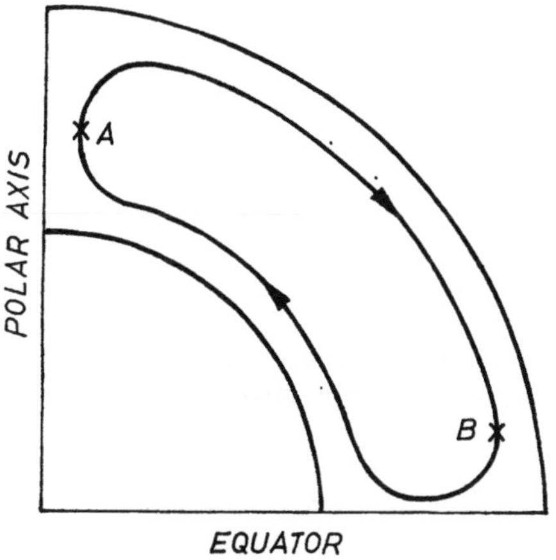

Fig. 2.2. The simplest form of meridional circulation in the SCZ. The selected streamline ('clockwise' flow) leads to the observed rotation law. On *AB* the flow gains energy from the centrifugal force.

they are directed towards the equator at the surface. Consequently, and we know this from Zöllner, the rotation rate at the equator increases. Immediately after this idea was published Steenbeck and Krause (1965) confirmed this by explicitly solving the differential equation with the appropriate boundary conditions. Kippenhahn demonstrated how the gas takes the energy it needs for streaming from the non-conservative centrifugal force (Fig. 2.2). In order to allow the circulation to move towards the equator at the surface, that is with the centrifugal force acquiring energy, the angular velocity has to increase with increasing distance (super-rotation). Super-rotation existed if the lateral Austausch dominated the radial one. For the observed rotation law, therefore, the turbulence must be more intense horizontally than vertically. This, however, seemed somewhat strange. Kippenhahn developed an iteration procedure for high viscosity and/or slow rotation and found that turbulence (alone!) gave rise to a radial rotation law, which then generated the required meridional circulation that, in turn, produced the acceleration of the equator. This approach gave a very large flow velocity at the surface, about 300 m/s, a value which is much too high compared with the observations. Kippenhahn therefore suggested that the full non-linear problem should be attacked numerically, to avoid the convergence difficulties of the analytical approach. The results of such calculations were presented in 1969 by Köhler in his doctoral thesis. He constructed 224 two-dimensional difference equations from the Navier-Stokes equations using the Wasiutynski tensor with $A_\phi = A_\theta$ and constant density. The solution of these equations yielded, for various assumed viscosities, the values shown in Table II.i (cf. Fig. 2.3).

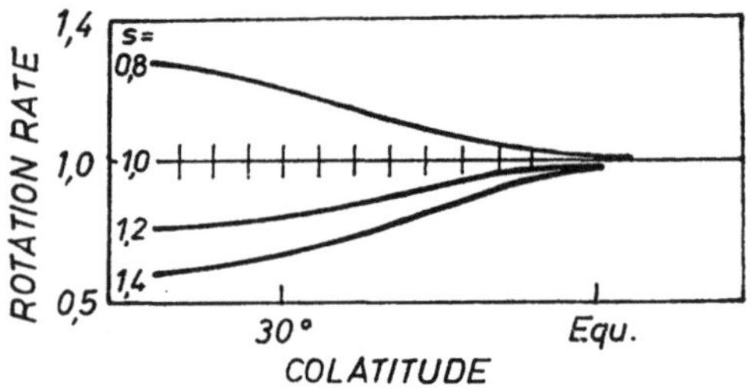

Fig. 2.3. The variation of angular velocity near the surface with Kippenhahn's anisotropy parameter s. For $s > 1$ the angular velocity increases towards the equator.

Table II.i. Anisotropic viscosity and differential rotation (Köhler, 1969)

A_r/ϱ	$s = A_\phi/A_r$	$\Omega_{eq}/\Omega_{pole}$	$\bar{u}_{\theta,max}$
$5 \cdot 10^{12}$ cm²/s	1.2	1.22	2 m/s
$5 \cdot 10^{13}$ cm²/s	1.2	1.13	10 m/s

Köhler also estimated the effect of the solar wind on the SCZ's rotation law. He found it to be small, the reason being that, because of the high eddy viscosity, even very small (radial) gradients of Ω are able to transport angular momentum which is then carried away by the solar wind and the embedded magnetic fields. The velocities found by Köhler are significantly less than those deduced by Kippenhahn and are of an acceptable order of magnitude. Velocities of a few m/s are sufficient to produce the observed rotation law. In Chapter 5 we will show that the velocity at the surface may even vanish completely.

The isolines of the angular velocity have a marked cylindrical symmetry because of the super-rotation (Fig. 2.4). Köhler also described the behavior of the angular velocity for fast circulation (or low viscosity), when the Coriolis force dominates friction and the angular momentum $\Omega\, r^2 \sin^2\theta$ becomes constant on the streamlines. Because its distance from the axis is large, the equator rotates very slowly, independently of any anisotropy.

Except for the assumption of constant density, Köhler's calculations contained the essential hydrodynamic ingredients of the heuristic Wasiutynski-ansatz. Further papers, such as those of Sakurai (1966) and Cocke (1967), completed the theory. Rüdiger (1974) then determined the viscosity tensor for simple classes of anisotropic turbulent fields. In particular, for reasons of simplicity, a horizontally two-dimensional turbulence was used as a model of an anisotropic field. Super-rotation resulted if an isotropic turbulence were superimposed on this two-dimensional motion. (The latter was wholly confined to planes perpendicular to the radial direction.) This agrees with Kippenhahn's findings: the lateral Austausch must dominate the radial Austausch.

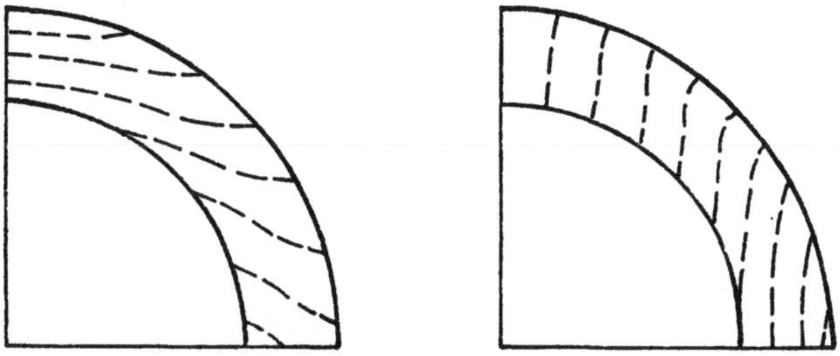

Fig. 2.4. Isolines of angular velocity in the convection zone. Right: cylindrical surfaces which were the rule in early theories of differential rotation. Left: disk-shaped isoplanes as are required, not only by dynamo theory, but also by modern theories of the solar differential rotation.

A final point remained to be settled. How important, in fact, was the assumption of constant density for these treatments? Köhler asserted that "it was no problem" to take into account the strong dependence of density on depth. From the hydrostatic equation

$$\varrho^{-1} \nabla p = \boldsymbol{g}^{\text{eff}} \equiv \boldsymbol{g} + \Omega^2 \boldsymbol{r} - (\boldsymbol{r}.\boldsymbol{\Omega})\,\boldsymbol{\Omega}, \tag{11}$$

one obtains

$$\nabla\times(\varrho^{-1}\nabla p) = \varrho^{-2}(\nabla p\times\nabla\varrho) = [\boldsymbol{g} + \Omega^2\boldsymbol{r} - (\boldsymbol{r}.\boldsymbol{\Omega})\boldsymbol{\Omega}]\times\nabla(\ln\varrho). \tag{12}$$

The ϕ-component of this expression represents a circulation which is called "baroclinic", this name being used whenever the isoplanes of pressure do not coincide with the isoplanes of density. All meridional circulations that do not originate from the non-conservative part of the centrifugal force are baroclinic. In consequence rigid rotation can drive meridional circulation as a "baroclinic flow".

Recently Moss and Vilhu (1983) have pointed out that these flows, which are formally similar to Eddington-Sweet circulations, are extremely sensitive to deviations from adiabaticity. Brief remarks on this question have also been made by Krause (1976) and by Mestel (1978). The latter summarizes the origins of meridional circulation starting from

$$-\varrho^{-1}\nabla p + \Omega^2\mathbf{s} + \nabla\psi, \tag{13}$$

where $s = r \sin\theta$ is the distance from the axis of rotation. The curl of this expression is

$$\varrho^{-2}(\nabla\varrho \times \nabla p) + s(d\Omega^2/\mathrm{d}z)\,\mathbf{1}_\phi. \tag{14}$$

Hence we have meridional circulation if (i) the isoplanes of pressure and density do not coincide and (ii) the rotation is not uniform. The baroclinic flow vanishes for constant density as well as for barotropic stratifications, $p = p(\varrho)$, but not for arbitrary dependencies $p(\varrho, T)$. Because of buoyancy, circulation arises even from "differential" density. A pole-equator difference in density in a rotating SCZ can only be determined, however, if the energy transport is considered simultaneously. The simplification used by Kippenhahn and Köhler ignores the circulations arising in non-adiabatic, non-polytropic stratifications where the real temperature gradient deviates from the adiabatic. This simplification decouples the system of hydro- and thermo-dynamic equations. In order to check the constraint for adiabaticity, attention has turned to the energy transport in rotating turbulent media. Yet, we still adhere to Zöllner's idea, that the observed differential rotation originates from a global flow. The causes for this streaming are becoming increasingly apparent.

In a short letter Weiss (1965) focussed on the influence of rotation on convective energy transport. Only the vertical component of the angular velocity vector, $\Omega\cos\theta$, would be relevant; the poles and equator would be affected differently. At the poles the super-adiabatic temperature gradient should be 3—4 times larger than at the equator, where the surface temperature would exceed that at the poles by 20 K. The corresponding circulation would be towards the equator, producing the correct rotational behavior. This is very similar to an argument put foward by Durney and Roxburgh (1971) who postulated a latitude-dependent turbulent thermal conductivity. This was related to the entropy gradient, ∇S, by

$$\boldsymbol{F}^{\mathrm{conv}} = -\varkappa T\nabla S, \tag{15}$$

with

$$\varkappa = \varkappa_T(1 + \varepsilon\cos^2\theta), \tag{16}$$

where ε denotes the normalized pole-equator difference in eddy conductivity and respresents the influence of rotation. With this ansatz they studied very thin rotating shells. For each hydro- and thermo-dynamic quantity a θ-dependence was found and, by a suitable choice of ε, the desired rotation law was produced. The equator turned out to be considerably warmer (by 70°K) than the poles, and the streaming velocity was found to be about

4 m/s. The turbulent Prandtl number, the ratio of the eddy viscosity to the turbulent conductivity, was not discussed in their paper.

Since thermodynamic relations had been introduced into their theory, additional boundary conditions were required. In particular, the significant, unwanted temperature differences at the solar surface required careful consideration. Durney and Roxburgh assumed that the energy flux at the lower boundary was both radial and independent of latitude. The same conditions were adopted by Belvedere and Paternó (1977b), whereas Spruit (1977), Durney (1981), Rüdiger (1982a) and Schmidt (1982) preferred to apply a radiation condition for the solar surface that incorporated the Stefan-Boltzmann law. This is, probably, justified, though there is still some lack of clarity about the lower boundary condition. The demand, which is used throughout, that all quantities be independent of θ, only prevents θ-dependencies from being smuggled into the theory via a boundary condition.

What is striking in the concept of anisotropic viscosity is the linearity of the Reynolds stresses in Ω. This peculiarity can only be justified if the period of rotation is long compared with the lifetime of the turbulent elements. The product of the angular velocity and the correlation time must be small. That this is really the case in the Sun was already questioned by Weiss in his short paper. Indeed, the mixing length theory provides lifetimes of 10—20 days for the deeper part of the SCZ.

The rotational terms in the turbulent energy transport are even functions of the angular velocity because "heat is not interested" in the sense in which the star rotates. Consequently, one must also ask whether the Λ-terms in the zonal momentum fluxes will depend on Ω^3. This is the question Iroshnikov addressed in 1969. He developed an approximation to higher orders in Ω, that is to faster rotations:

$$\begin{aligned} Q_{r\phi} &= \dots \sin\theta\, r\partial\Omega/\partial r + \dots \sin\theta\, \Omega + \dots \sin\theta \cos^2\theta\, \Omega^3, \\ Q_{\theta\phi} &= \dots \sin\theta\, \partial\Omega/\partial\theta + \dots \cos\theta \sin^2\theta\, \Omega^3. \end{aligned} \tag{17}$$

The last term in the second expansion was discussed in great detail but was not given explicitly. Comparing (17) with the angular momentum transport by the mean flow (see Chapter 4),

$$\bar{u}_r \bar{u}_\phi \sim \Omega \sin\theta \cos^2\theta, \tag{18}$$

one recognizes decisive similarities in the θ-dependencies. Since meridional circulations lead to latitude-dependent angular velocities, via the functions $\sin\theta\cos^2\theta$, it would not be surprising if the Ω^3-terms in (17) could do the same thing. Iroshnikov derived from these Reynolds stresses a rotation law of the form

$$\Omega \sim 1 + \tau_c^2 \Omega_\odot^2 \sin^2\theta. \tag{19}$$

He recognized that not only meridional circulation but also the Ω^3-effect (that is "fast" rotation) could produce an equatorial acceleration. He found super-rotation ($\partial\Omega/\partial r > 0$) and surface velocities of about 5—10 m/s.

Clearly, Iroshnikov had opened up a new area for investigation. It was no longer Zöllner's circulations alone that could produce the observed differential rotation: the Reynolds stresses could also do it, by themselves. Obviously, the correlation tensor Q_{ij} had to be investigated for rotating turbulence as carefully as possible. The necessary Reynolds stresses could, it was found, arise from the Λ-effect,

$$Q_{r\phi}^{(r)} = \Lambda_V \Omega \sin\theta \equiv \nu_T V \Omega \sin\theta ,$$
$$Q_{\theta\phi}^{(r)} = \Lambda_H \Omega \cos\theta \equiv \nu_T H \Omega \cos\theta , \tag{20}$$

even when the rotation is rigid. Here the superscript (r) on Q indicates that the tensor is evaluated for a rigid rotation. Clearly the structure of (20) suggests that "the Λ-effect is the α-effect of hydrodynamics". The quantities ν_T and Λ have the same physical dimensions, and we therefore expect them to have the same order of magnitude. For a rotation that is not too slow, the Λ itself has a θ-dependence of the form

$$V = \sum_{l=0} V^{(l)} \sin^{2l}\theta , \qquad H = \sum_{l=1} H^{(l)} \sin^{2l}\theta , \tag{21}$$

and a θ-dependent angular velocity of rotation is a direct consequence. Thus one finds, for the surface of a convection zone of fractional thickness d, the rotation law (Rüdiger, 1982b)

$$\Omega \sim \text{constant} + \frac{1}{2}\sum_{l=1} (\mathrm{d}\, V^{(l)} + l^{-l} H^{(l)}) \sin^{2l}\theta . \tag{22}$$

Low powers of $\sin\theta$ in Λ therefore produce low powers of $\sin\theta$ in the rotation law, whereas high powers in the former produce high powers in the latter, leading to zones of equatorial acceleration that are much too narrow. The observations only fit theories with $l = 1$, or at most $l = 2$, because terms of higher order than $\sin^4\theta$ do not occur with significant amplitude in the observed rotation law.

It can also be seen from (22) that any Λ-mode with non-vanishing l generates latitudinally-dependent differential rotation. Positive $V^{(1)}$ and $H^{(1)}$ directly lead to equatorial acceleration. The pole-equator difference in the angular velocity is proportional to the constants themselves, and in addition varies with the depth d of the convection zone. The fundamental constant $V^{(0)}$ does not appear in (22). It does not lead to latitudinally varying rotation, but it is the main factor determining the radial variation of the rotation, through the relation

$$d(\ln\Omega)/\mathrm{d}(\ln r) \simeq V^{(0)} .$$

Positive $V^{(0)}$ therefore leads to super-rotation, and negative $V^{(0)}$ to sub-rotation. The latter, for example, obviously corresponds to the case when the vertical Austausch (i. e. the coefficient relating stress to rate of strain) dominates or, equivalently, when Kippenhahn's s is less than unity.

Homogeneous isotropic turbulence contributes nothing to the Λ-effect. The first non-trivial result arises for two-dimensional turbulence in the horizontal plane: $V^{(0)}$ is then found to be necessarily positive, $V^{(1)}$ positive for small molecular viscosity (and negative for large viscosity or decreasing turbulence spectra) and $H^{(1)}$ zero (Rüdiger, 1977a). A two-dimensional model was chosen only because of its simple structure, not for any physical reasons (cf. Nickel, 1969). Later this restriction was lifted, and the form of the cells was allowed to be freely adjustable in the radial direction. This led to nonzero values for the coefficient $H^{(1)}$. When the turbulence has a more complicated form, $V^{(1)}$ and $H^{(1)}$ no longer have transparent expressions, but $V^{(0)}$ remains simple: it changes from positive to negative values only when the vertical intensity of the turbulence is strongly increased. In order to produce the observed differential rotation, the correlation time for the turbulence must, it has been estimated, be approximately half the rotation time, i. e. about 10 days. Durney (1983) coined the apt phrase "giant granulation" for the situation. A granulation of this kind is definitely not found in the upper layers of the convection zone, and one is faced with the question: how is it possible that angular momentum transport takes place mainly in the depths of the SCZ but differential rotation is still seen on the solar surface?

The majority of turbulence models have been, except for rotational effects, isotropic in the horizontal plane but, at the expense of complicating the formulae, this simplication too can be removed. The first authors to do this formally were Ruzmaikin and Vainshtein (1978), Durney and Spruit (1979) and Rüdiger (1980). Yet one has to be careful: in order to avoid singularities at the poles the coefficient $H^{(0)}$ must vanish there. In a certain sense Drobyshevski's (1977) investigation is also relevant in this context. He showed that, if two preferred directions exist in the turbulence field, the correlations $Q_{r\phi}$ and $Q_{\theta\phi}$ do not in general vanish.

For some special turbulence models, Gough (1978) determined the angular momentum coefficient $V^{(0)}$ which turned out to be, as was also shown by Durney and Spruit, negative everywhere. The fundamental mode $V^{(0)}$ was also calculated by Gailitis and Rüdiger (1982) for another class of models in radially stratified regions; again $V^{(0)}$ was found to be negative. Negative $V^{(0)}$ is equivalent to $A_\phi < A_r$ in (2.8). The consequence of $V^{(0)} < 0$ is that the vertical Austausch predominates and, contrary to Kippenhahn's premise, this leads to $\partial\Omega/\partial r < 0$ (sub-rotation). This conforms with both the observations (see Chapter 3) and the demands of dynamo theory. For sub-rotation the isoplanes of the angular velocity are rather more disk-like than cylindrical (Fig. 2.4).

2.3. Instability Theory: Convection and Rotation

Obviously, the concepts we have described require that assumptions be made about the structure of the prevailing turbulence. The question arises of how these assumptions should be influenced, if at all, by the instability arguments commonly applied to the Bénard problem. If these arguments have force, the theory of the solar convection zone would essentially require only information about the Sun's geometry and energy production. The pure Bénard problem deals with the origin of laminar flow patterns in a plane liquid layer which is heated at the bottom, cooled at the top, and is under the influence of gravity. As early as 1961, in the well-known book of Chandrasekhar, such a theory was expounded for rotating spheres of fluid, heated from within. According to the so-called Boussinesq approximation [which, incidentally was first used by Oberbeck (1879)], density deviations only appear in the buoyancy term. For certain regimes of flow (wave-number) the related Rayleigh number *Ra* (temperature difference) was determined for the steady marginal state ($\partial/\partial t = 0$). The flow pattern associated with the smallest possible (i. e. the "critical") Rayleigh number was chosen, this being considered to be the most probable regime of flow, the so-called "preferred mode".

At first, only axisymmetric ($\partial/\partial\phi = 0$) flows were studied. Bisshopp (1958), and Chandrasekhar also, calculated for each marginal mode the critical Rayleigh number as a function of the Taylor number, *Ta*, this being the appropriate non-dimensional measure of the given angular velocity. This critical Rayleigh number, which must be exceeded for the onset of convection, increases with increasing angular velocity. Rotation obviously has a stabilising, turbulence-suppressing effect, but when Roberts in 1968 investigated for the first time non-axisymmetric convection patterns in which variables were proportional to exp ($im\phi$), he unearthed a surprising result. He found that, for sufficiently fast rotations, the critical Rayleigh number *Ra* is a minimum when $m = 2$ and not $m = 0$ (Fig. 2.5). The relation between the critical Rayleigh number and the longitudinal wave-number, m, when $Ta = 10{,}000$ is, for rigid boundaries,

m	0	1	2	3
10^{-4} *Ra*	1.64	1.80	1.58	1.86

Paradoxical as it may sound, the motion most likely to develop is a non-axisymmetric one. *At first sight, it seems most unlikely that a radial buoyancy force, together with the axially symmetric effects of rotation, could together result in asymmetric motions.* This phenomenon is associated with the major topic of the formation of global structures (synergetics). An example often cited is the existence of Jacobi ellipsoids when masses of self-gravitating fluids rotate sufficiently rapidly. Magnetic stars, too, are in principle non-axisymmetric. The curvature of the spherical surface is an important factor in the preference for non-axisymmetric convection. At a plane surface, streams could emerge that are independent of the coordinate parallel to Ω, and be unaffected by rotation.

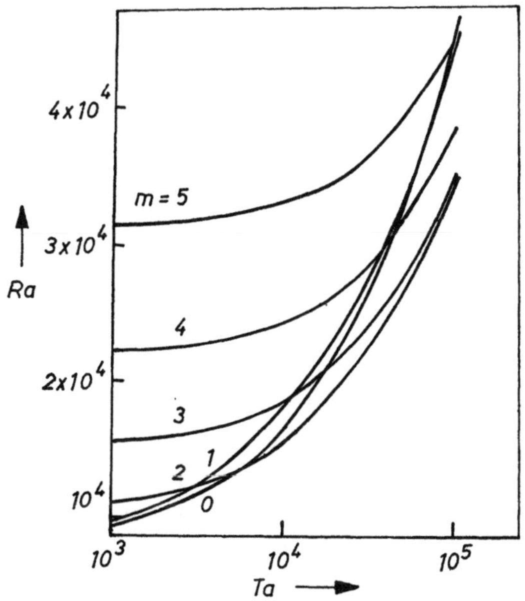

Fig. 2.5. The variation of the critical Rayleigh number with the Taylor number. For sufficiently fast rotations it is not axisymmetric convection ($m = 0$) but that with $m = 2$ which is the easiest to excite (Roberts, 1968).

Roberts's results influenced all subsequent investigations. In particular there were three important publications, published in 1970 and 1971, which should be discussed. Earlier results on axisymmetric convection had by then lost their significance. The Boussinesq approximation, however, continued to be used for thermal convection in rotating spheres. Busse (1970) found that the flow at the lowest Rayleigh number depended on the horizontal coordinates as $P_l^l(\theta) \exp(il\phi)$, i. e. it had identical longitudinal and latitudinal wave-numbers. The Legendre polynomial P_l^l is proportional to $\sin^l \theta$, i. e. it has a maximum at the equator. The eigenvalue l was about π/d ($\mathrm{d} = 1 - x_i$, $\varkappa_i$ being the normalized radius of the base of the convecting shell), i. e. about 10. The differential rotation is the axisymmetric part of the solution of the Reynolds equation, when averaged over ϕ. As expected, the resulting rotation law takes the form $\Omega \sim \sin^{2l} \theta$. According to observation,

l should be equal to 1 or 2, thus there are still discrepancies. In Busse's theory, as well as in similar theories by Durney (1970) and by Yoshimura and Kato (1971), differential rotation, or more precisely $\partial\Omega/\partial\theta$, depends linearly on the global angular velocity Ω. This is a pure form of the Lebedinski — Biermann effect, with convection existing mainly at the equator and therefore transporting more angular momentum there than elsewhere: inhomogeneous turbulence under the influence of slow rotation. Iroshnikov's higher orders of Ω, tantamount to faster rotation, were ignored. The duration of the convective elements, which in reality is very long, and the strong equatorial concentration of the resulting rotation law ($\sim \sin^{20}\theta$), do not favor these almost linear theories.

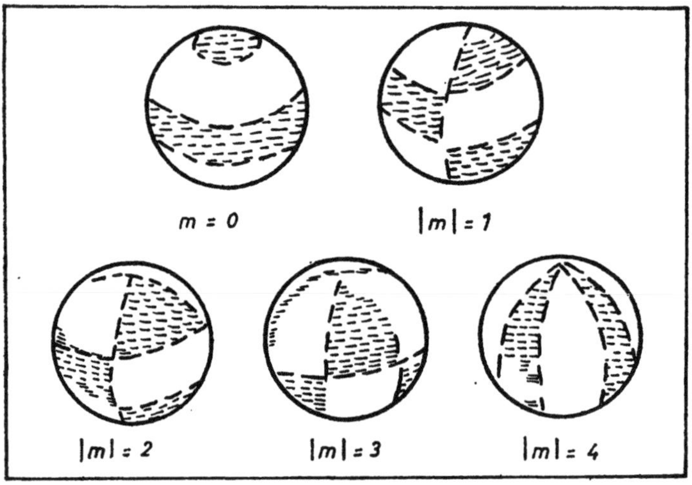

Fig. 2.6. Illustration of the spherical functions $Y_4m(\theta, \phi)$: $m = 0$ describes axial symmetry, $m = 4$ reflects the "banana (sectorial) structure" in convection theory; m defines the zonal structure (after Unno *et al.*, 1979).

In his paper of 1970, Durney also dealt with rotating convection, neglecting the nonlinear terms in the equations governing the fluctuations ("Herring's approximation"). He formulated an initial-value problem and found that over long periods only the sectorial mode $l = m = 10$ survives. The shape of the corresponding convection cells are reminiscent of bananas extending from pole to pole (Fig. 2.6). They transport angular momentum preferentially towards the equator, and it was therefore claimed that this explained the equatorial acceleration. But (in direct contradiction to the observations) the convective heat flux also had a pronounced maximum at the equator. Durney wrote: "If differential rotation is produced by the interaction of large convective cells with rotation, we would also expect that the pole-equator difference in temperature (or flux) would develop... both differential rotation and the pole-equator difference in temperature have their origins inside the convection zone, and as the local Taylor number becomes small near the surface, both effects are smoothed out. This smoothing is greater for ΔF than for the differential rotation because the Prandtl number is small," (see also Durney, 1971).

The sector ("banana") structure of the cells was not surprising. It reflected a correlation length hypertrophied through rotation in the North-South direction (N-S). This was exactly the substance of the theorem stated by G. I. Taylor in 1921. In the steady, slow motion of a rotating, inviscid uniform medium

$$(\boldsymbol{\Omega} \cdot \nabla)\, \boldsymbol{u} = 0. \tag{23}$$

The turbulence is thus two-dimensional, the flow does not depend on the coordinate parallel to $\boldsymbol{\Omega}$, i. e. on z. This Taylor-Proudman theorem (see Proudman, 1916) holds under real conditions only for a sufficiently rapid rotation. It guarantees that the large Coriolis force that then exists can be balanced by a pressure gradient: a turbulent flow of this form is uninfluenced by rotation. In experimental terms, the force of this theorem is manifest in the two main properties of rotating turbulence as stated by Tritton (1978):

(i) the components of the turbulence intensity are not decoupled,

(ii) rotation leads to a considerable increase in the correlation length parallel to the axis of rotation, as well as to a moderate increase in directions perpendicular to it.

Figure 1.2 of Greenspan's (1969) book is an impressive demonstration of the Taylor-Proudman effect.

Only a few years later Busse (1973) presented a more complete form of his theory. He considered a coupled system of equations for differential rotation, meridional circulation and heat transport. For the modes with

$$l = m = \pi \mathsf{d}/(1 - \mathsf{d}), \tag{24}$$

a horizontal scale much larger than supergranules, the Reynolds stress and turbulent heat flux were determined using the Boussinesq approximation. The equations lacked only the contributions made by the centrifugal force. From a desire to make the theory complete, in particular to include the balance between the Reynolds stresses and the meridional flow, the calculations were very complex. Explicit solutions were given for the velocity field, but not for the heat flux. The rotationally generated temperature differences at the boundaries of the SCZ were required to vanish. The inner surface was permitted to rotate either freely or rigidly, but in both cases the angular velocity proved to be a maximum at the equator. For rigid inner rotation, however, the angular velocity was a minimum, not at the poles, but at mid-latitudes.

The theory would have been acceptable had it been correct to assume that (i) such a theoretical structure for convection really corresponded to the giant cells and (ii) small scales may be summarily parameterised through an "eddy viscosity". Interesting too are Busse's remarks on the influence of the meridional circulation. He finds it to be unnecessary for maintaining differential rotation. Its influence is minimal in shallow zones with a stress-free lower surface, and (for small Prandtl numbers at least) it does not play any other role either. It is solely through the Reynolds stresses that the rotationally induced N-S direction of the convection cells produce the equatorial acceleration, though it should be noted that it is the shape (or rather the anisotropy) of the cells, and not merely their existence, that is the important factor. In the final analysis, it is the nonlinear interaction of large-scale convection near the equator that excites the differential rotation.

The Coriolis force has two principal effects on convection: suppression and deformation. They occur side by side. In high latitudes, suppression prevails, whereas near the equator it is deformation which dominates, as may be seen from the N-S alignment. Deformation seems to have been seen on the Sun as a N-S coherence of motions. Suppression has not yet been observationally detected. Yoshimura and Kato (1971) considered motions of very large horizontal dimensions ($K_r \gg K_\phi$). In applying this assumption to the observed data, they were the first to refer to the observational results of Ward (1965) who had discovered a systematic correlation between the longitudinal and latitudinal proper motions of sunspot groups. He had found that the more rapidly rotating groups moved towards the equator while the more slowly rotating groups moved towards the poles. The cross-correlation, $Q_{\theta\phi}$, in mid-latitudes amounted to about 0.1 (deg/day)^2 (see Chapter 3). Ward spoke of global whirls or waves carrying angular momentum to the equator. This would directly lead to an increased equatorial angular velocity. To think, he along with other authors, almost believed that the overall problem of "differential rotation" had been solved. Each Newtonian fluid element exhibited a finite horizontal cross-correlation for a non-uniform rotation; but, in a case where the equator rotated more rapidly than high latitudes, the correlation had a *negative* sign. Apart from an unpublished criticism by Leighton, Ward's findings have never been questioned. In the present book we show that they have as great a significance for the theory of the SCZ as the differential rotation itself.

To return to Yoshimura and Kato, these authors compared the results of their Boussinesq approximation, again with $l = m$, with the Ward-profile for horizontal motions. The resulting correlation, $Q_{\theta\phi}$, was presented separately for the cases of rigid and differential rotation. This provided an opportunity to estimate the eddy viscosity. It turned out that it hardly influenced the results at all: the correlations were only slightly reduced. The main contribution came from the Λ_H-effect (Fig. 2.7). As m was increased the correlation decreased, and the smaller scales transported progressively less angular momentum. The numerical estimates used the values $u' = 5 \cdot 10^3$ cm/s and $\nu_T = \chi_T = 5 \cdot 10^{12}$ cm^2/s. This velocity was that found by

Bumba (1967) for giant cells whose horizontal dimension is substantially larger than that of the supergranules. There were good reasons now for assuming that m was substantially less than Durney's value of 10: m might be 2 or 3. The procedure for calculating the influence of rotation on convection/turbulence has hardly changed since that time: small-scale motions are still parameterised by an enhanced "viscosity", while larger-scale modes are treated explicitly, either numerically or analytically, with $m = 0.1, \dots 20$ (cf. Stix, 1983).

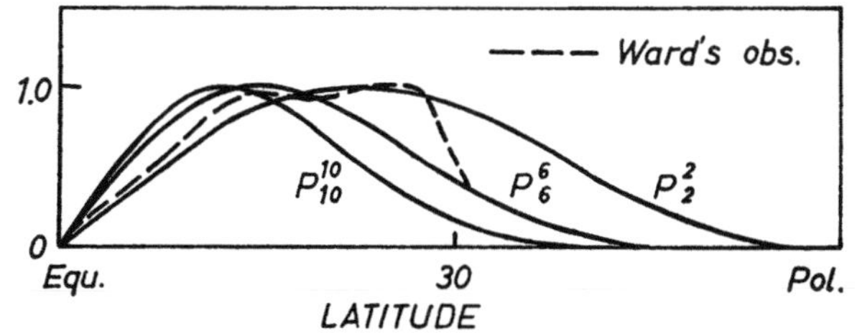

Fig. 2.7. The latitudinal profile of the cross-correlation $Q_{\theta\phi}$ in the theory of "slow rotation, inhomogeneous turbulence" after Yoshimura and Kato (1971). The broken line shows the observational results of F. Ward (1965).

A fundamental difficulty is to be found in all papers on "convection and rotation" up to the present day. The characteristic time of 10^7 s (about 100 days), based on the velocity of $5 \cdot 10^3$ cm/s given above and on the solar radius, is very long compared with the rotation period of the Sun. Or, to put it another way, the angular velocity of the global rotation is not small compared with the "eddy frequency", $2\pi/\tau_c$. (The other possible time scale, R^2/ν_T, is even larger.) This means that the influence of rotation on convection surely *cannot* be described by a theory that is linear in Ω. This objection to the most widely used method is in fact more serious than had been originally anticipated. Numerical investigations of a rotating shallow SCZ at small Taylor numbers, i. e. at slow rotation, do *not* lead to a θ-dependent angular velocity (Gilman and Foukal, 1979). An analytical theory linear in Ω can never provide a differential rotation without the help of a meridional circulation (Kippenhahn, 1963; Rüdiger, 1983a). To put it more forcefully: the fundamental discovery that the observed differential rotation reflects the strong, nonlinear influence of rotation on the solar convection zone is concealed by the theories under discussion. Irrespective of their pioneering nature and background, it is possible to doubt whether such theories are relevant to the Sun.

A big step towards overcoming the drawbacks of the Boussinesq approximation was taken by Vandakurov in 1975. With the inclusion of finite dissipation, he determined the convective motions under the two assumptions that (i) the temperature stratification is only weakly nonadiabatic, and (ii) the modes grow only slowly. Arbitrary density stratifications were allowed, and the rotation was assumed to be slow. Vandakurov determined the Reynolds stresses and referred to an "azimuthally averaged azimuthal force". This was calculated both for rigid and differential rotation. Therefore, in addition to the Λ-term, an eddy viscosity was also present. This turned out to be negative and perhaps in consequence the author refrained from addressing the question of whether the pole or the equator should rotate faster in his theory.

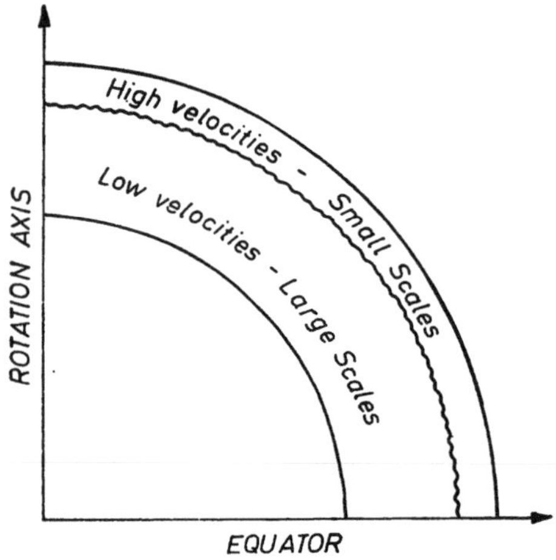

Fig. 2.8. Schematic structure of the SCZ. Near the surface there is small-scale, almost isotropic convection (granulation, etc.) which is only weakly influenced by rotation ("viscosity"). In the deeper layers there is assumed to be large-scale, anisotropic convection which is strongly influenced by rotation.

It is convenient to say something here about negative eddy viscosities. In 1974 Krause and Rüdiger, and in 1976 Kraichnan also but by a different method, calculated turbulent viscosities in "two and three dimensions". For 3D-fields the viscosity was always positive, as had been anticipated. For 2D-turbulence, however, Kraichnan obtained a negative answer, while Krause and Rüdiger determined a value that could in some circumstances be negative, but only exceptionally. Today we are rather inclined to think that for a real, naturally-occurring turbulence, one can never speak of an effective viscosity that is negative, and we shall consider it here only in one exotic case.

Further progress on the instability theory of rotating spherical shells could only be made using numerical methods. First of all we must mention the extensive work of Gilman, and the more recent contribution of Glatzmaier, on differential rotation, and of Hathaway (1984) and Hathaway *et al.*

(1980) on the Λ-effect. Nor must we neglect Gilman's (1983) substantial discussion of his own and Glatzmaier's models. (To Gilman we also owe the impessive Figures 2.8 and 2.9.) He writes,

"Let me give a few more detailed results from the latest model calculations for global convection and differential rotation. These results are based on two models: my own, and that of Glatzmaier. The physics of these models is rather similar, but the mathematical solution techniques are different. Both models are compressible, and typically contain several pressure scale heights. Both calculations begin significantly below the photosphere (10% of the solar radius) because they are incapable

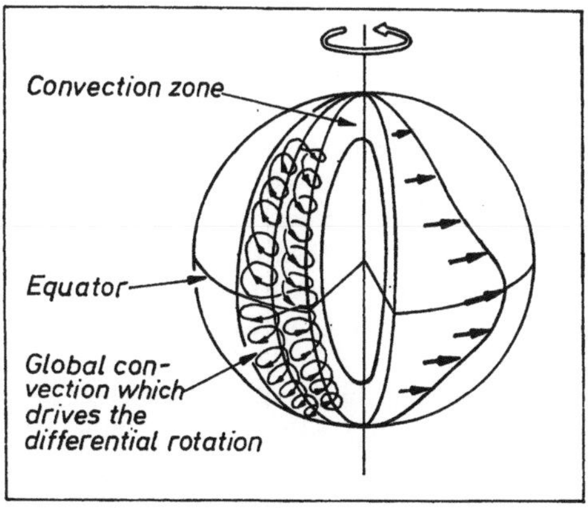

Fig. 2.9. Representation of global convection and differential rotation after Gilman (1981). Left-handed helicity is preferred. The axis of the helix is not coplanar with the rotation and corresponding equatorial axes.

of resolving supergranule scale convection. Both parameterize the effects of small scale (subgrid scale) eddy transports of momentum and energy quite crudely. Among the physical differences between the models are the treatment of the lower boundary. My model has a rigid lower boundary above which the stratification is all convectively unstable. Glatzmaier's model also uses a spherical harmonic expansion in the horizontal dimension that allows better treatment of the poles than does mine, which uses a Fourier expansion in longitude, coupled with a finite difference grid in latitude and radius.

"The differences between these models seem to be relatively unimportant in that the main results are quite similar. The fundamental conclusion from calculations using both these models of relevance to the Sun is that, starting from a solar luminosity, a solar rotation rate, and plausible, though crude, representation of the eddy diffusivities of momentum and entropy, they generate a broad spectrum of convection that in turn drives an equatorial acceleration of about the correct magnitude and profile. When one looks in a little more detail at the solution, some departures from solar observations do show up, but they do not negate the main agreement, suggesting we may be not too far from the 'correct' answer for the Sun.

"A number of additional results from these models should also be mentioned. For both, the angular velocity decreases with depth and is roughly constant on cylinders concentric with the axis of rotation. Convection in low latitudes is also predominantly in the form of convective rolls with a north-south axis. These results are both quite similar to that of the incompressible case studied earlier in a number of papers by me [see, e. g. Gilman (1979), (1980), and earlier references cited therein], and arise from the strong influence of rotation upon the flow.

"In general, the spectrum of convection (measured, say, by the kinetic energy in each longitudinal wave number m) is quite broad. In my calculations, the kinetic energy drops by a factor of 10 compared to the spectrum peak (somewhere in the range $m = 10$ to 15), by $m = 30$ to 40. At certain times, however, there are exceptions, when a single longitudinal wave number contains a substantial fraction of the total convective kinetic energy. Glatzmaier found a case when wave number 14 contained 34% of the kinetic energy of the convection for a period of 2 months. No other wave number had more than 10% of the energy...

"Two possible discrepancies with solar observations that show up are that the amplitude of the horizontal flow in the convection near the top of the model convection zone (but still 10% below the surface) is a factor of 2—3 larger than the upper limits estimated from the Mt. Wilson Doppler velocity measurements by LaBonte *et al.* (1981). If both observations and theory are correct, then the global convection must be attenuated by the factor of 2—3 in the upper most 10% of the convection zone. Simple estimates of the amount of this attenuation to be expected (Stix, 1981) are not encouraging.

"Another difficulty is that we get better agreement in detail with the solar differential rotation if the convection zone is deep, extending, either as an unstable layer (my model), or through penetration (Glatzmaier model) to a depth of 40% or so of the radius. Shallower layers give a narrower equatorial acceleration, and some tendency to spin up the poles (my model). A convection zone depth or a depth of convective mixing of 40% conflict with estimates of burning of Li and Be (Vauclair *et al.*, 1978) and with observed five minute oscillation frequencies." (Gilman, 1983).

Evidence for the large-scale N-S cylindrical convection structures he mentions have not yet been convincingly provided. La Bonte *et al.* (1981) rule out velocities above 12 m/s. Cram *et al.* (1983) made a search in higher latitudes and found evidence for areas with dimensions 10^{10} cm within which the velocities were about 25 m/s. The focal point in all the theories is the variation of angular velocity with depth. In most cases equatorial acceleration could be achieved only at the expense of a simultaneous superrotation. The isoplanes $\Omega =$ constant are then cylindrical. Unfortunately, the radial rotation law can only be measured indirectly, though we now know three independent methods by which it may be inferred. Radial differential rotation is manifested in:

(i) the varying rotation of the photosphere (Doppler observations) compared with the probably more deep-seated "tracers" such as sunspots,
(ii) the behavior of the migration of the toroidal activity belt of the solar dynamo,
(iii) the mode structure of the 5-min oscillation (Deubner, 1975).

All these require sub-rotation ($\partial\Omega/\partial r < 0$) in the SCZ, which means not cylindrical but disk-shaped isoplanes. Of course, arguments (i) and (iii) are of uncertain weight. Perhaps the dynamo argument is the most convincing. All in all, sub-rotation is favored: an inwardly increasing angular velocity. Later on we shall include "$\partial\Omega/\partial r < 0$" at the surface as an added "observation" to build into the theory of differential rotation.

The depth of the SCZ is a second, important quantity that can only be measured indirectly. Similar statements are made here but only on the basis of mixing length theory. The "20%" of the old days has developed into 30% and this value must be regarded as the upper limit if one does not want to conflict with other observations. The solar Li abundance can only be understood if the temperature at the base of the SCZ is at most $2.5 \cdot 10^6$ K. Hence it cannot be too deep.

2.4. Determination of Λ: Plane Layer and Oblique Rotation

Can one say that, irrespective of the shortcomings of mean field theory, irrespective of rough approximations such as the second order correlation approximation (SOCA), the existence of the Λ-effect has been irrefutably established? Are reliable quantitative statements possible? Following the numerical investigations of Hathaway *et al.* (1979, 1980), the answers to both these questions appears to be in the affirmative. They examined, numerically and in depth, convection in a rotating plane layer, even in cases where gravity did not coincide with the axis of rotation. Reference to such "oblique" configurations was made by Chandrasekhar as early as 1961 in his celebrated book, but the developments we require were not undertaken at that time. Recalling Biermann's remark quoted in § 2.2, it is quite sufficient for our purposes to consider thermal instability in a layer that is heated from below, cooled from above, and inclined to the axis of rotation (Figure 2.10), and it was precisely this model that was investigated by Hathaway *et al.*

The model studied in their first paper consisted of a rigidly rotating layer of *ideal* Boussinesq fluid. When the rotation was sufficiently rapid, roll-shaped structures of high horizontal wave-number and with axes in the N-S direction, i. e. parallel to $\mathbf{\Omega}$, were the fastest to grow. This indicated that these modes are preferred, and may be expected to emerge from an evolving convection pattern. This result fits in well with the "banana" structures derived from the $m = l$ calculations. The preponderance of N-S rolls applies at low latitudes whereas, near the poles, the differences between the growth rates of modes of different horizontal structure (but the same wave-number) vanish, as they should. The zonal momentum fluxes, $Q_{r\phi}$ and $Q_{\theta\phi}$, could also be determined from these solutions. Since, apart from the convection, the layer is in rigid rotation, these correlations refer directly to

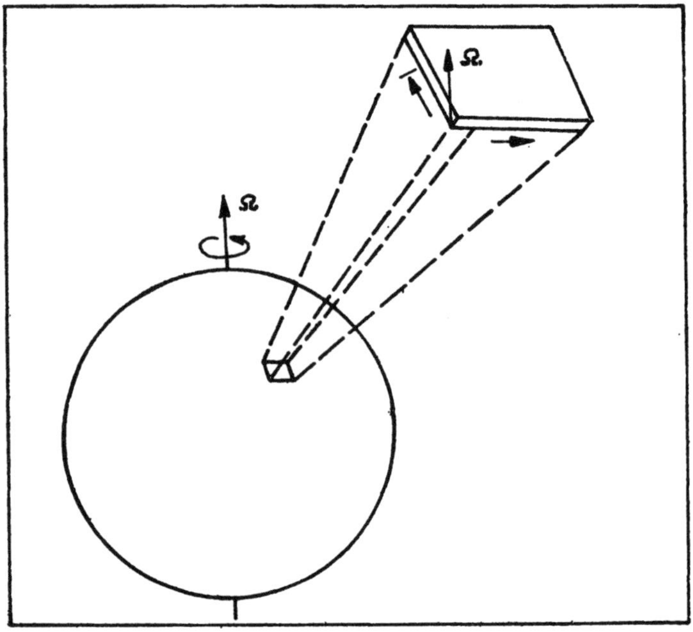

Fig. 2.10. The geometry used in the papers by Hathaway *et al.* The flow takes place in a plane layers. The associated rotation vector has two nonvanishing components.

the Λ-effect. For N-S rolls it follows that

$$\Lambda_V = 0, \qquad \Lambda_H \sim \sigma\tau_c/K^2, \tag{25}$$

where $\tau_c = (\alpha g|\mathrm{d}T/\mathrm{d}z|)^{1/2}$ is the turnover time and σ the dimensionless growth rate of the mode. The turbulent angular momentum transport is directed towards the equator and vanishes for large K.

In the following year Hathaway *et al.* (1980) repeated calculations for real, viscous and thermally-conducting gases. A Prandtl number of unity was used, the surfaces were considered rigid, and the mean density was taken to be constant. The model was now being used only to determine which convective modes are most easily excited. The Rayleigh number, temperature gradient, Taylor number, Prandtl number and latitude appeared parametrically, and the flow was characterised by the horizontal wave-numbers k and l. The corresponding growth rates were determined, and the modes with the largest growth rate were regarded as preferred. Unlike their earlier model employing ideal fluids, the disturbances of small scale now have only a small growth rate. Wave-numbers of between 5 and 10 developed fastest. Over the whole range of Taylor and Rayleigh numbers, the N-S rolls dominated the E-W rolls at $\theta = 45°$. The effective measure of rotation proved to be not so much the Taylor number as the dimensionless quantity

$$\Omega^* = 2\tau_c\Omega\,. \tag{26}$$

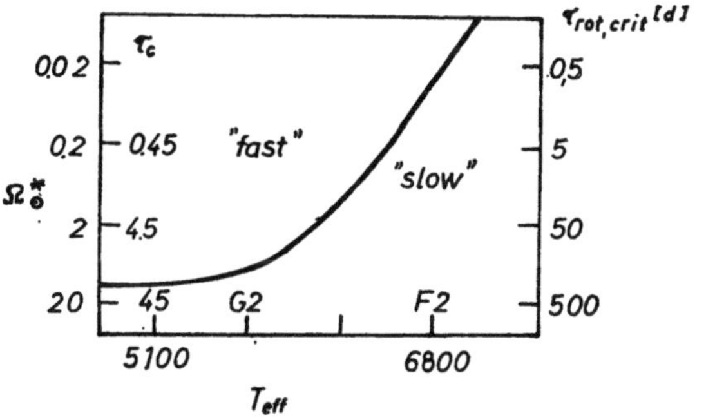

Fig. 2.11. Turnover times $\tau = L/u'$[d] near the bottom of the convection zones along the main-sequence (after Durney and Latour, 1978), where $\Omega^* = 2\tau_c\Omega$, and $\tau_{\text{rot, crit}}$ is the rotation period for which $\Omega^* = 1$. The actual solar period lies on the "fast" side of the line which separates states where rotation influences the turbulence strongly from states where that influence is weak.

$\Omega^* \gg 1 (\ll 1)$ implies that the influence of rotation is strong (weak). It is worth recording that Ω^* is of order 10^{-4} for granules, 10^{-1} for supergranules, and between 1 and 10 for giant cells. One therefore expects the supergranulation to be weakly influenced by rotation in the upper SCZ, but strongly influenced in the lower. Using mixing length calculations, Durney and Latour (1978) estimated the turnover time, $\tau_c = L/u'$, at a distance of half a mixing length from the base of the zone, to be about 14 days, corresponding to $\Omega^* = 6$ (Figure 2.11).

The Hathaway *et al.* publication of 1980 very impressively depicted the growth rates at various latitudes (Figure 2.12). Whereas near the poles ($\theta = 15°$) hardly any difference existed between the different convection

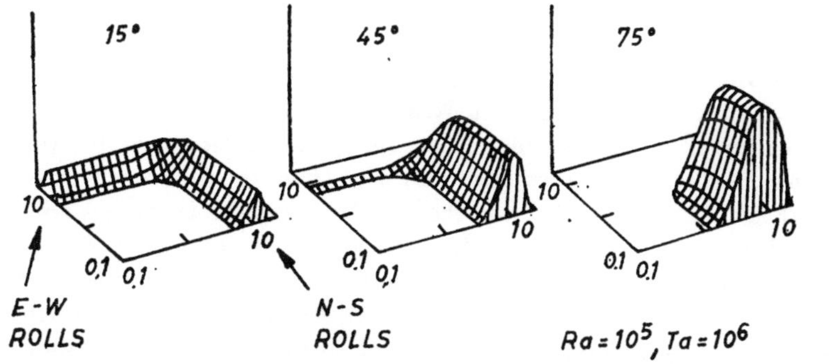

Fig. 2.12. The growth rates for convection in rotating plane layers at varying distances from the poles. Near the equator (RHS), rolls with axes in the N-S direction are clearly preferred whereas, near the poles (LHS), almost no differences exist. The results of Hathaway, Toomre and Gilman have a high degree of generality. They can also be interpreted as a demonstration of the Λ-effect.

patterns, the N-S rolls are definitely preferred near the equator ($\theta = 75°$). The inference seems to be clear: with the exception of the poles, roll structures form with axes parallel to $\mathbf{\Omega}$. The turbulent momentum transport is directed towards the equator, Λ_H is positive, and Λ_V vanishes. The effect of the very strong density gradient of the Sun remains open. In 1983, Hathaway and Somerville established a three-dimensional, time-dependent model, with 38,400 grid points, in order to investigate the influence of rotation on convection. They still neglected surface curvature, and density gradient, retaining the Boussinesq approximation. Again they assumed rigid boundaries, and a Prandtl number of one; they used a Rayleigh number of 10,000.

Table II.ii. Numerical results for the zonal fluxes of angular momentum at $\theta = 75°$ (Hathaway and Somerville, 1983).

Ta	Ω^*	Rotation	$Q_{r\phi}/(u_r u_\phi)$	V	$Q_{\theta\phi}/(u_\theta u_\phi)$	H
10^4	1	modest	−0.3	−0.62	0.16	1.2
10^5	3	fast	−0.015	−0.01	0.80	2.0

Table II.iii. Radial Λ-effect at the equator after Busse and Cuong (1977)

$d = 1 - x_i$:		0.4			1	
x	0.6	0.8	1	0.2	0.6	1
$Q_{r\phi}$	0.7	−0.5	1	−0.8	−1	−1

Particularly important to them was the change in the inclination of the axis of rotation necessary to excite various patterns of convection. Because of the advective terms, $u'u'$ and $u'T'$, the equations they solved were strongly nonlinear.

Their results for the critical Rayleigh numbers and for the correlations $Q_{r\phi}$ and $Q_{\theta\phi}$ are of dominating interest. Near the equator the N-S rolls again proved to be maximally unstable. Near the poles distinctions between North, South, East and West disappear and with them the differences in the critical Rayleigh numbers. Only near the equator were the turbulent zonal momentum fluxes calculated (Table II.ii; the velocities are r. m. s. values).

Clearly, the flux of angular momentum is directed inwards and towards the equator. Nevertheless, such inwardly directed and subrotation-producing radial flux arises very rarely, the most recent instance being Hathaway's (1982) paper. The great programs for studying rotating spherical shells (Gilman, starting in 1977; Busse and Cuong, 1977) often produce an outwardly directed flux, $Q_{r\phi} > 0$. Table II.iii gives normalized values at the equator, using the model of Busse and Cuong. The first row of the table refers first to a shell occupying the outer 40% of the sphere, and then to the full sphere. One sees here negative values of $Q_{r\phi}$ at the equator. The authors state: "The s-component of the angular momentum flux is mostly negative with positive values near the outer boundary". (Here s is the distance from the polar axis.) They also discovered the way in which "polar" and "equatorial" regions actually differ in the context of convection theory. Polar volumes are those with $s < x_i$, and equatorial ones are those with $s > x_i$. While in polar regions the motion resembles that of convection in a plane layer rotating about a vertical axis, the motion becomes columnar in equatorial regions, resembling convection in a plane layer rotating about a horizontal axis.

The results of Hathaway and Somerville summarized in Table II.ii contain information beyond merely that of sign. As the angular velocity increases, $Q_{r\phi}$ decreases and $Q_{\theta\phi}$ increases. To estimate the dimensionless fluxes V and H we use (20), (26) and $\nu_T \sim \tau_c(\overline{u'^2_r}\,\overline{u'^2_\phi})^{1/2}$, from which it immediately follows that

$$\begin{aligned} V &= 2Q_{r\phi}/(\overline{u'^2_r}\,\overline{u'^2_\phi})^{1/2}\Omega^* \sin\theta\,, \\ H &= 2Q_{\theta\phi}/(\overline{u'^2_r}\,\overline{u'^2_\phi})^{1/2}\Omega^* \cos\theta\,. \end{aligned} \tag{27}$$

Whether one may really set $\overline{u'^2_r} = \overline{u'^2_\theta}$, cannot be judged at this stage. The values obtained from (27) are also listed in Table II.ii. They are of order unity. Unfortunately, the published details are insufficient to allow us to determine $V^{(0)}$, $V^{(1)}$ and $H^{(1)}$ more accurately. This information is necessary before conclusions about the resulting rotation laws can be reached. Nor do the mean flows, as determined by the authors, offer any further help in this respect. In the upper part of the layer the mean flow is directed from North-West to South-East, while in the lower part it is in the opposite direction. Sub-rotation is easily seen despite the lack of curvature of the boundaries. Fast rotation drastically restricts the flow. To be useful in our context, it would be necessary to repeat the calculations of Hathaway and Somerville with different inclinations of the axis of rotation. Busse (1982) added some specific remarks concerning the type of instability that they found. He analyzed their unexpected finding that, in a layer of constant height, mean zonal flows are generated. Their origin must lie in the inclination of the rotation axis in conjunction with a special type of instability.

Chapter 3. Current Observations

Imagine there is no heaven ...
above us only sky.
J. Lennon, 1971

In this Chapter we will give an account of modern observations that bear on our topic. Our aim is not to survey the entire literature, and readers wishing to cover the field more thouroughly should refer to the excellent reviews of Howard (1978, 1984), Paternò (1978), Beckers (1981), Gilman (1985) and Schröter (1985). Indeed a sentence or two from these surveys may even appear in our account of the subject.

3.1. Convective Modes

The photosphere, defined as the visible surface of the Sun, is extremely thin. Defining its upper surface as the level of minimum temperature and its lower surface to be at an optical depth, τ, of unity, one finds that its thickness is only 500 km. It is a stable region, in contrast to the zone beneath it, where the prevailing temperature gradient exceeds the adiabatic gradient: $-\mathrm{d}T/\mathrm{d}r > g/C_p$. Only through the overshooting of the motions into the photosphere can the velocity pattern at the top of the convection zone be seen, more or less exactly. Our knowledge of the SCZ relies on mixing length calculations. Recent models, which are frequently used and even more frequently quoted, are those of Baker and Temesváry (1966), Spruit (1974) and Staude (1977). They paint a consistent picture of the SCZ. Characteristic of this consistency are the similar values they give of the density and temperature at the top and bottom of the zone. The values of the main variables are summarized in the Tables III.i and III.ii. (Unless other-

Table III.i. Surface values in SCZ models.

	T_e	$\varrho_\odot$	$T_\odot$	u'	L	Lu'	L/u' (min)
Baker and Temesvàry	5783	$3.9 \cdot 10^{-7}$	6237	$2.3 \cdot 10^5$	$2.9 \cdot 10^7$	$7 \cdot 10^{12}$	2.1
Spruit	5770	$3.3 \cdot 10^{-7}$	6390	$1.1 \cdot 10^5$	$1.3 \cdot 10^7$	$1 \cdot 10^{12}$	2.0
Staude	5780	$8.3 \cdot 10^{-7}$	6442	$3.0 \cdot 10^5$	$1.6 \cdot 10^7$	$5 \cdot 10^{12}$	0.9

wise stated, all values in this book are given in c.g.s. units, and the intensity of the turbulence is always given as an r.m.s. velocity.) In these tables T_e is the effective temperature, $\varrho_\odot$ and $T_\odot$ are the density and temperature at the top of the convection zone, and u' and L refer to the depth at which u' is a maximum. The turnover time, L/u', proves to be very small, in contrast to the diffusivity Lu' which is large.

Table III.ii. Values at the base of the SCZ.

	x_i	D(km)	ϱ_i	T_i	u_i	L_i	$L_i u_i'$	L_i/u' (days)
Baker and Temesvàry	0.72	190,000	0.21	$2.1 \cdot 10^6$	2000	$1.5 \cdot 10^{10}$	$3.0 \cdot 10^{13}$	87
Spruit	0.71	198,000	0.20	$2.2 \cdot 10^6$	2500	$1.0 \cdot 10^{10}$	$2.4 \cdot 10^{13}$	46
Staude	0.72	193,000	0.09	$1.7 \cdot 10^6$	2100	$1.9 \cdot 10^{10}$	$4.0 \cdot 10^{13}$	100

At the bottom of the SCZ the interesting quantities are given in Table III.ii, in which D is the thickness of the convecting domain. It should be noted that the diffusivities, Lu', do not differ greatly from the value given in Table III.i (cf. Fig. 2.1). This is because the mixing length rises strongly with depth. The turnover time increases inwards dramatically, and hence the Coriolis parameter Ω^*, defined in (2.26), takes values near 20 (cf. Fig. 2.11). At the top of the SCZ, however, Ω^* has a very small value. Thus, with respect to the angular momentum transport, the diffusion parameter, Lu', remains almost constant while non-diffusive effects ($\sim\Omega^*$) are potent only in the deeper parts of the SCZ. This conclusion should be relevant to any theoretical discussion of the mean-field hydrodynamics of the Sun.

At a depth of 1000 km the models have already become nearly adiabatic, i.e. the real temperature gradient is almost the same as the adiabatic gradient. A slight difference between them suffices to produce the necessary convective energy transport because the diffusion coefficient, Lu', is large. Only the upper surfaces of the convective cells can be seen in the form of granules. At any time there are millions of these. They typically have a diameter of 1000 km = 1 Mm. Their lifetime is about 8 min and the mean distance between the centers of adjacent cells is nearly 1500 km. Individual granules may live for 30 minutes which, however, is still too short for them to be of use as tracers for rotation observations. Their spatial configuration can best be depicted by the intensity power spectrum (Fig. 3.1). This shows a maximum at a wave-number of 3.5 Mm^{-1} and even a fine structure, with three peaks. Brightness and velocity are strongly correlated: the hot granules rise and the cool intergranules sink. The eddy heat flux, $\overline{u'_r T'}$, is therefore positive. We are mainly interested in the observed velocities and we later present these together with other numerical values in Table III.iv. As it is a small-scale phenomenon, granulation can be expected to be rather isotropic. The r.m.s. vertical and horizontal velocities may therefore be roughly equal.

Quite a different situation arises for the supergranulation which are associated with the tops of large convective cells. Since no significant horizontal temperature difference corresponds to these, supergranulation is nearly invisible in white light. It is mainly seen through its large-scale velocity pattern. The cells are irregular in shape with a mean diameter of about 32,000 km and are thus slightly larger than active regions. Material rises at the centers of the cells, moves horizontally outwards, and sinks at their edges. Typical velocities are some hundreds of m/s, i.e., higher than those in strong gales on Earth. The depth of the cells may be 10,000 km. They

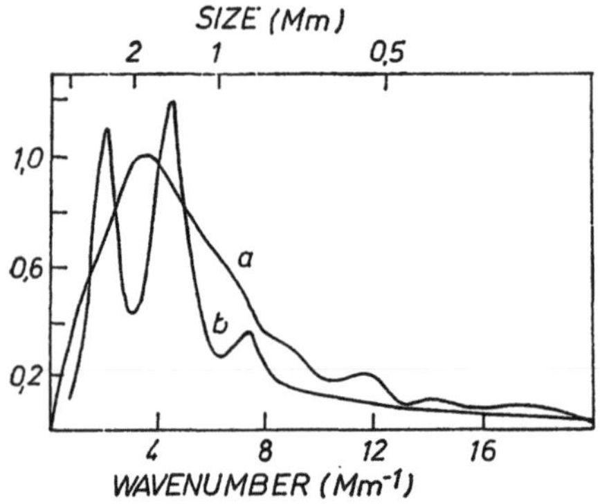

Fig. 3.1. Power spectrum of granular brightness fluctuations (after Wittmann, 1979): (a) curve combining several published spectra, (b) recent observations which even reveal a fine structure.

have lifetimes of 1—2 days though some of them can be seen for several days longer. Although essentially convective, supergranulation appears to produce no noticible variations in heat-flux. We could continue by reporting on further properties of supergranulation and (say) magnetism. The reader is, however, referred to the reviews mentioned above, which should also be consulted for details about cellular mesogranulation whose scale is intermediate between granulation and supergranulation. This mode is characterized by small velocities, only about 60 m/s.

We are especially interested in the "giant cells" which were predicted in 1968 by Simon and Weiss. It was suggested that giant cells extend across the entire depth of the SCZ and should therefore have large horizontal scales also. If this is correct they should create large-scale modulations in the solar rotation. Analyses have generally produced similar results (Table III.iii). It may be worth commenting on Kuhn's result. His values follow from an analysis of the rotation signal based on features having lifetimes of

Table III.iii. Observed parameters for giant cells.

	u(m/s)	horizontal scale (Mm)
Kubicela and Karabin (1977)	32	
Beckers (1978)	35	300
Howard (1979)	40	200
Kuhn (1983)	2	100–1000

several days. In the resulting diagnostic diagram, he did not find any large-scale structures at low frequencies.

From their Ca II network observations, Schröter and Wöhl (1976) deduced the existence of chromospheric cells with a longitudinal extension of 1000 Mm and a flow velocity of 80 m/s. The velocity pattern was antisymmetric with respect to the equator (Fig. 3.2). Subsequently, further

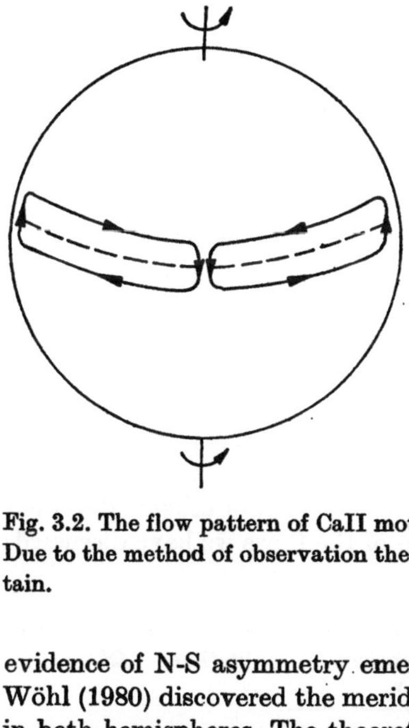

Fig. 3.2. The flow pattern of CaII mottles as derived for 1975 by Schröter and Wöhl. Due to the method of observation the latitudinal dimension of the giant cell is uncertain.

evidence of N-S asymmetry emerged, as for example when Balthasar and Wöhl (1980) discovered the meridional motion of sunspots to be southwards in both hemispheres. The theoretical implications of such findings were to be of dramatic importance. They indicated that the angular polynomials in θ relevant to the theory had precisely the opposite parity to those that had usually been employed. We may, for the moment, defer consideration of such equatorially-asymmetric meridional motion.

There have been suggestions that large-scale patterns of magnetic field are related to giant cells (Bumba, 1967). La Bonte *et al.* (1981), however,

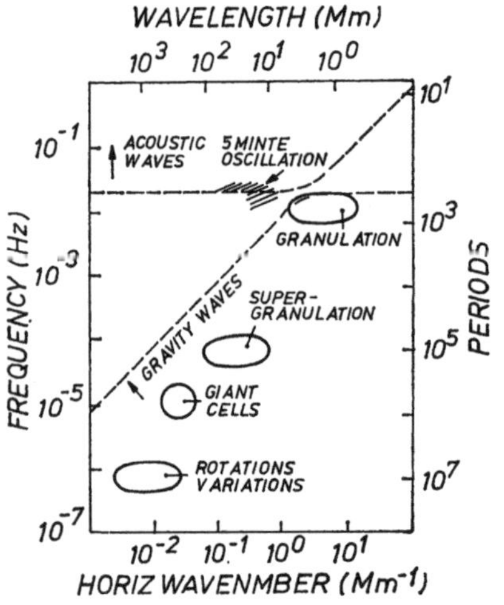

Fig. 3.3. Diagnostic diagram of convective modes (Beckers, 1981). All the modes lie in the "gravity region", i.e. buoyancy dominates pressure in the balance of forces.

were not successful in isolating any long-lived large-scale patterns. Cram *et al.* (1983) observed patterns of about 100 Mm which in the polar region rotated with a period of about 30 days (as compared with 35 days for the polar supergranulation). They could not state whether this new mode belonged to a "giant granulation" or to the tail of the supergranules distribution. La Bonte *et al.* (1981) were unable to detect cells with velocities greater than 3–12 m/s.

Even a cursory inspection of the data leads to the conclusion that we are a long way from the truth about slow and large-scale motions. This is unsatisfying insofar as the largest modes naturally (i) are the most anisotropic ones and (ii) possess the largest Ω^*: Both features are important to the theory of the differential rotation. The present state of knowledge about giant cells is, however, insufficient to form the basis of a theory. Beckers (1981) has produced an impressive diagnostic diagram of the flows we have discussed;

Table III.iv. The solar velocity modes.

mode	velocity (m/s) (peak to peak)		dimensions (Mm)		lifetime	Ω^*
	horiz.	vert.	horiz.	vert.		
Merid. flow	30					
Giant cells	40		100	200	0.5 yr	10^2
Supergranulation	200	40	32	10	1.5 day	0.6
Granulation	1300	650	1.5		8 min	$2 \cdot 10^{-3}$
5 min-oscillation		570	20		5 min	

it is reproduced in Fig. 3.3. Another form of presentation is Table III.iv, which also includes our current knowledge of the velocities. Care must be taken to ensure that values refer to one and the same level. Again we follow Beckers and choose the depth $\tau = 0.1$ (140 km above $\tau = 1$). We find that the Coriolis parameter, Ω^*, for giant cells is much too large for any form of linearisation or series expansion to be justified.

3.2. Rotation Laws

At present there are two basic methods for measuring solar rotation. One of these utilizes the Doppler shifts of the radiating plasma; the other relies on the motion of useable tracers. Unfortunately, tracers are (except for the granulation which, however, is too short-lived) always associated with magnetism. The true plasma rotation can be seen only by using the spectroscopic method. The resulting rotation law relates directly to solar hydrodynamics, our principal object of study in this book. In contrast, the rotation laws derived from tracers must be used with caution, since we must expect that the magnetic fields will have introduced deviations away from the true rotation law.

The most frequently used representations of the angular velocity profile on a surface are

$$\begin{aligned}\Omega &= A + B\cos^2\theta + C\cos^4\theta \qquad A > 0,\\ &= A + B + C - (B + 2C)\sin^2\theta + C\sin^4\theta,\\ &= A + \frac{B}{5} + \frac{3C}{35} + \frac{2}{15}\left(B + \frac{2}{3}C\right)\frac{P_3^1}{\sin\theta} + \frac{8C}{315}\frac{P_5^1}{\sin\theta},\end{aligned} \tag{1}$$

where the final form, which uses spherical harmonics, is of particular importance for the theory. When temporal variations due to cyclic magnetic phenomena are included, representation $(1)_1$ no longer makes sense. Observers would be well advised to favor the more physical formulation $(1)_3$. The profile (1) possesses extrema at poles and equator. In mid-latitudes, the profile is monotonic if either $BC > 0$ or $|B| > 2|C|$. As the first condition is satisfied by solar values, it is quite impossible to describe, in addition, such a small equatorial deceleration as seems to appear in the observations of La Bonte and Howard (1982).

(a) Plasma rotation

Using their Doppler measurements, Howard and Harvey (1970) found the mean values $A = 13.76$, $B = -1.74$ and $C = -2.19$ (all in deg/day) for the rotation of the solar photosphere. The unit 1 deg/day corresponds to 0.202 μ rad/s or to 140 m/s (at the equator). Thus the rotation law can be written in any of the forms

$$\Omega = 2.78\,(1 - 0.13\cos^2\theta - 0.16\cos^4\theta)\ \mu\ \text{rad/s},$$
$$\Omega = 1.99\,(1 + 0.62\sin^2\theta - 0.22\sin^4\theta)\ \mu\ \text{rad/s}, \tag{2}$$
$$\Omega = 2.67\,(1 - 0.033\,P_3{}^1/\sin\theta - 0.0043\,P_5{}^1/\sin\theta)\ \mu\ \text{rad/s}.$$

From the final expression we see that, in the general representation,

$$\Omega = \Omega_\odot \sum_{n=1,3,\dots} \omega_{n-1} P_n{}^1/\sin\theta, \tag{3}$$

the coefficients decrease progressively:

$$\omega_0 = 1, \qquad \omega_2 = -0.033, \qquad \omega_4 = -0.0043. \tag{4}$$

More recently, Howard *et al.* (1983) published results covering an observation period of 14 years which included data from 188 rotations. The mean rotation law

$$\Omega = 2.87\,(1 - 0.12\cos^2\theta - 0.17\cos^4\theta)\ \mu\ \text{rad/s} \tag{5}$$

resulted. The new equatorial velocity proves to be significantly larger than that of $(2)_1$.

Instead of (3) we shall also use the modal representation $(2)_2$, i.e.

$$\Omega = \Omega^{(0)} \sum_{l=0,1,\dots} \omega^{(l)} \sin^{2l}\theta, \tag{6}$$

which is useful from the observational point of view. Here $\Omega^{(0)}$ is the angular velocity of the poles. According to Howard and Harvey, the fractional overall pole-equator difference in the angular velocities,

$$\delta\Omega = (\Omega_{\text{pole}} - \Omega_{\text{eq}})/\Omega_{\text{eq}}, \tag{7}$$

is about 30%. The actual mean rotation law is monotonic with no very dramatic equatorial increase; as a maximum we have

$$|\Omega^{-1} \sin\theta\,\partial\Omega/\partial\theta| \approx 0.4, \tag{8}$$

which occurs at $\theta \approx 30°$. We must also mention the measurements recorded by the Stanford Solar Observatory. Scherrer *et al.* (1980) found that the rotation of the solar plasma since 1976 was distinctly greater than that given in the Mt. Wilson result (2). Daily Doppler observations of the photospheric rotation showed no temporal variations of Ω. What was to be even more important was the discovery that the observed rotation rate is the same as that of sunspots and of the large-scale magnetic field structure (see below). In their paper Howard *et al.* (1984) contrasted these results to those usually quoted.

(b) Spot rotation

As was to be expected, the rotation of tracers differs slightly, but significantly, from the "true" profile (2). Sunspots generally rotate faster than the surrounding plasma, by about 4% or 80 m/s at each latitude (cf. Fig. 3.4).

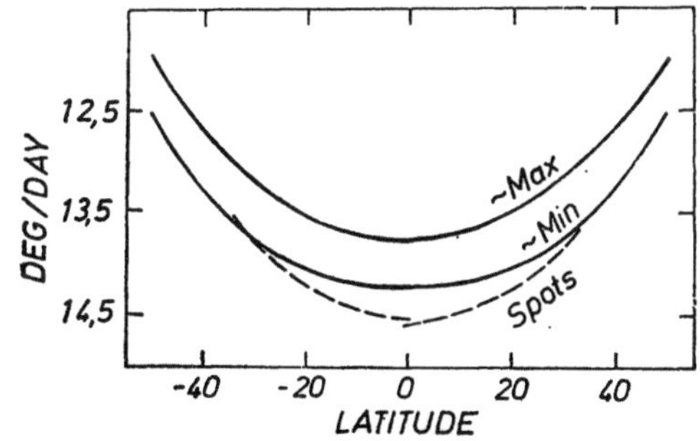

Fig. 3.4. Solar rotation laws: the two upper lines represent photospheric rotation near the activity minima and maxima (Howard, 1978) and the dotted line represents sunspot rotation (Ward, 1965). In these figures we found it preferable to use the latitude $B = 90° - \theta$ instead of the colatitude θ.

The plasma in the spot moves with the "spot phenomenon"; the two are linked (Beckers, 1977). The result most often quoted for sunspots is due to Newton and Nunn (1951):

$$\Omega = 2.90\,(1 - 0.19\cos^2\theta)\ \mu\ \text{rad/s}, \tag{9}$$

which was derived from the analysis of individual long-lived sunspots. In comparison to the 1865 result of Faye, only the "gradient" has been reduced, from 0.22 to 0.19.

The most recent result is that of Howard *et al.* (1984), who analyzed the Mt. Wilson white light plate collection (1921—1982) and found

$$\Omega = 2.934\,(1 - 0.196\cos^2\theta)\ \mu\ \text{rad/s}. \tag{10}$$

Note that the sunspot rotation laws are slightly steeper than the plasma rotation laws. An analysis of about 5,000 single spots, as well as of groups of spots, over a period of nearly 35 years led Lustig (1983) to propose a rotation law similar to that of Newton and Nunn. He asserted, however, that the latitudinal gradients of Ω in the northern and southern hemispheres are remarkably different.

There are also significant differences between the rotation laws for various types of spots:

(i) small asymmetric spots move round the Sun 2% faster than large symmetric spots,
(ii) individual spots are 1% faster than groups,
(iii) small groups are faster than large groups.

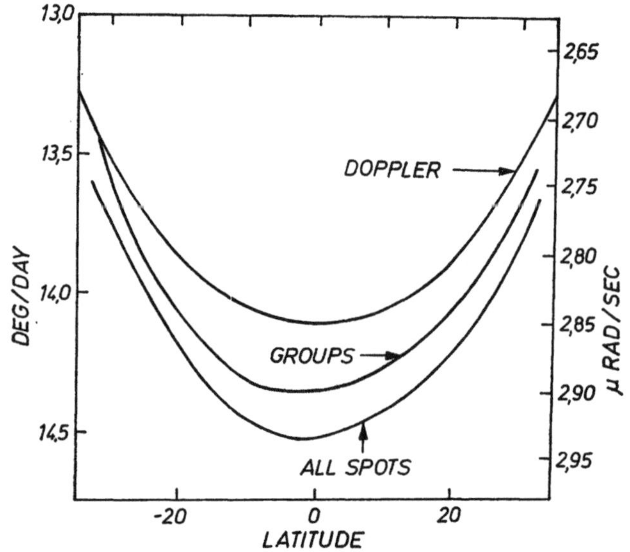

Fig. 3.5. A graphical illustration that spots and spot groups possess different rotation profiles.

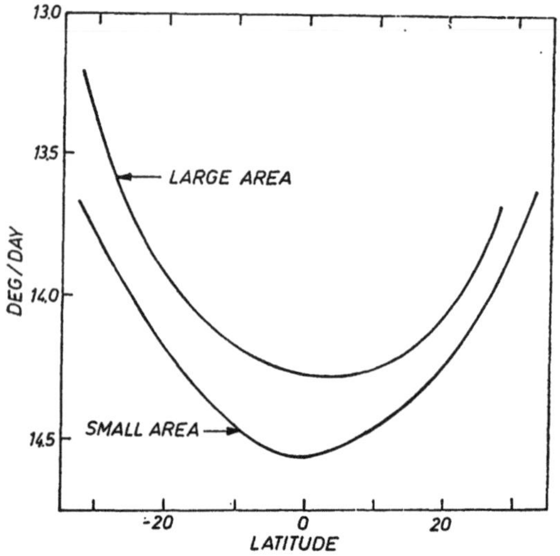

Fig. 3.6. An illustration of the fact that groups of different sizes rotate at different rates.

(Ward 1965, 1966; Howard *et al.* (1984); see Fig. 3.5 and 3.6). The angular velocity for tracers increases along the following sequence: largest spots, groups, medium-sized spots, small spots. Groups rotate at the same rate as large spots (but not as rapidly as the largest). There is also a striking age-dependence in the rotational behavior. Schröter and Wöhl (1978) mention that old spots (defined to be those whose recurrence number exceeded 5) yield flatter rotation profiles than young spots, while Godoli and Mazzucconi (1978) found that the rotation rate is even faster if younger spot groups alone are considered. They also reported that the θ-profile of spots is steeper in their early life.

Balthasar *et al.* (1982) found that the rotation law for old spots approached that of the plasma. During their lifetime the velocity of the spots decreased by about 5 m/s per day. Again the θ-profile was sharper for young spots, a phenomenon which has also been seen in the photospheric magnetic fields. Regions of strong magnetic fields outside the sunspot belt rotate faster than weak field regions (Foukal, 1976). Many other tracers have been observed, and the resulting rotation laws have been reviewed in great detail by

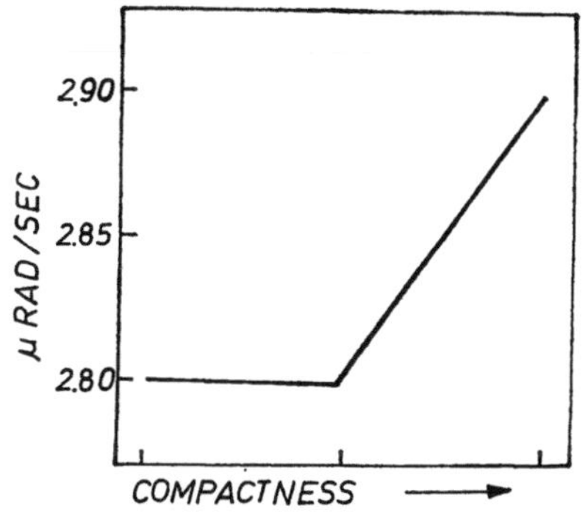

Fig. 3.7. The equatorial rotation of K facule with different compactness (Belvedere *et al.*, 1978). Very compact faculae rotate like spots, while the more diffusive structures rotate like the plasma.

Howard (1978) and Paternò (1978), as well as by Beckers (1981). Laws similar to those governing spots emerged, for example, from investigations of CaII emission and solar faculae.

Reporting their observations of chromospheric faculae in the H and K lines over a period of 10 years, Belvedere *et al.* (1978) discussed the rotation law for differently structured tracers. The inclusion of objects of large area removed the differences between the equatorial rotation of spots and plasma. The most compact faculae rotate like spots, in contrast to the less compact ones which approximately follow the plasma rotation. In order to demonstrate the influence of the facular structure on rotation, the constant A is plotted versus the compactness of the faculae in Fig. 3.7. No similar results hold for the "gradient" of the angular velocity [corresponding to B and C in (1)]. Older measurements of white-light faculae were summarized by Schröter and Wöhl (1978). Here there were marked deviations in the high-latitude rotation from one investigation to another. Near the equator faculae seem to rotate even faster than sunspots.

Interestingly enough, in a study published in 1978, Golub and Vaiana established both rotation laws, i.e. the law for the plasma and the law for the spots, from one and the same tracer class. They measured the small X-ray emissions caused by eruptions of the magnetic field. There are both long-lived and relatively short-lived emissions, and it appeared that the latter followed the plasma law, while the longer-lived events followed the spot law. Clearly, the higher rotation rate of the sunspots has something to do with scales of magnetic field that are neither too small nor too large. The smallest

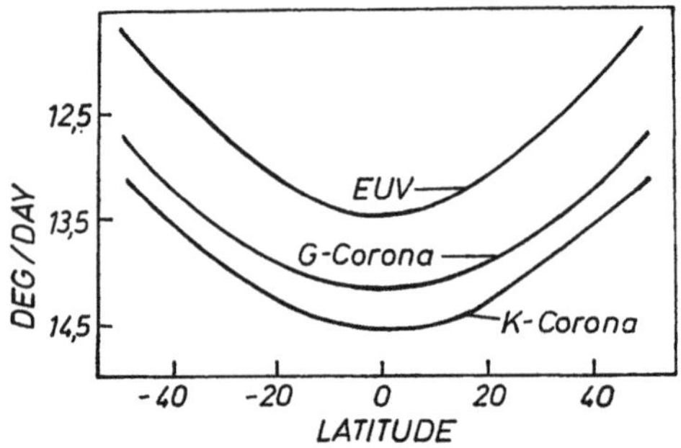

Fig. 3.8. Chromospheric and coronal rotation: Skylab EUV observations reveal the slow rotation of chromosphere. It is not quite certain whether the difference between the K corona and the green line corona is real. The profiles are somewhat flatter than those for plasma or spots.

scales rotate at the same rate as the photopheric gas. Another important tracer exists in the form of CaII mottles. These are fine structures of only a few thousand km in length and appear in the light of the line λ 3933 Å. They correspond to weak magnetic fields so the magnetic coupling should not be too strong. More than 75% of the mottles can still be identified after 4 hours. They are uniformly distributed over the solar surface and are part of the chromosphere.

The rotation law for the CaII network derived by Schröter and Wöhl (1976) is intermediate between laws (2) and (9). Chromospheric rotation appeared to be significantly slower than sunspot rotation, and nearly coincident with the plasma law (2). Occasionally the angular velocity in the northern hemisphere at $\theta \leqq 60°$ was almost the same as that near the equator; occasionally the rotation was rigid. A similar situation had already been observed by Hence and Dupree (1973) in the chromospheric Lyman continuum.

The coronal tracers, all of which are magnetic, also appear to rotate differentially. As expected, there is agreement with the spot results at equatorial latitudes, while at higher latitudes the decrease in Ω is somewhat smaller. The green line corona rotates even faster than sunspots. In contrast, the EUV spectro-heliograms produced in the Skylab corona program revealed a steep rotation profile but with a slow equatorial velocity [$A = 2.73\ \mu$ rad/s, Dupree and Hence (1972)]. No dependence on height has been found. Fig. 3.8 summarizes these results. It is probable that the interpretation of the measurements may even have over-emphasized magnetic effects because time-scales of more than one rotation period were used.

(c) Rigid rotation

Searches undertaken by Wilcox and Howard (1970), as well as by Wilcox *et al.* (1970), revealed that the photospheric magnetic fields rotate like sunspots, and at the equator they share nearly the same high velocity. For magnetic patterns of longer duration, however, the rotation becomes increasingly uniform, and phenomena which recur for more than five successive rotations have a nearly rigid-body motion. Rigid rotation appears to be a third profile of Ω which exists in the Sun.

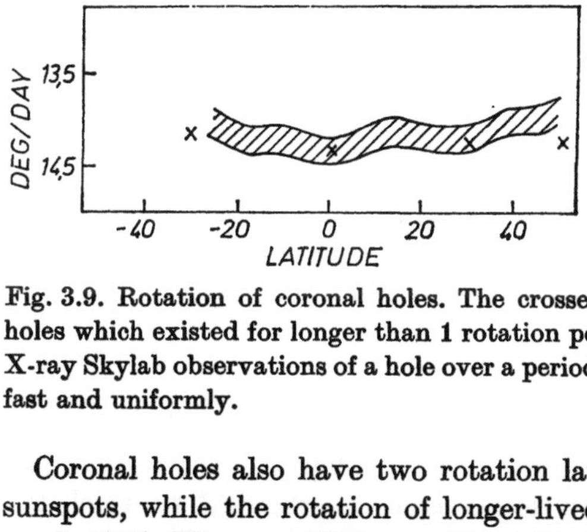

Fig. 3.9. Rotation of coronal holes. The crosses represent EUV measurements for holes which existed for longer than 1 rotation period. The other region concerns soft X-ray Skylab observations of a hole over a period of 5 months. Holes rotate relatively fast and uniformly.

Coronal holes also have two rotation laws. Short-lived holes rotate like sunspots, while the rotation of longer-lived holes is nearly rigid (Timothy *et al.* 1975; Wagner, 1975; see Fig. 3.9). As coronal holes are intimately related to solar magnetism, the dual character of tracer rotation can best be stated as follows: individual elements rotate in accordance with the sunspot law (9) while large-scale long-lived features rotate rigidly at a rate somewhat *slower* than sunspots at the equator, but still faster than the plasma.

A peaceful coexistence between rigid and differential rotation in the solar atmosphere has been suggested by Antonucci (1978). Under special conditions, rigid rotation can be seen at coronal, chromospheric and even photospheric levels. The key feature is the lifetime of the rotation tracers, and this varies throughout the solar activity cycle. In the corona, features exist with either rigid (EUV coronal holes, green line corona) or differential (compact X-ray regions) rotation. The whole atmosphere may be organized in a very large-scale pattern (over a period of more than a month) which does not show differential rotation. The phenomena which do exhibit differential rotation have, in general, a smaller size and shorter duration. Antonucci points out that coronal rotation measurements normally require tracers of long duration, hence rigid rotation may dominate the results. In order to detect differential rotation in the corona, one must use tracers with lifetimes of less than one rotation period. Ca-line events do, indeed, appear to rotate differentially:

$$\Omega = 2.96\,(1 - 0.19\cos^2\theta)\ \mu\ \mathrm{rad/s}. \tag{11}$$

The coexistence of this rotation law with the rigid rotation of the long-lived Ca emission is demonstrated in Fig. 3.10.

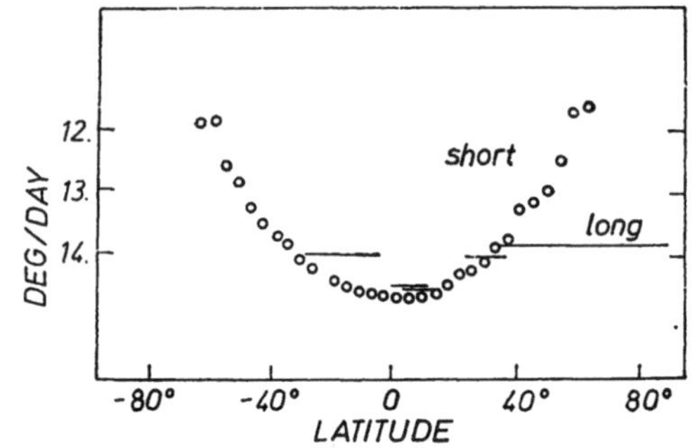

Fig. 3.10. The coexistence of two rotation laws in the Ca chromosphere. The long-lived pattern ($\tau_c > 27$d) rotates almost rigidly, the short-lived pattern ($\tau_c < 1$d) rotates non-uniformly (derived from Antonucci (1978)).

(d) Radial rotation law

One is tempted to put the various observed rotation laws in order. No comment is needed on the hydrodynamic surface law (2). The sunspot law (9) is often assumed to belong to deeper layers of the SCZ to which the magnetic tracers are "anchored". This may be correct, provided they can slip easily through the overlying layers.

As Foukal (1972) pointed out, the magnetic Reynolds number, $R_m = \mu_0 \sigma L u'$, of the outer layers of the Sun is very large ($\simeq 10^5$). This suggests that there is no slip between the magnetic field and the fluid flow; i.e., that they move in phase. This estimate ignores the turbulently-generated decrease of electrical conductivity but, assuming that the picture is basically correct, we may ask next, "Which dominates: field or flow?". A preliminary answer can be obtained immediately from the ratio of magnetic to kinetic energies (or between the Alfvén and turbulent velocities). If this ratio is large then the magnetic field determines the motion; if it is small, the opposite is true. Foukal argues that in the chromosphere the ratio is very large, and is even large in the photosphere. In these regions, therefore, the flow strictly follows the field configuration unless it is forcefree. Neither in the chromosphere nor the photosphere can the motions distort the fields. Only at a depth of about 5000 km does the ratio become unity and, at deeper levels, the field follows the motion. In other words, the magnetic field is anchored to this sub-photospheric layer and the rotation of magnetic tracers reflects the rotation of that level. Consequently, the fact that the sunspots rotate faster than the plasma at the surface may allow us to infer that the visible layers rotate more slowly than do the deeper regions. However, if computed with a characteristic length of 5000 km, the gradient would be too large. For a difference between the spots and the plasma of only 0.1 μ rad/s, one obtains

$$\mathrm{d}(\ln \Omega)/\mathrm{d}(\ln x) \simeq -5. \tag{12}$$

Duvall (1980) gave what may possibly be a confirmation of the idea, and may also have provided an escape from (12). By measuring the rotation of the equator he could extract the supergranulation as a tracer. In this way he eliminated the complications of magnetic tracers. Supergranule cells are formed by nonmagnetic plasma. Nevertheless, their rotation is characteristic of that of the sunspots. The inferred rotation rate was 3% faster than the rate obtained from the Doppler effect in consecutive periods. As supergranulation has some connection with layers at greater depth (say 10^4 km), it appears likely that its rotation also carries information about those layers. When re-computed with a characteristic length of 10^4 km, the logarithmic gradient (12) is reduced to -2.

We must bear in mind, however, that there exists neither a theoretical basis for Foukal's mechanism nor statistics that include the magnetic field strengths of the spots. Only the phenomenologial classes "young", "old", "large", "small" have been used. "Young and small" move faster than "old and large". Can we be sure that the former are anchored more deeply than the latter? It is also important to know whether the slowly-rotating class "old and large" really refers to one and the same layer in the Sun. Or does one, e.g. "old", belong to the innermost region and the other, i.e. "large", to the outermost region? If this were the case, layers at two different depths in the Sun would rotate slowly and a maximum of Ω would lie between them. Indeed the fact that the very oldest members of any tracer class exhibit rigid rotation and are likely to belong to the deepest layers (see below) could be a hint that the slowly-rotating class, "old and large", is not homogeneous with respect to depth. If this is the case, the differential rotation of the oldest spots should be less. In any event, to complete the study of phenomenological features, measurements of the magnetic flux are necessary.

Another open question concerns the implications of the rigid-body rotation law derived from the long-lived magnetic tracers. Its (uniform) angular velocity lies between the values of A for spots and the values of A for the plasma. Compared with the latter, the longitudinal drift of largescale magnetic tracers, such as coronal holes, is about 150 deg/yr. The latitudinal drift in the butterfly diagram is about 3 deg/yr (1.2 m/s). According to Stix (1974, 1976a), rigid rotation might be understood in terms of the sectorial structure of a solar dynamo model, i.e. of non-axisymmetric magnetic modes. Such modes always have wave characteristics, with phase speeds that may be related to the drift in longitude. There are dynamo models for the Sun in which the two-sector mode S1 (S means symmetric with respect to the equator; 1 means associated with $m = 1$ dependence on φ) is more easily excited than the normal axisymmetric dipole. Indeed, an analysis of the 1973 observations of magnetic fields and coronal holes revealed that the modes S1 and S3 dominate (Stix, 1977). Yet the theoretical drift rates of about 25 deg/yr for such modes are too small. The question arises whether the remaining difference reflects the rotation rate of rather deep layers of the SCZ, indicating that this is higher than the rotation rate for the surface but lower than that for spots. As the rotation is rigid, this suggests that the deepest layers are close to the base of the SCZ. So the conclusion appears to be that the latter rotates more rapidly than the surface but more slowly than the region to which the sunspots are anchored. If all this is correct, then we have an indication that *the radial Ω-profile has a slight maximum within the SCZ.*

There is, however, some reason to question this appeal to dynamo theory. Following the arguments and extensive model calculations of Rädler (most recently those of 1986), it seems extremely unlikely that the solar sectorial structure can be thought of in traditional mean-field terms. Differential rotation and non-axisymmetric magnetism appear to be completely incompatible, as the two magnetic parities are thoroughly mixed. It is thus hard to believe that a mean-field sector mode is actually more easily excited than the normal axisymmetric dipole.

A rotation faster than the photospheric has been observed in the chromosphere. In H_α, Livingston (1969) found this greater chromospheric rate, and showed that its latitudinal profile is relatively flat. As we have already stressed, the observational results obtained from tracers confirm both these findings. There is, however, a more recent method of investigating the rotation of the Sun's interior. In principle, the inner rotation profile can be deduced from frequency measurements of the global p-mode oscillations. Such oscillations exhibit two essential features: the basic rotation splits the frequencies, and the location of the associated point in the Sun selects the mode number of the oscillation. The rotational splitting of the eigen-frequency into $2l + 1$ levels (symmetrically about the value for $\Omega = 0$) is reminiscent of to the Zeeman effect. Normally the shift of frequencies is proportional to the product $m\Omega$ (where m is the longitudinal mode number), and hence the frequency difference for the two sectorial harmonics with $m = l$ (retrograde) and $m = -l$ (prograde) should provide the most accurate measure. Taken together, the two features highlighted above allow the general rotation law $\Omega = \Omega(x)$ to be determined, but only if a sequence of suitable modes is available for observation. High-degree p-modes are related to the outer layers of any solar model and low-degree modes are concerned with the deep interior (see Cox, 1980).

Deubner *et al.* (1979) initiated the search by probing the inner rotation with high-degree p-modes. They observed an inward increase in the linear velocity of about 80 m/s for the first 15,000 km,

$$d(\ln \Omega)/d(\ln x) \simeq -2: \tag{13}$$

Their observations, as well as those of Claverie *et al.* (1981) for low-degree modes — which exist throughout the Sun — have not yet been finally confirmed. That the solar core rotates about 5 times faster than the surface was, at that time, a characteristic statement. The rotation profiles presented today are much flatter, with a slight minimum somewhere below the base of the convection zone. In particular, no great differences appears to exist between the rotation rates at $x = 1$ and $x = x_i$ (Duvall *et al.*, 1984).

Additional evidence for a sub-rotating SCZ has recently been given by Hill *et al.* (1984). They analyzed the rotational splitting of non-radial oscillation modes through measurements of the diameter. For the equatorial rotation rate of the SCZ they obtained

$$d(\ln \Omega)/d(\ln x) \simeq -1.5, \tag{14}$$

implying $V^{(0)} \simeq -1.5$. In contrast to (13), such a profile is flat enough to fulfill the Solberg-Rayleigh criterion,

$$d(s^2\Omega)/ds > 0, \tag{15}$$

for stability in the presence of axisymmetric disturbances [for details, see the monograph by Tassoul (1978)].

3.3. Horizontal Cross-Correlation. Meridional Flow. Mean Temperature

There are other mean quantities whose distribution over the solar surface can be determined by direct observation. They are also related to the rotation problem in more than one way. Our present knowledge of their true values is, however, unsatisfactory. The non-vanishing components of the surface velocity both lie completely within the horizontal layer. We have already discussed the mean longitudinal part, $\Omega R \sin \theta$. Now we turn to the latitudinal component $\bar{u}_\theta$ (the meridional flow) and the correlation between longitudinal and latitudinal fluctuations, i.e. $Q_{\theta\phi} = \overline{u_\theta' u_\phi'}$. The latter is also known as "the horizontal Reynolds stress", "the horizontal cross-correlation", "the covariance" or "the lateral turbulent flux of angular momentum".

In order to introduce a suitable normalization we have to keep in mind that $Q_{\theta\phi}$

(i) vanishes unless the star rotates; hence it must vanish with Ω,

(ii) is an axial expression (which changes sign if $\Omega \to -\Omega$); hence it must be odd in Ω,
(iii) must vanish at the equator if, as we assume, the turbulence pattern has equatorial symmetry (the "symmetric mode"),
(iv) must vanish at the poles if, as in our case, the mean quantities are assumed to be axisymmetric.

Taking these properties into account, it seems reasonable to introduce a function $w(\theta)$, defined by

$$Q_{\theta\phi} = \nu_T \Omega_\odot \cos\theta \sin^2\theta \, w(\theta), \tag{16}$$

which we are tempted to call the "Ward profile", after its first observer, F. Ward. The product $\nu_T\Omega_\odot$ is the simplest scalar quantity that has both the correct dimensions and the correct symmetries.

A modal expansion of w is useful:

$$w = w^{(1)} + w^{(2)} \sin^2\theta + \cdots = \sum_{l=1} w^{(l)} \sin^{2l-2}\theta. \tag{17}$$

In order to prove statement (iv), we consider the axis of rotation and introduce Cartesian coordinates (x, y), which are perpendicular to the axis of rotation. If the mean quantities are symmetric with respect to that axis, we have

$$\overline{u_x'^2} = \overline{u_y'^2}, \qquad \overline{u_x' u_y'} = 0. \tag{18}$$

If spherical components u_θ' and u_ϕ' are defined by

$$u_\theta' = u_x' \cos\phi + u_y' \sin\phi, \qquad u_\phi' = -u_x' \sin\phi + u_y' \cos\phi, \tag{19}$$

then, at the poles, we have

$$\overline{u_\theta' u_\phi'} = (\overline{u_y'^2} - \overline{u_x'^2}) \sin\phi\cos\phi + \overline{u_x' u_y'} (\cos^2\phi - \sin^2\phi). \tag{20}$$

This expression vanishes as a consequence of (18). In Chapter 6 we shall demonstrate that the turbulent flux of angular momentum which creates the non-uniform rotation, $\Omega(\theta)$, also produces the profile $w(\theta)$.

The Reynolds equation also requires that $Q_{\theta\phi}$ vanishes on the rotation axis. Unless $Q_{\theta\phi} \sim \sin^2\theta\cos\theta$, the tensor divergence for Q_{ij} would not exist as $\theta \to 0$. This is also the reason why the coefficient $H^{(0)}$ does not appear in the series expansion (2.21). The sum $\sum H^{(l)} \sin^{2l}\theta$ can only start from $l = 1$. By comparing the two expressions for the cross-correlation, namely (17) and (2.21), we immediately find that

$$w^{(1)} = -2\omega^{(1)} + H^{(1)} \approx -1.2 + H^{(1)}. \tag{21}$$

Here we have used the value $\omega^{(1)} = 0.62$ for the solar differential rotation deduced from the observations. If the Λ-effect did not exist, i.e. if $H^{(1)} = 0$, the "natural" value for $w^{(1)}$ would have been $w^{(1)} \approx -1$. The observed dramatic departure from this value should be taken as an indication that the turbulent momentum fluxes contain non-diffusive parts. This is one of the reasons why we infer the presence of the non-diffusive, non-Boussinesq terms, Λ_V and Λ_H.

What is the actual empirical result? Is a (say) faster longitudinal motion in the northern hemisphere correlated with a poleward ($Q_{\theta\phi} < 0$) or with an equatorward ($Q_{\theta\phi} > 0$) motion? Ward (1965) was the first to provide clear unequivocal evidence for a *positive* correlation, namely

$$Q_{\theta\phi} \simeq 0.1\ (\text{deg/day})^2 \simeq 2 \cdot 10^7\ \text{cm}^2\,\text{s}^{-2}, \tag{22}$$

(Fig. 3.11). He considered the proper motion of sunspot groups, the faster of which tend to move towards the equator. Profile (17) for w could be described approximately by

$$w^{(1)} \approx 2, \qquad w^{(2)} \approx 0. \tag{23}$$

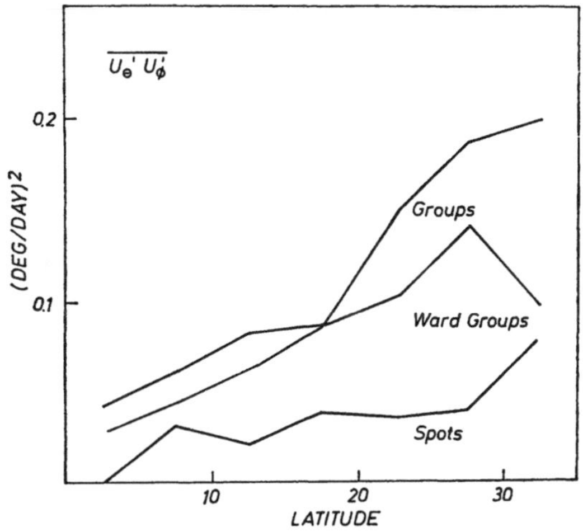

Fig. 3.11. The correlation $Q_{\theta\phi}$ of longitudinal and latitudinal velocity (northern hemisphere). The TMT is clearly equatorwards. New results by Gilman and Howard (1984b) confirm the findings of Ward (1965). Note the difference between individual spots and groups: spots appear to tend towards the "natural" negative sign of $Q_{\theta\phi}$.

This result has a strong implication in the theory. It indicates the existence of a horizontal Λ-effect, even at the surface of the Sun. There, however, we expected Λ to be small because of the weak influence of rotation on the turbulence in the surface layers of the SCZ.

In an unpublished manuscript, Leighton is reported to have argued that the observed correlation might arise from some systematic features of the spots themselves, and might not reflect the true motional state of the plasma. A brief resumé of his argument has been presented by Gilman and Howard (1984b) who themselves extracted the positions of individual spots and groups of spots from a 62 year period, using data accumulated in the Mt. Wilson plate collection. They obtained a cross-correlation with the same sign and of similar magnitude as Ward's (Fig. 3.11). This in no way confirms Leighton's argument that the properties of the spots, rather than those of the entire fluid, are responsible. However, for the small solitary spots, which should especially be almost passive tracers, the new result is considerably less (see also Ribes, 1986).

Spot-groups probably portray the physical state of the deeper layers better than individual, isolated spots. Short-lived, completely passive tracers are expected to behave as the simple Boussinesq relation (2.3) implies, i.e. $w^{(1)} \approx -1$. In this context, it is relevant to mention the measurements made by Belvedere *et al.* (1976), Schröter and Wöhl (1976) and Tuominen (1982). Belvedere *et al.* used the motion of solar faculae in order to determine the correlation, $Q_{\theta\phi}$. Faculae may be even more suitable than spots for this purpose, because of their weaker coupling to the magnetic field. They obtained correlations which had the same sign as Ward's but were larger by more than an order of magnitude. No latitudinal profile could, however, be derived. The end products of the investigation by Schröter and Wöhl, who measured proper motions of CaII plages, were the same. They obtained for $Q_{\theta\phi}$ the value $4 \cdot 10^7$ cm^2/s^2. In contrast, Tuominen's analysis of the motion of sunspot groups yielded relatively small values for this correlation.

It is important to emphasize that the actual magnitude of this horizontal Reynolds stress is today so uncertain as to be an open question. How unfortunate this lack of information is! For, were this stress known, its impact on the theory would be as profound as the knowledge of the rotation law has been to theory in the past.

With regard to the latitudinal component of the mean motion, $\overline{u}_\theta$, the observations show some uniformity only when restricted to Doppler measurements, which have led to the following results:

Beckers	(1978)	42 m/s
Duvall	(1979)	20 m/s
Howard	(1979)	20 m/s.

In each case, the motion was directed polewards.

Howard has also identified a pattern of upward movement near the equator. It is a difficult task to isolate such a flow since there are many factors that complicate the motion. Furthermore the observing sessions, during which measurements could be made, were in all cases relatively short. Another report of meridional flow (towards the equator, 30 m/s) was made by Perez-Garde *et al.* (1981). All empirical results, of course, need confirmation.

Most recently, La Bonte and Howard (1982) have uncovered a poleward flow of 16 m/s by re-analyzing some Doppler data and removing the apparent velocity patterns. They advanced a new concept — the longterm average presence of a radial inflow in magnetic regions. The measured value of this meridional flow, averaged over the years 1967—1980, was only 6 m/s but, when corrections for the inflow were taken into account, this was amended to 16 m/s. Their study also suggests that magnetism influences the slow velocity patterns.

In summary, the claim can now be made with some confidence, that a photospheric mean *poleward* flow exists of 10—20 m/s. The last word on this topic is, however, by no means yet written (Ribes, 1986).

This is also true of the investigations using tracers such as sunspots or CaII mottles. Statistics of sunspot groups observed at Greenwich (Tuominen, 1941, 1961; Becker, 1954; Richardson and Schwarzschild, 1953: see also Coffey and Gilman, 1969) suggest that there is a mean equatorward motion at low latitudes and a poleward motion at higher latitudes. Some authors (Ward, 1965, 1973; Balthasar and Wöhl, 1980) contested the statistical significance of this conclusion. In any case, the order of magnitude of the flow is not more than a few m/s. For deeper layers, we conclude that there is, in contrast to the photosphere, no clear evidence of a mean meridional flow. Ward (1973) favored a mean meridional velocity of less than 1 m/s. Such data, he concluded, cannot be used to infer latitudinal flows, but may be invoked as an argument against theories which rely on the meridional circulation to generate the equatorial acceleration.

Among the observations of the chromospheric CaII network, those of Schröter *et al.* (1978) are of particular interest. These authors compared results from two periods of observation. During the first they found a significant poleward drift of 30 ± 9 m/s in both hemispheres, but this result was not confirmed in the second. The authors noted that during the first period the normal differential rotation was observed, whereas in the second the rotation was more rigid.

Fortunately, a much clearer situation exists for the "differential temperature". Of course, the question arises as to whether the influence of rotation also creates latitude-dependent thermodynamic quantities, such as mean temperature, pressure and density. Any pole-equator difference in the mean temperature or heat-flux at the solar surface would be of particular interest. Strikingly enough, all modern investigations have reached the same conclusion. Pole-equator differences in the mean temperature, if present at all, are very small. Values of the temperature differences reported in the literature are:

Appenzeller and Schröter (1967)	< 6 K
Plaskett (1970)	$\approx 5\%$
Caccin *et al.* (1970)	$< 1\%$
Altrock and Canfield (1972)	< 1.5 K
Rutten (1973)	< 3 K
Noyes *et al.* (1973)	< 7 K
Falciani *et al.* (1974)	≈ 0 K.

Thus, any successful model of the dynamics of the convection zone must account for the extremely flat profile in latitudinal temperature.

To sum up, we can interpret the reported observations at two levels. To start with, there are the results of which we are certain. We can be sure that

(i) the turbulent angular momentum transport is *not* directed polewards ($Q_{\theta\phi} \cos\theta \geqq 0$),
(ii) the meridional flow is *not* directed equatorwards ($\overline{u}_\theta \cos\theta \leqq 0$),
(iii) the mean temperature is *not* "differential", i.e. it does not depend on θ.

The second level is more positive but also relies more on interpretation:

(i) the angular momentum transport is equatorward,
(ii) the meridional flow is small and poleward.

It must be stressed, however, that there is a striking weakness in the observations. Seldom do the results of different observers coincide. No extended periods of simultaneous observations have been made by two or more observatories. Reviewing the material, it almost seems as though a hidden parameter influences the velocity pattern. Whether this parameter is simply the phase of the activity cycle is an important question.

3.4. Temporal Variations

It is only recently that evidence has accumulated, indicating that the mean velocity field at the solar surface varies with time. And it is the magnetic solar cycle that plays the dominant role. One at once suspects that the variation in a velocity component during the cycle must be caused by Lorentz force.

La Bonte and Howard (1982) analyzed Doppler measurements dating back to 1967 (Fig. 3.12) and showed that the coefficients B and C are periodic, repeating every 11 years; B and C are in antiphase and, at one magnetic minimum (1976), B was maximal and C minimal. Hence, at a minimum the rotation profile is flatter at low latitudes and steeper at high latitudes. At an activity minimum lower latitudes rotate faster, while near its maximum the higher latitudes rotate faster. The superiority of representation $(1)_3$ is here demonstrated, for it provides a unified picture, while $(1)_1$ and $(1)_2$ do not. The two coefficients $\omega_2(1)$ and $\omega_4(1)$ are in phase, and both are minimal at an activity minimum (Snodgrass and Howard, 1985).

A similar result was obtained by Lustig (1983) who found that the rotation during minimum activity was always faster and more rigid than around maximum. Of course, since he analyzed the motion of sunspots, his conclusion only applies at low latitudes, where it agrees with the results of La Bonte and Howard. When high latitudes are considered the rotation profile is steeper. Also Tuominen and Kyröläinen (1982) reported variations in the

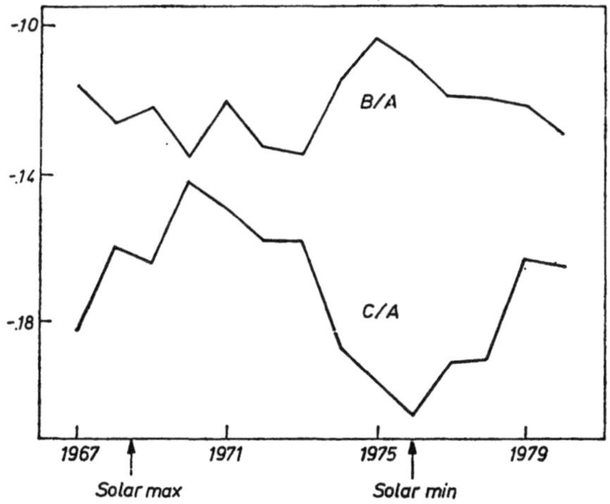

Fig. 3.12. The variation of B and C during an activity cycle (after La Bonte and Howard, 1982). B and C are in antiphase. High latitudes rotate faster during the spot maxima.

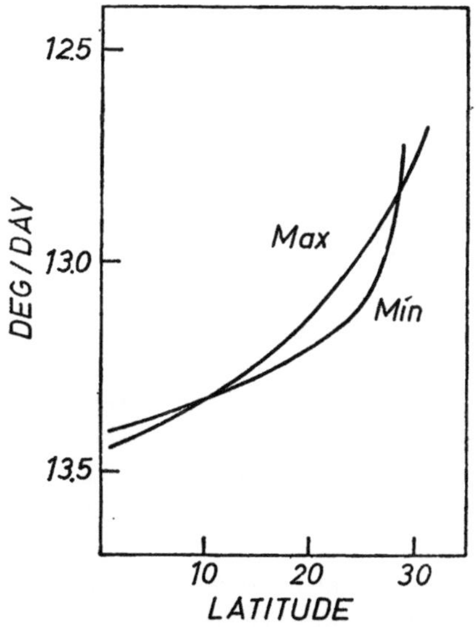

Fig. 3.13. Sunspot rotation at activity maxima and minima (Tuominen and Kyröläinen, 1982). At a minimum the low-latitude rotation is less sheared than at a maximum. Since only a few spots appear in the equatorial region during a minimum, the results are less reliable for mimima than for maxima.

angular velocity in the sunspot zone. At a minimum, the rotation is more rigid at low latitudes and more sheared at high latitudes (Fig. 3.13).

In the same year Godoli and Mazzucconi, as well as Tuominen and Virtanen (1984), proposed a division of the sunspot zone into two parts. The line of maximal activity in the butterfly diagram defines two regions whose common border, *A*, migrates equatorwards. At a given latitude the "lower" spots (between the equator *E* and the activity line *A*) rotate with an above average angular velocity. It is very tempting to interpret this result as observational support for the "torsional oscillation" which had been report-

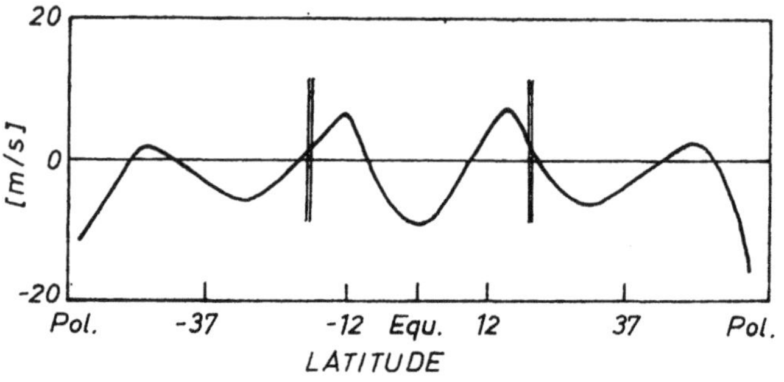

Fig. 3.14. The torsional oscillation pattern near an activity maximum. Two fast and two slow belts exist in each hemisphere. During the cycle the pattern migrates equatorwards and the equator is always decelerated. The vertical bars indicate the active latitudes.

ed earlier, in 1980, by Howard and La Bonte. At any epoch, some latitudinal bands of the Sun are rotating faster than they do on average, and some are rotating more slowly. At a fixed latitude the velocity oscillates between eastward and westward directions (relative to the mean) with a period of about 11 years. In each hemisphere there are two zones which rotate more slowly and two which rotate more rapidly than their average (Fig. 3.14). Over a period of 22 years the whole pattern migrates at about 2 m/s towards the equator. The most rapidly rotating zone is always on the equatorial side of the activity belt, and the maximum activity appears close to the zero line (the "shear zone") of the velocity oscillation.

These results concern the variable *profile* of solar rotation. The question remains whether the equatorial velocity, *A*, also varies during the cycle. As early as 1976 Howard pointed to evidence that *A* is higher during a magnetic minimum (cf. Fig. 3.4). Other data also tended to confirm this, as, for instance, that of Livingston and Duvall (1979), who had already found the polar deceleration at minimum mentioned above. Balthasar and Wöhl (1980) studied data, collected over 30 years, on the proper motion of sunspot groups and obtained an indication of a maximum *A* during minimum activity.

Lustig (1983), also using sunspot observations, found A to be larger during minimum than at maximum. Yet the differences are rather small (see Table III.v.).

Table III.v. Long-time peak-to-peak variations of the Sun's equatorial rotation, A

	Observatory	Variation (deg/day)	period
Howard	Mt. Wilson	0.3	1967—1975
Livingston and Duvall	Kitt Peak	0.4	1966—1978
Balthasar and Wöhl	Greenwich	0.2	1940—1968
Lustig	Kanzelhöhe	0.1	1947—1981
Belvedere *et al.*	Catania	0.4	1967—1977

More evidence has recently come from sunspot observations made at Mt. Wilson. Howard *et al.* (1984) and Gilman and Howard (1984a) investigated the daily solar disk photographs taken between 1921 and 1982. They found that the angular velocity was maximal near an activity minimum, though the amplitude itself varied from cycle to cycle. There were also slight differences between the hemispheres and the latitudes concerned, these differences possibly being due to the chaotic character of the solar cycle. The authors found a mean peak-to-peak variation of 0.16 deg/day (Fig. 3.15); for the latest cycle the value is 0.20. The variations derived from Doppler observations are similar but about twice as large. It should also be mentioned that an additional lower peak may occur shortly after a maximum.

The observations of the variable solar rotation described in this section are summarized in Fig. 3.16. The rotation law is different at solar maximum

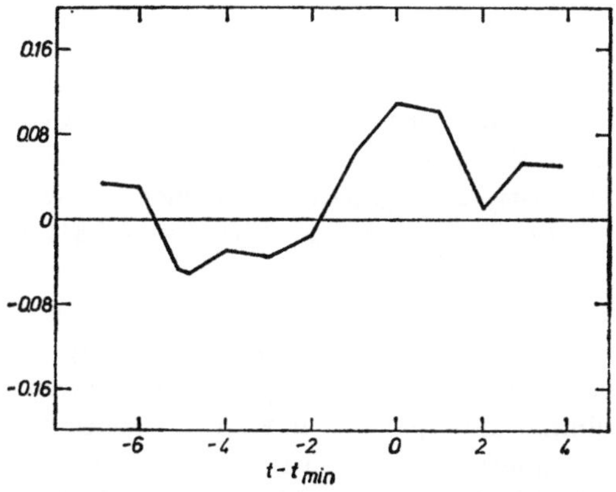

Fig. 3.15. The variation of the equatorial rotation rate during an activity cycle (after Gilman and Howard, 1984a). Rotation is highest at a magnetic minimum. During a maximum there is a second small peak. These results are based on sunspot statistics for 1921—1982.

and minimum: at a minimum the slope is flatter for lower latitudes than at a maximum. The opposite is true at higher latitudes; there the slope is steeper. The higher order terms (the torsional oscillation) introduce a wavy component to the profile.

Recently, the old discrepancy concerning the meridional motion of sunspot groups has also become more definite. In fact, Tuominen and Kyröläinen (1982) have discovered a distinct difference in the latitudinal distribution of the meridional drift for the activity maxima and minima, using Greenwich sunspot group observations. Tuominen *et al.* (1983) detected the presence of a pattern in the meridional flow. One stream is on the poleward side of an activity zone, flows towards the poles, and apparently rotates more slowly.

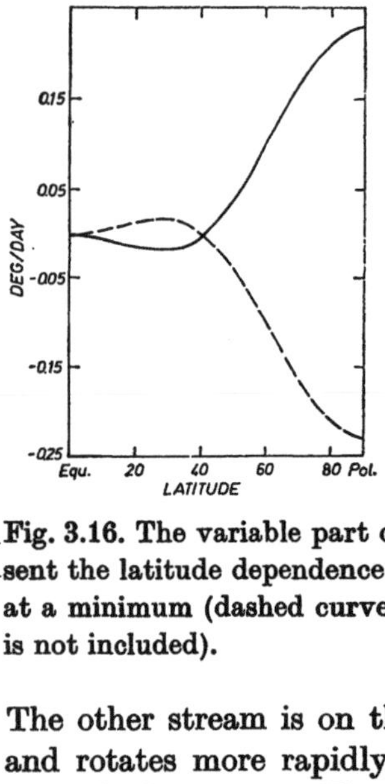

Fig. 3.16. The variable part of the solar differential rotation. The two curves represent the latitude dependence of $B + C$ at a solar maximum (continuous curve) and at a minimum (dashed curve). The zero-point on the ordinate scale is arbitrary (A is not included).

The other stream is on the equatorward side, flows towards the equator, and rotates more rapidly. The zero line between the two streams drifts towards the equator. The velocity of the streams is a few m/s, in latitude as well as in longitude. The most statistically significant evidence for cyclic meridional motion was obtained by Tuominen and Virtanen (1984) by relating the meridional drift directly to the butterfly diagram. In all the cycles between 1874 and 1935 the mean motion was towards the poles on the poleward side of the line of maximal activity and towards the equator on the equatorward side. Combining results from all the cycles, they obtained a mean velocity of 3.5 m/s. A similar flow pattern having *two* zones with poleward flow and *two* zones with equatorward flow can also be seen in the paper by La Bonte and Howard (1982). Thus, it would seem that the time-dependent part of the meridional circulation is larger than, or at least comparable with, the steady part — in sharp contrast to the differential rotation.

All the variations mentioned follow the solar cycle with a mean period of 11 years. They also share the non-harmonic nature of the cycle: the time from minimum to maximum is shorter than from maximum to minimum. Even the presence of daily variations in the equatorial plasma velocity have been reported. Indeed, the fundamental paper by Howard and Harvey (1970) described day-to-day changes of several hundreds of m/s. On some days the equator was even seen to rotate more slowly than higher latitudes, and physical speculations in favor of giant cells have been based on this. Using the linear photodiode array RETICON, Perez Garde *et al.* (1981) found daily variations in the equatorial velocity of up to 30 m/s, but measurements using the newer, two-dimensional diode array RETICON 100×100 reduced the significant variations to a maximum of 10 m/s from an averaging time of several hours (Küveler and Wöhl, 1983). So it would seem to follow that, the better the instrument the smaller the variation.

Chapter 4. Transporters of Angular Momentum

Our thinking becomes calm on realizing that a puzzling situation is merely a special version of the general.
Pauli, 1921

4.1. The Conservation Law

The basic equation for any hydrodynamical problem is the Navier-Stokes equation which, for a Newtonian fluid in an inertial frame of reference, is

$$\varrho[\partial \boldsymbol{u}/\partial t + (\boldsymbol{u} \cdot \nabla)\, \boldsymbol{u}] = -\nabla p + \varrho \boldsymbol{g} + \nabla \cdot \pi + \mu^{-1} \nabla \times \boldsymbol{B} \times \boldsymbol{B}. \tag{1}$$

The standard notation has been used: ϱ is the density, p the pressure, $\boldsymbol{B}$ the magnetic field, $\boldsymbol{g}$ the acceleration due to gravity, $\boldsymbol{u}$ the velocity and π_{ij} the viscosity tensor, where

$$\pi_{ij}/\varrho = \nu(u_{i,j} + u_{j,i}) + (\zeta - 2\nu/3)\, u_{m,m}\delta_{ij}. \tag{2}$$

Here ν is the kinematic (shear) viscosity; ζ, the kinematic bulk viscosity, is always neglected in the following analysis.

For future reference, we give the azimuthal (ϕ) component of the non-magnetic form of (1) in spherical coordinates (r, θ, ϕ):

$$\varrho\left(\frac{\partial u_\phi}{\partial t} + u_r \frac{\partial u_\phi}{\partial r} + \frac{u_\theta}{r}\frac{\partial u_\phi}{\partial \theta} + \frac{u_r u_\phi}{r} + \frac{\cot\theta}{r}\, u_\theta u_\phi\right)$$
$$= \frac{1}{r^3}\frac{\partial}{\partial r}(r^3 \pi_{\phi r}) + \frac{1}{r \sin^2\theta}\frac{\partial}{\partial \theta}(\sin^2\theta\, \pi_{\theta\phi}), \tag{3}$$

where, since $u_\phi = r\Omega \sin\theta$,

$$\begin{aligned} \pi_{\phi r} &= \pi_{r\phi} = \varrho\nu r \sin\theta\, \partial\Omega/\partial r, \\ \pi_{\phi\theta} &= \pi_{\theta\phi} = \varrho\nu \sin\theta\, \partial\Omega/\partial\theta. \end{aligned} \tag{4}$$

Thus the right-hand side of (3) is simply

$$\frac{\sin\theta}{r^3}\frac{\partial}{\partial r}\left(\varrho\nu r^4 \frac{\partial\Omega}{\partial r}\right) + \frac{1}{r \sin^2\theta}\frac{\partial}{\partial\theta}\left(\varrho\nu \sin^3\theta \frac{\partial\Omega}{\partial\theta}\right). \tag{5}$$

The mass conservation law,

$$\partial\varrho/\partial t + \nabla \cdot (\varrho \boldsymbol{u}) = 0, \tag{6}$$

becomes, in spherical coordinates,

$$\partial\varrho/\partial t + r^{-2}\partial(r^2\varrho u_r)/\partial r + (r \sin\theta)^{-1}\, \partial(\varrho \sin\theta u_\theta)/\partial\theta = 0. \tag{7}$$

We shall assume axial symmetry ($\partial/\partial\phi = 0$) throughout. For any situation where rotational forces are important, the azimuthal component of the Navier-Stokes equation is particularly relevant for, when the stress tensor is symmetric (as it is in our case), it implies conservation of angular momentum. To make the connection more transparent, the following considerations may be helpful. The basic equation of magnetohydrodynamics is

$$\partial(\varrho u_i)/\partial t + T_{ik,k} = \varrho g_i, \tag{8}$$

where the stress tensor, T_{ik}, is given by

$$T_{ik} = \varrho u_i u_k + \left(p + \frac{2}{3}\,\varrho\nu\nabla \cdot \boldsymbol{u} + \frac{1}{2}\,\mu^{-1}B^2\right)\delta_{ik} - \varrho\nu(u_{i,k} + u_{k,i}) - \mu^{-1}B_iB_k. \tag{9}$$

Let $\boldsymbol{e}$ be the vector field whose components in spherical and Cartesian coordinates are, respectively,

$$\boldsymbol{e} = (0, 0, r\sin\theta), \qquad \boldsymbol{e} = (-y, x, 0).$$

Multiplying (8) scalarly by $\boldsymbol{e}$, we obtain

$$\partial(\varrho\boldsymbol{e} \cdot \boldsymbol{u})/\partial t + (e_iT_{ik})_{,k} = e_{i,k}T_{ik} + \varrho\boldsymbol{g} \cdot \boldsymbol{e}. \tag{10}$$

Obviously, the crucial term in this equation is the first term on the right-hand side. Written out in detail, it is

$$T_{xx}e_{x,x} + T_{xy}e_{x,y} + T_{yx}e_{y,x} + T_{yy}e_{y,y} \equiv T_{yx} - T_{xy},$$

where the second form follows from the definition of $\boldsymbol{e}$. Thus, if $g_\phi = 0$ and the stress tensor T is symmetric ($T_{xy} = T_{yx}$), we have the conservation law,

$$\partial(\varrho r u_\phi \sin\theta)/\partial t + \nabla \cdot \boldsymbol{t} = 0, \tag{11}$$

where

$$t_i = r\sin\theta\, T_{i\phi}. \tag{12}$$

The meridional components of this vector are

$$\begin{aligned} t_r &= r\sin\theta(\varrho u_r u_\phi - \varrho\nu r\sin\theta \cdot \partial\Omega/\partial r - \mu^{-1}B_rB_\phi), \\ t_\theta &= r\sin\theta(\varrho u_\theta u_\phi - \varrho\nu\sin\theta \cdot \partial\Omega/\partial\theta - \mu^{-1}B_\theta B_\phi), \end{aligned} \tag{13}$$

and we can write

$$\boldsymbol{t} = \varrho r^2\sin^2\theta \cdot [\Omega\boldsymbol{u} - \nu\nabla\Omega - (\mu r\varrho\sin\theta)^{-1}B_\phi\boldsymbol{B}]. \tag{14}$$

The basic equation of angular momentum transport is therefore

$$\frac{\partial}{\partial t}(\varrho r^2 \sin^2\theta\Omega) + \nabla\cdot\left(\varrho r^2 \sin^2\theta\Omega\boldsymbol{u} - \varrho r^2 \sin^2\theta\nu\nabla\Omega - \frac{r\sin\theta}{\mu}B_\phi\boldsymbol{B}\right) = 0. \quad (15)$$

In a laminar flow of a Newtonian fluid there are thus three transporters of angular momentum: the meridional flow, the molecular viscosity and the magnetic field. We consider them in turn. With $g_\phi = 0$ we excluded the angular momentum transport by non-axisymmetric gravity, as it appears, for example, in spiral configurations. It can be included if the term $(\psi_{,i}\psi_{,k} - \psi_{,j}\psi_{,j}\delta_{ik}/2)/4\pi G$ is added to (9) when ψ is the gravity potential and G the gravity constant.

4.2. Viscosity

If neither meridional circulation nor magnetic forces are present, any large-scale non-uniform rotation is always smoothed out in a characteristic time of $\tau_{\text{dec}} \approx R^2/\nu$. When the molecular viscosity and solar radius are used, τ_{dec} is very large:

$$\tau_{\text{dec}} \approx 50 \cdot 10^{20} \cdot 10^{-3}\, s \approx 10^{11}\ \text{yrs}. \quad (16)$$

Were this the whole story, no special explanation of differential rotation would be called for, since it could be thought of as a fossil from the birth of the Sun. The fact, however, that the northern and southern hemispheres have identical rotation laws already makes the fossil hypothesis seem unlikely. Moreover, no other long-lived, hydrodynamical phenomenon has been observed on the Sun. The "true" viscosity may therefore be much larger than the molecular viscosity. Indeed, if the scales of the granules (1,000 km, 10 min) are combined together as an "eddy viscosity", as was done in (2.5), a value of 10^{12}—10^{13} cm^2/s is obtained.

Turbulent diffusivities are larger than the molecular diffusivities by many orders of magnitude. In passing, we note that 10^{13} cm^2/s is also the magnitude required for the magnetic diffusivity if the relatively short decay times of sunspots are to be understood. This is a strong argument in favor of the eddy character of all diffusivities in the Sun. Using $\nu = 10^{13}$ cm^2/s, the decay time of the differential rotation is reduced to about 15 years, and the fossil rotation idea is no longer tenable. The highly intensified small-scale dynamics does not lend itself to the creation of large-scale flow structures unless there is a mechanism to generate them. Fascinatingly, such a mechanism is intrinsic to turbulence, but its understanding requires a more sophisticated insight to turbulent processes than a mere appreciation of the concept of an eddy viscosity.

From the sign of the momentum flux (13), it is evident that angular momentum is transported outwards for sub-rotation ($\partial\Omega/\partial r < 0$) and inwards for super-rotation ($\partial\Omega/\partial r > 0$).

The eddy viscosity is computed for various turbulence models in Chapter 16. Its possible anisotropy, and the implications of anisotropy for rotation theory, are discussed in Chapter 7. An eddy viscosity owes its existence to the Reynolds stresses, and can only be understood after (15) has been reformulated for turbulent media. Laboratory measurements, as well as elementary considerations, show that the ratio ν_T/ν is proportional to Re/Re_{crit}. Here Re_{crit} denotes the value of the Reynolds number at the onset of turbulence.

4.3. Meridional Flow. Lorentz Force

If the angular momentum fluxes from the friction and meridional parts of (15) are to balance, we must have

$$\nabla \cdot (\varrho r^2 \sin^2 \theta \cdot \nu \nabla \Omega) = \varrho \boldsymbol{u} \cdot \nabla (r^2 \Omega \sin^2 \theta), \tag{17}$$

where the stationary form of the continuity equation (6) has been applied. A detailed analysis of (17) is contained in the next Chapter. To obtain a qualitative picture, we suppose that the density and viscosity are constants and assume the rotation law

$$\Omega = \Omega_\odot + \Omega_2 \sin^2 \theta, \tag{18}$$

on and beneath the surface (Ω = constant, Ω_2 = constant). The radial flow may be neglected as it vanishes at the surface, and the meridional flow must vanish at poles and equator, suggesting that

$$u_\theta \simeq U_\odot \sin \theta \cos \theta . \tag{19}$$

The amplitude, $U_\odot$, of the flow may be assumed to be latitude-independent, so that it follows immediately from (17) that

$$5\nu\Omega_2 = U_\odot R \Omega_\odot . \tag{20}$$

Thus, a positive $U_\odot$ (equatorward flow) leads to a positive Ω_2, i.e. to equatorial acceleration. The normalized pole-equator difference of the angular velocity defined in (3.7), is

$$\delta\Omega = -U_\odot R / 5\nu . \tag{21}$$

An equatorward flow of 10 m/s will give $\delta\Omega \approx -2$, but only when ν is given a value, $5 \cdot 10^{12}$ cm^2/s, appropriate to an eddy viscosity. (Had a molecular value been used, an almost arbitrarily small equatorward drift would have sufficed to create the observed equatorial acceleration.)

The argument is quite similar when magnetic fields are used to transport the angular momentum. The equation corresponding to (17) is then

$$\nabla \cdot (\varrho r^2 \sin^2 \theta \nu \nabla \Omega) = -\mu^{-1} \boldsymbol{B} \cdot \nabla (r \sin \theta \, B_\phi). \tag{22}$$

As is customary, the magnetic field is written as the sum of toroidal and poloidal parts.

$$\boldsymbol{B} = \boldsymbol{B}_\phi + \nabla \times \boldsymbol{A}, \tag{23}$$

where both of the vectors $\boldsymbol{A}$ and $\boldsymbol{B}_\phi$ are purely zonal. The most representative field (the dipole) can be described by

$$A_\phi \simeq \check{A} \sin\theta, \qquad B_\phi \simeq \check{B} \sin\theta \cos\theta, \tag{24}$$

which becomes, in spherical coordinates,

$$\boldsymbol{B} \simeq (\check{B}_r \cos\theta, \qquad \check{B}_\theta \sin\theta, \qquad \check{B} \sin\theta \cos\theta). \tag{25}$$

The toroidal field $\check{B}$ vanishes at the solar surface. After making the same simplifications with respect to density, viscosity and angular velocity as we made earlier, and applying (22) to the surface layer, we obtain

$$\nabla \cdot (\varrho r^2 \sin^2 \theta \nu \nabla \Omega) = -\mu^{-1} \check{B}_r R \cos^2\theta \sin^2\theta \, \mathrm{d}\check{B}_\phi/\mathrm{d}r, \tag{26}$$

and therefore

$$10\nu\Omega_2 = -(\varrho\mu)^{-1} \check{B}_r R \cdot \mathrm{d}\check{B}_\phi/\mathrm{d}r. \tag{27}$$

It follows that

$$\delta\Omega = \frac{R\check{B}_r \, \mathrm{d}\check{B}_\phi/\mathrm{d}r}{10\varrho\nu\mu\Omega_\odot} \approx -10^{-2} \cdot \mathrm{sign}\,(B_r B_\phi), \tag{28}$$

where the product $B_r \, \mathrm{d}B_\phi/\mathrm{d}r$ represents the Lorentz force. Equatorial acceleration appears for a negative Lorentz force or, what is the same here, for a positive $B_r B_\phi$. Our numerical estimate is based on the surface values $B_r \approx 1$ Gauss, $\varrho \approx 10^{-3}$ g/cm^3, $\nu \approx 10^{13}$ cm^2/s and $B_\phi \approx 100$ Gauss at a depth of 10^4 km (Schüssler, 1981).

We conclude that the Lorentz force also plays an active role in causing non-uniform rotation but not (at least for the Sun) with the same efficacy as the meridional flow does. A warning should be given, however, not to oversimplify the estimates, since they are sensitive to the density profile, as well as to that of the toroidal magnetic field, and these are steep near the surface.

A detailed analysis of the role of the Lorentz force will be presented in Chapter 12, and there applied to the solar dynamo and the observed torsional oscillations of the Sun. With the channel flow model developed there, we will be able to reach some conclusions concerning a suspicion often voiced by solar physicists: Is it really true that the strong increase of density with depth will prevent a flow pattern from being generated magnetically in the bulk of the SCZ? The implicit belief is that, at all depths, $u\varrho =$ constant, where u is the flow amplitude. Simple calculations do not support this idea. The kinematic viscosity prevents the formation of vertical profiles that are too steep. We will discover that when the density gradient is large $u(x_i)\varrho(x_i)$

$\approx 10^4 u(1)\varrho(1)$. This result should be relevant to the theory of torsional oscillations. The observed magnetically-generated flow pattern cannot exist in the outermost layers without involving the deeper layers of the SCZ also.

4.4. The Main Transporter: Reynolds Stress, Turbulence

Transparently, equation (15) does not to include any eddy diffusivities. The question arises as to why the term "eddy viscosity", as it is used in turbulence theory, should be relevant. The actual reason is, of course, the presence of velocity fluctuations in the Sun's outer layers, i.e. in the SCZ. The temperature decreases more rapidly with r in the SCZ than does the adiabatic gradient, and convective instability sets in. Cowling (1951) generalized the Schwarzschild criterion to allow for the presence of a basic rotation. He found from linear perturbation theory that a mode, characterized by the wave-numbers l, m, n (corresponding to the direction radially from the axis of rotation, the azimuthal ϕ-direction and the z-direction parallel to $\boldsymbol{\Omega}$, respectively) becomes unstable if (the negative of) the temperature gradient exceeds a modified adiabatic gradient, or more precisely if

$$-\nabla T \geqq (\varrho C_p)^{-1} \nabla p + 4[m^2 + (l \cos\theta - n \sin\theta)^2]^{-1} \Omega^2 T n^2 \boldsymbol{g}, \tag{29}$$

where T is the temperature. Thus, a mode with small n, that is one for which $\partial/\partial z \approx 0$, should be the first to become unstable. We should recall here that the layer as a whole will be unstable with respect to convection when any single mode is linearly unstable in this way. At sufficiently large super-adiabaticity, nonlinear interactions excite a whole spectrum of modes. The non-stationary character of the velocity field at the solar surface indicates that the SCZ is, indeed, in just such a turbulent state (Stix, 1983). Actually, two distinct questions should be considered. One is whether the condition for the onset of turbulence is altered by the presence of rotation, and the second is whether turbulence, once established in a rotating star, is substantially different in its general character from the turbulence that would occur in a similar, but non-rotating, star. Only the second query is considered in this book.

In a turbulent medium all field quantities vary irregularly in space and time. Let F be such a fluctuating field, considered as a random function. The corresponding mean field, $\overline{F}$, is then defined as the expectation value of F in an ensemble of identical systems, and the fluctuating part, F', of F is defined to be the difference between F and $\overline{F}$:

$$F = \overline{F} + F'. \tag{30}$$

The following relations, often called *the Reynolds rules*, hold

$$\begin{aligned} &\overline{\overline{F}} = \overline{F}, \qquad \overline{F'} = 0, \\ &\overline{F + G} = \overline{F} + \overline{G}, \qquad \overline{\overline{F} + \overline{G}} = \overline{F} + \overline{G}, \qquad \overline{F'\overline{G}} = 0, \end{aligned} \tag{31}$$

where G is a second fluctuating field. The averaging operator commutes with the differentiation and integration operators in both space and time.

These rules hold exactly in two practically important cases. The first is when the motion is taken to be a random variable and the averages are the corresponding expectation values. The motion has then widely been called "turbulence" and the language of turbulence theory has been used. The second case is when averages with respect to space and/or time can be used. In particular, averaging with respect to the longitude, ϕ, proves to be of especial value. The mean of a field then becomes synonymous with its axisymmetric part, while its "fluctuating part" becomes its asymmetric part. For example, because of its independence from ϕ, the global angular velocity, Ω, is (apart from the factor $r \sin \theta$) identical to the axisymmetric part of the longitudinal flow. Such a definition of the mean is basic to the theory of the nearly-symmetric dynamo (see Braginsky, 1976), as well as to the non-axisymmetric numerical models of the solar differential rotation developed by Gilman, starting in 1972. Gilman was able to calculate explicitly the interaction of the modes which produced the observed differential rotation. Naturally, the motion was called "convection", and the terminology of convection theory was used throughout this work.

The crucial question is whether there are *essential* differences between these two ways of treating nonlinear mode interactions. In principle, the relationship between the statistical and the space/time averages is given by the ergodic theorem, about which much has been written. According to this theorem, the two averages are equal if the flow is stationary and the averaging intervals are sufficiently large. For naturally occurring processes we can, of course, expect only a pseudo-stationality, with respect to time or space scales that are large compared with the scales of the turbulence. Intuitively we may expect that real observations will, for the most part, be concerned with temporal averages rather than with statistical ones (see Kröner, 1971). Nevertheless, our treatment will be analytical, and consequently we shall prefer to use the language of turbulence theory; we must appeal, therefore, to the ergodic theorem.

Let us now identify the fields, $\boldsymbol{F}$ and G, we introduced above with the (vector) fields $\boldsymbol{u}$ and $\boldsymbol{B}$:

$$\boldsymbol{u} = \overline{\boldsymbol{u}} + \boldsymbol{u}', \qquad \boldsymbol{B} = \overline{\boldsymbol{B}} + \boldsymbol{B}', \tag{32}$$

density fluctuations being ignored. Then equation (1) becomes

$$\varrho[\partial\overline{\boldsymbol{u}}/\partial t + (\overline{\boldsymbol{u}} \cdot \nabla)\, \overline{\boldsymbol{u}}] = -\nabla \cdot (\varrho Q) - \nabla \overline{p} + \varrho \boldsymbol{g} + \nabla \cdot \pi + \overline{\boldsymbol{j}} \times \overline{\boldsymbol{B}} + \overline{\boldsymbol{j}' \times \boldsymbol{B}'}, \tag{33}$$

where Q_{ij} denotes the one-point correlation tensor

$$Q_{ij} = \overline{u_i'(\boldsymbol{x}, t)\, u_j'(\boldsymbol{x}, t)}\,. \tag{34}$$

The additional stresses, $-\varrho Q_{ij}$, appearing in (33) originate from the turbulence and are known as "Reynolds stresses"; they are determined, essentially, by the correlation tensor of the turbulence. This tensor is, by definition, symmetric with respect to its indices i and j: there is no place in turbulence theory for choices of Q_{ij} that are not symmetric. The Reynolds equation (33) is identical to the usual Navier-Stokes equation with all quantities replaced by their mean values, except for the additional Reynolds stresses

$$\pi_{ij}^{(R)} = -\varrho Q_{ij} \tag{35}$$

(and, of course, the turbulent Maxwellian stress $\pi_{ij}^{(M)}$). The tensor (35) provides the mean momentum transport by the fluctuating motions, its divergence, $\nabla \cdot \pi^{(R)}$, being the actual momentum exchanged by the turbulence and the mean flow. Whenever turbulent motion exists in a star, the viscous stress π_{ij} is negligible compared with the Reynolds stress $\pi_{ij}^{(R)}$.

The central problem of this formulation lies in the fact that the three scalar equations (33) introduce new, unknown quantities, namely the six components of Q_{ij}. Following Boussinesq and others, it is usually supposed that the turbulent stresses should take the same form as the viscous stresses, but with the molecular viscosity replaced by an eddy viscosity (2.5). Historically, the tensorial aspect at first played only a minor role. The Boussinesq relation was originally written only in a Cartesian form such as (2.4), but including the eddy viscosity (2.5) of free turbulence. The mixing length was usually taken to be proportional to the size of the larger eddies. In stellar interiors the mixing length is generally assumed to be of the same order as the density or pressure scale-heights. As such a scale-height increases downwards while the r.m.s. velocity correspondingly decreases, the product of the two is approximately constant throughout the SCZ. Without exception we shall, in all our discussions below, assume that the eddy viscosity is constant. In summary, it is customarily supposed that the correlations, Q_{ij}, are functions of the mean flow, and are given by the relation (2.2). Equations (2.3) provide two important examples of this relation for spherical geometry.

If these ideas are applied directly to the conservation law (15) for the angular momentum we immediately discover that

$$\partial(\varrho r^2 \sin^2\theta \cdot \Omega)/\partial t + \nabla \cdot \Big\{ r \sin\theta \cdot [\varrho r \sin\theta \cdot \Omega \bar{\boldsymbol{u}} + \varrho\, \overline{u_\phi' \boldsymbol{u}'} - \mu^{-1}(\overline{B_\phi' \boldsymbol{B}'} + \bar{B}_\phi \bar{\boldsymbol{B}})] \Big\} = 0 . \tag{36}$$

Only two components of the correlation tensor appear in this equation, namely its off-diagonal ("cross") components, $Q_{r\phi}$ and $Q_{\theta\phi}$, which represent the radial and latitudinal transport of turbulent angular momentum. It is a deep and far from trivial result that, because of the effect of the Coriolis force on a prevailing turbulence that is non-isotropic, these quantities do not, in general, vanish. It should particularly be borne in mind that, on the one hand, an upward (downward) motion is deflected to the west (east), so that the product $u_r'u_\phi'$ is negative; on the other hand, an eastward (westward) velocity acquires an upward (downward) component from the Coriolis force, so that $u_r'u_\phi'$ is positive. Therefore, provided the original horizontal and vertical motions are statistically different, as will be the case for anisotropic turbulence, the fluctuating velocity components $u_r'u_\phi'$ will be correlated and their product will have a non-zero average: $Q_{r\phi} \neq 0$. This situation even holds when the mean motion is one of solid-body rotation!

4.5. Turbulent Angular Momentum Transport (TMT)

We look next for a convective pattern which implies a non-vanishing turbulent angular momentum transport (TMT). Initially, our argument will be geometrical, rather than physical. Let us analyse the situation at any point, P, within the star, and let $\mathring{g} = OP/|OP|$ be the unit radial vector, where O is the center of the star. Denoting an arbitrary unit vector by $\mathring{e}$, we may express the velocity fluctuation as

$$u' = a[\mathring{e} \times (\mathring{e} \times \mathring{g})] + b(\mathring{e} \times \mathring{g}) + c\mathring{e}, \tag{37}$$

where a, b, c are arbitrary and uncorrelated. It is then easy to write down the components (in spherical coordinates):

$$\begin{aligned} u_r' &= a[(\mathring{e}\cdot\mathring{g})^2 - 1] + c\mathring{e}\cdot\mathring{g}, \\ \sin\theta\cdot u_\theta' &= a(\mathring{e}\cdot\mathring{g})\,[(\mathring{g}\cdot\mathring{\Omega})\,(\mathring{e}\cdot\mathring{g}) - \mathring{e}\cdot\mathring{\Omega}] + b[\mathring{e}\cdot(\mathring{\Omega}\times\mathring{g})] \\ &\quad + c[(\mathring{e}\cdot\mathring{g})\,(\mathring{g}\cdot\mathring{\Omega}) - \mathring{e}\cdot\mathring{\Omega}], \\ \sin\theta\cdot u_\phi' &= a(\mathring{e}\cdot\mathring{g})\,[\mathring{e}\cdot(\mathring{\Omega}\times\mathring{g})] + b[\mathring{e}\cdot\mathring{\Omega} - (\mathring{e}\cdot\mathring{g})\,(\mathring{g}\cdot\mathring{\Omega})] \\ &\quad + c[\mathring{e}\cdot(\mathring{\Omega}\times\mathring{g})]. \end{aligned} \tag{38}$$

(Here $\mathring{\Omega}$ is a unit vector in the z-direction). A short calculation yields

$$\begin{aligned} \sin\theta\,\overline{u_r'u_\phi'} &= (\mathring{e}\cdot\mathring{g})\,[\mathring{e}\cdot(\mathring{\Omega}\times\mathring{g})]\,\{\overline{c^2} - [1 - (\mathring{e}\cdot\mathring{g})^2]\,\overline{a^2}\}, \\ \sin^2\theta\cdot\overline{u_\theta'u_\phi'} &= [\mathring{e}\cdot(\mathring{\Omega}\times\mathring{g})]\,[(\mathring{e}\cdot\mathring{g})\,(\mathring{g}\cdot\mathring{\Omega}) - \mathring{e}\cdot\mathring{\Omega}]\,[(\mathring{e}\cdot\mathring{g})^2\overline{a^2} - \overline{b^2} + \overline{c^2}]. \end{aligned} \tag{39}$$

Thus, for non-vanishing cross-correlations it is required that

$$\mathring{e}\cdot\mathring{g} \neq 0, \qquad \mathring{e}\cdot(\mathring{g}\times\mathring{\Omega}) \neq 0, \tag{40}$$

i.e. the characteristic vector $\mathring{e}$ is neither perpendicular to $\mathring{g}$, nor co-planar with $\mathring{g}$ and $\mathring{\Omega}$. Obviously, it must lie somewhere between the meridional plane and the azimuthal direction. What role does $\mathring{e}$ play? Of course, it reflects the anisotropy of the turbulence, since completely isotropic fields never possess cross-correlations. If we choose, for example $a = b = 0$, the field under consideration is a special one with motion in only one direction, namely $\mathring{e}$. This exotic case leads to

$$\sin\theta Q_{r\phi} = -(\mathring{e}\cdot\mathring{g})\,[\mathring{e}\cdot(\mathring{g}\times\mathring{\Omega})]\,\overline{c^2},$$
$$\sin^2\theta Q_{\theta\phi} = -[\mathring{e}\cdot(\mathring{g}\times\mathring{\Omega})]\,\big((\mathring{e}\cdot\mathring{g})\cos\theta - (\mathring{e}\cdot\mathring{\Omega})\big)\,\overline{c^2}. \tag{41}$$

The direction of $\mathring{e}$ may not lie in the meridional plane or be perpendicular to it, and it must be sufficiently oblique to ensure non-vanishing zonal momentum fluxes. This result, however, is almost trivial since otherwise the field considered would have no ϕ-component.

The case $c = 0$ may be more instructive. The turbulent motion is then forced to lie in a plane perpendicular to $\mathring{e}$, which therefore provides an "axis" for the motion. The correlations sought are derived from (39) by substituting $c = 0$, i.e.

$$\sin\theta Q_{r\phi} = (\mathring{e}\cdot\mathring{g})\,[\mathring{e}\cdot(\mathring{\Omega}\times\mathring{g})]\,[(\mathring{e}\cdot\mathring{g})^2 - 1]\,\overline{a^2},$$
$$\sin^2\theta Q_{\theta\phi} = [\mathring{e}\cdot(\mathring{\Omega}\times\mathring{g})]\,[(\mathring{e}\cdot\mathring{g})\cos\theta - (\mathring{e}\cdot\mathring{\Omega})]\,[(\mathring{e}\cdot\mathring{g})^2\overline{a^2} - \overline{b^2}]. \tag{42}$$

Thus, again, it is essential that the "axis", $\mathring{e}$, deviates from the meridional plane, as well as from the azimuthal direction. It may be seen from these expressions that anisotropic fields with $\mathring{e}\cdot\boldsymbol{u}' = 0$ produce TMT if their axes are sufficiently tilted in the sense of (40). This finding corresponds to the picture given by Gilman (1981) in which the planes of motion are, indeed, clearly aligned (see Fig. 2.9).

The tilt of the "axis" of the convective elements necessary for finite TMT cannot exist in a medium that has only one preferred direction (e.g. a non-rotating star). An additional preferred direction (e.g. $\boldsymbol{\Omega}$) must be present. Only then will the convective motions be "sufficiently complex". It is not easy to provide a three dimensional representation of the streamlines but, for the simpler case of incompressible flow, two dimensions suffice to give an

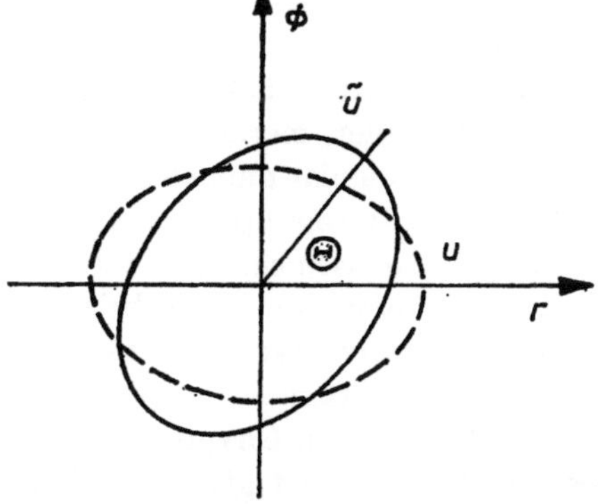

Fig. 4.1. Streamlines for an anisotropic eddy in the (say) $r\phi$-plane. The aligned configuration (broken line) does not possess a cross-correlation. The tilted cell, however, possesses correlations $Q_{r\phi}$ in agreement with (45).

impression. Fig. 4.1 shows a streamline in the $r\phi$-plane which is tilted at an angle Θ to the axis. The coefficients a and b used above are now correlated and $c = 0$. The tilted pattern $\tilde{\boldsymbol{u}}$ can be expressed in terms of the original pattern, $\boldsymbol{u}$, through the transformation

$$\begin{pmatrix} \tilde{u}_r \\ \tilde{u}_\phi \end{pmatrix} = \begin{pmatrix} \cos\Theta & \sin\Theta \\ -\sin\Theta & \cos\Theta \end{pmatrix} \begin{pmatrix} u_r \\ u_\phi \end{pmatrix}. \tag{43}$$

Among the correspondingly transformed correlations there is, for example, the relation

$$\tilde{Q}_{r\phi} = \frac{1}{2}\sin 2\Theta \cdot (Q_{\phi\phi} - Q_{rr}) + \cos 2\Theta \cdot Q_{r\phi}, \tag{44}$$

so that, even if the original pattern $\boldsymbol{u}$ has no cross-correlations, the tilted pattern may have one:

$$\tilde{Q}_{r\phi} = \frac{1}{2}\sin 2\Theta \cdot (Q_{\phi\phi} - Q_{rr}). \tag{45}$$

Hence, sufficiently anisotropic flow systems ($Q_{rr} \neq Q_{\phi\phi}$) produce TMT provided they are not aligned with the coordinate axes. The tilt Θ needed can be thought of as being a consequence of the basic rotation.

One should not forget, however, that the mere existence of TMT does not by itself give the differential rotation, but only its non-diffusive part, i.e. the Λ-effect. Finally, it should be noted that differences between related quantities appear in expression $(42)_2$. The horizontal cross-correlation may therefore be sensitive to the turbulence model chosen.

We turn now to physical aspects of TMT. At first sight, it may seem that the nonzero TMT provided by the Boussinesq relations (2.3) can generate the non-uniform rotation with ease. At a deeper level, this observation adds little to our understanding of differential rotation: to be sure the viscosity is effectively enhanced and, therefore, an Ω-profile is smoothed out more rapidly than if no turbulence were present. The existence of non-uniform rotation requires angular momentum transporters that do not vanish when the rotation is uniform. As was demonstrated earlier, meridional flow is one of these. In contrast to viscosity, it transports angular momentum even for a rigid rotation and for this reason it necessarily sets up a differential rotation. The Reynolds stress *in its Boussinesq formulation* cannot create the necessary non-diffusive part of the TMT. It has therefore come to be believed for a considerable time that Reynolds stresses generally have only a restricted significance for the theory of the non-uniform rotation.

But this is not true. As we have already described in Chapter 2, Biermann (1951) argued that the Boussinesq relations do not give the whole truth for *anisotropic* turbulence. For a quantitative estimate of the angular momentum flux he constructed a spectral function for the turbulence with two parts. The first part represented a completely isotropic motion, and the second motion is solely in the direction of gravity, $\boldsymbol{g}$. The relative sizes of these two contributions reflect the degree of anisotropy of the turbulence. In the equatorial plane, to which Biermann restricted his argument, the angular momentum flux due to the isotropic part is

$$-A_1 r^2 \, \mathrm{d}\Omega/\mathrm{d}r, \tag{46}$$

where A_1 is an eddy viscosity. The eddies of the second part of the turbulence displace the angular momentum along the radius r, so creating the momentum flux

$$-A_2 \, \mathrm{d}(r^2\Omega)/\mathrm{d}r \equiv -A_2 r^2 \, \mathrm{d}\Omega/\mathrm{d}r - 2A_2 r\Omega, \tag{47}$$

where A_2 is the analogue of A_1 associated with the anisotropy. The assumption of vanishing net flux, therefore, yields directly

$$(A_1 + A_2) \, \mathrm{d}\Omega/\mathrm{d}r = -2A_2\Omega/r, \tag{48}$$

which is equivalent to equation (2.10). It is the last term on the right-hand side of (47) that creates the non-uniformity of the angular velosity Ω. And it is also this term which survives even when uniform rotation is assumed. It is similar, in some respects, to the α-effect of mean-field MHD where, for certain kinds of turbulence, correlations survive even when the prevailing magnetic field is uniform

$$(\overline{\boldsymbol{u}' \times \boldsymbol{B}'})_i = \cdots \alpha_{ik}\overline{B}_k: \tag{49}$$

It is this analogy which prompted us to call the property

$$Q_{ij} = \cdots + \Lambda_{ijk}\Omega_k \tag{50}$$

the Λ-effect. This name reflects the Longitudinal character of the effect (ϕ-correlations; ϕ-component of the Reynolds equation) and it also serves to remind us of Lebedinski, who was the first to recognize the effect via the formal concept of an anisotropic eddy viscosity (cf. the historical outline in Chapter 2). We already used this notation in our first publication in this field (Krause and Rüdiger, 1974a).

4.6. Structure of the Λ-Effect

The tensor Λ_{ijk} must be symmetric in its indices i and j. Since Q_{ij} is polar, we see from (50) that Λ_{ijk} must be axial, i.e. it must change sign if the (say) right-handed coordinate system is replaced by a left-handed one. In other words, Λ_{ijk} reverses sign under an improper orthogonal transformation while Q_{ij}, being polar, does not. As a preferred direction (anisotropy!) we simply choose the radial direction $\mathring{\boldsymbol{g}}$. In this case only $\mathring{\boldsymbol{g}}$, δ_{ij} and ε_{ijk} can be used to construct the tensor Λ_{ijk}. It follows that

$$\Lambda_{ijk} = \Lambda_V(\varepsilon_{ipk}\mathring{g}_j + \varepsilon_{jpk}\mathring{g}_i)\,\mathring{g}_p. \tag{51}$$

No other tensorial form exists with the required symmetry with respect to i and j. In particular, it should be noted that Λ_{ijk} can be formed only with an even number of g factors. The mean features of those turbulence models that maintain differential rotation are, therefore, unchanged under the transformation $\boldsymbol{g} \to -\boldsymbol{g}$. More generally, only that part of the spectrum, which does not distinguish between "up" and "down", is effective in accelerating the solar equator.

To obtain the required cross-correlations using the result (51), we substitute it into (50) and then multiply by $\hat{g}_i \hat{e}_j$ and $\varepsilon_{imn} \hat{e}_m \hat{g}_n \hat{e}_j$ in turn, remembering that

$$Q_{r\phi} \equiv \hat{g}_i \hat{e}_j Q_{ij}, \qquad Q_{\theta\phi} \equiv \varepsilon_{imn} \hat{e}_m \hat{e}_j \hat{g}_n Q_{ij}. \tag{52}$$

Now $\hat{\boldsymbol{e}}$ is the unit vector in the eastward direction. By following this procedure we obtain the simple but, nevertheless, quite general expressions

$$Q_{r\phi} = \sin\theta\,(-\nu_{vv} r \partial\Omega/\partial r + \Lambda_V \Omega), \qquad Q_{\theta\phi} = -\nu_{hh} \sin\theta\, \partial\Omega/\partial\theta, \tag{53}$$

which are characterized by three free parameters. For presentational purposes, a matrix representation in spherical polar coordinates may be useful:

$$Q_{ij} = \begin{pmatrix} \overline{u_r'^2} & 0 & 0 \\ 0 & \overline{u_\perp'^2} & 0 \\ 0 & 0 & \overline{u_\perp'^2} \end{pmatrix} - \sin\theta \begin{pmatrix} 0 & 0 & \nu_{vv} r \partial\Omega/\partial r \\ 0 & 0 & \nu_{hh} \partial\Omega/\partial\theta \\ \nu_{vv} r \partial\Omega/\partial r & \nu_{hh} \partial\Omega/\partial\theta & 0 \end{pmatrix}$$
$$+ \sin\theta \cdot \Omega \begin{pmatrix} 0 & 0 & \Lambda_V \\ 0 & 0 & 0 \\ \Lambda_V & 0 & 0 \end{pmatrix}. \tag{54}$$

This may be written in the more symbolic form

$$Q_{ij} = Q_{ij}^{(0)} + Q_{ij}^{(d)} + Q_{ij}^{(r)}, \tag{55}$$

where the superscript (o) indicates "original", (d) "differential" and (r) "rigid". Only the final terms, $Q_{ij}^{(d)} + Q_{ij}^{(r)}$, influence the rotation, and only $Q_{ij}^{(r)}$ contains the pure Λ-effect (cf. Kichatinov, 1986, 1987).

The consequences of this effect for the theory of differential rotation are developed in Chapter 6, while Chapters 14 and 15 provide detailed calculations of Λ.

Equation (51) may be regarded as the first term in an expansion for small Ω. We may add successively higher power of Ω to Λ_{ijk}, complicating the corresponding correlation Q_{ij} [given by the ansatz (50)] as we do so. A systematic look at all the possibilities that can contribute to TMT helps to clarify the situation. The only tensor Λ_{ijk} that can be formed without any $\boldsymbol{\Omega}$ has already been given by (51). At the next level of the approximation where one $\boldsymbol{\Omega}$ is included in Λ_{ijk}, two possibilities arise:

$$\hat{g}_i \hat{g}_j \Omega_k, \qquad (\hat{g}_i \Omega_j + \hat{g}_j \Omega_i)\, \hat{g}_k. \tag{56}$$

Both these tensors are symmetric in i and j and "axial". Neither of the forms (56), however, produces a nonzero $Q_{r\phi}$, so we may ignore them.

The next step is to include two $\boldsymbol{\Omega}$ vectors. Again we are only interested in contributions to $Q_{r\phi}$. The possible tensor combinations are

$$\Omega^2(\mathring{g}_i\varepsilon_{jpk} + \mathring{g}_j\varepsilon_{ipk})\,\mathring{g}_p\,, \qquad (\mathring{\boldsymbol{g}}\cdot\boldsymbol{\Omega})^2\,(\mathring{g}_i\varepsilon_{jpk} + \mathring{g}_j\varepsilon_{ipk})\,\mathring{g}_p\,,$$
$$(\mathring{\boldsymbol{g}}\cdot\boldsymbol{\Omega})\,(\Omega_i\varepsilon_{jpk} + \Omega_j\varepsilon_{ipk})\,\mathring{g}_p, \tag{57}$$

and $(57)_3$ introduces a qualitatively new feature.

We have shown that, up to third order in the angular velocity, the correlation tensor (50) is

$$Q_{ij} = \cdots + \{[A + B\Omega^2 + (\mathring{\boldsymbol{g}}\cdot\boldsymbol{\Omega})^2 C]\,(\mathring{g}_i\varepsilon_{jpk} + \mathring{g}_j\varepsilon_{ipk})$$
$$+ (\mathring{\boldsymbol{g}}\cdot\boldsymbol{\Omega})\,D(\varepsilon_{ipk}\Omega_j + \varepsilon_{jpk}\Omega_i)\}\,\mathring{g}_p\Omega_k\,, \tag{58}$$

and the consequences of this form become apparent when the related TMT are evaluated. One obtains

$$Q_{r\phi}^{(r)} = -[(A + B\Omega^2 + C\Omega^2\cos^2\theta)\sin\theta + D\Omega^2\sin\theta\cos^2\theta]\,\Omega\,,$$
$$Q_{\theta\phi}^{(r)} = D\Omega^3\sin^2\theta\cos\theta\,. \tag{59}$$

It can be seen that the new tensor appearing in (57), involving the term in D, produces $Q_{\theta\phi}$-correlations, while the other terms involving Ω^3 only modify the radial TMT. With the definitions (2.20) of the vertical and horizontal flux quantities Λ_V and Λ_H, we are able finally to write

$$\Lambda_V = -\,[A + B\Omega^2 + (C + D)\,\Omega^2\cos^2\theta]\,,$$
$$\Lambda_H = D\Omega^2\sin^2\theta\,. \tag{60}$$

Note that Λ_H is of higher order in Ω than Λ_V and that it vanishes at the poles. The quantity Λ_V varies from poles to equator so that the notation of (2.21) makes sense.

The terms which appear when higher orders in Ω are included in the expansion of Q_{ij} do not essentially influence the results. They only add to Λ_V and Λ_H powers of $\sin\theta$ and $\cos\theta$ higher than those appearing in (60).

4.7. Boundary Conditions

We cannot close this Chapter without examining the conditions that have to be applied at the top and bottom boundaries of the convection zone. In all our considerations we shall assume that no angular momentum can enter the zone and, for reasons of stationarity, that no angular momentum can leave it either. These requirements partially fulfil the stress-free conditions at the spherical surfaces of the SCZ. According to (11),

$$t_r = 0 \quad \text{at} \quad x = x_i, 1\,, \tag{61}$$

hence

$$\bar{u}_r\bar{u}_\phi + Q_{r\phi} - \nu r \sin\theta \cdot \partial\Omega/\partial r - (\mu\varrho)^{-1}(\bar{B}_r\bar{B}_\phi + \overline{B_r'B_\phi'}) = 0. \tag{62}$$

These conditions can be simplified for two reasons. First, u_r vanishes at $x = x_i$ and at $x = 1$. Second, we may remember that the turbulent viscosity is included in the Reynolds stresses Q_{ij}, so that the ν in (62) refers only to the molecular viscosity, which may generally [though not in deriving (66) below] be assumed negligible compared with the turbulent viscosity. Thus, assuming the SCZ is turbulent right up to its boundaries, (62) reduces to

$$Q_{r\phi} = (\mu\varrho)^{-1}(\bar{B}_r\bar{B}_\phi + \overline{B_r'B_\phi'}), \tag{63}$$

and, if magnetic effects are also neglected, the condition becomes

$$Q_{r\phi} = 0 \quad \text{at} \quad x = x_i, 1. \tag{64}$$

The lateral stress must also vanish, and in the same approximation,

$$Q_{r\theta} = 0 \quad \text{at} \quad x = x_i, 1. \tag{65}$$

The fact that the lower boundary can be assumed to be stress-free depends on the relatively low molecular and radiative viscosities beneath the SCZ. Thus no torque is exerted on the convection zone by the stellar interior. To drop this assumption and to apply a condition different from (64) is tantamount to allowing angular momentum to be carried through the lower boundary of the SCZ. Most frequently, the condition applied instead of (64) is $\Omega(x_i) = \text{constant}$. Models of this kind are called "rigid-free" in contrast to the "free-free" model defined by (64). If there is no turbulence in the layers adjacent to the boundaries, conditions (64) and (65) are replaced by the traditional relations

$$\partial\Omega/\partial r = \partial(r^{-1}\bar{u}_\theta)/\partial r = 0 \quad \text{at} \quad x = x_i, 1. \tag{66}$$

[The first of these is easily obtained, in the non-magnetic case where (66) applies, by setting $Q_{r\phi} = 0$ in (62) and by recalling that $\bar{u}_r = 0$ on the boundaries. For the second condition see e.g. $(5.55)_4$.]

Part II
Macroscopy

In this, the main part of the book, the different transporters of angular momentum are analysed in detail. No difficulties are encountered in Chapter 5, where we examine the transport of angular momentum by meridional flow. Similarly, when in Chapter 12 we investigate transport by the mean magnetic field we encounter no serious obstacle, partly because we there exclude the role played by the fluctuating magnetic fields. Matters are not so simple when we consider the Reynolds stress transporter. We shall appeal to the Λ-effect mechanism to which, because of its role as the non-diffusive part of the turbulent angular momentum transport (TMT), is intimately linked the idea of an eddy viscosity. The only known situation in which eddy diffusivities can be employed is when the scale of fluctuating and mean quantities are clearly separated. Thus we cannot regard the giant cells as being "just another kind of eddy" in the turbulence of the solar convection zone (SCZ). They are simply too big; the required clear separation between their scales and those of the differential rotation does not exist. We can, however, picture the turbulence as an agglomoration of convectively-produced fluctuations on all scales smaller than those of some "giant granulation", which we characterise by the mixing length and turnover time evaluated at the bottom of the SCZ. On the one hand, current observations of the connection that exists between the activity and the rotation of solar-type stars encourage this view. They reveal that the maximum turnover time in the convection zone of such a star is one of the most crucial factors in determining, for example, the period of its cyclic chromospheric activity (Noyes *et al.*, 1984). On the other hand, we shall find that the strong outward decrease of eddy scales implied by mixing length theory are not consistent with our models.

Taking due note of the two scales on which motion exists, we shall have to construct the tensors necessary to describe the effects of turbulence. These tensors, which for example comprise the Reynolds stress tensor, include coefficients whose values are, at this stage, unknown. We shall call them "free parameters". We must emphasize that ignorance of these parameters in no way invalidates the tensor forms we construct. And ultimately they can be legitimately determined by comparing the predictions of the theory with reality. This balance between theory and observation is in harmony with the attitude of modern nonlinear mechanics: the effort that would have to be expended to calculate these coefficients from first principles would be simply too great to be seriously contemplated. Our approach is, then, semi-empirical: the general form of the tensors demands a secure theoretical foundation in principles of invariance, but to particularise them to phenomena on the Sun requires that they be fitted to the observations.

To summarize the method we employ: first, the overall structure of the important tensors is determined, and then the unknown coefficients they introduce are determined by applying the theory to particular situations for which theoretical results can be compared with observable features of the Sun. For example, in our final model for the solar differential rotation (Chapters 10 and 11), some of the free parameters must be given special values before the theory can account for the slowness of the meridional drift seen on the surface of the Sun. We could find no clear reason, other than this choice of coefficients, why the meridional circulation should be so inhibited. Only one particular set of free parameters makes the drift slow. A parameter set, chosen in such a way, may have other implications that provide a consistency check. In the example just quoted, the fit to the meridional motion implies something about the radial profile of the differential rotation, and this might some day be empirically available, so providing an *a posteriori* check of our theory.

When we consider the turbulent transport of angular momentum in a convection zone for which the basic rotation is not slow, we immediately discover that three free parameters arise. When we wish also to study the effect of rotation on the turbulent transport of heat (Chapters 8 and 9), we find that two further coefficients enter. Thus, generally, five free parameters arise in any heuristic theory of mean-field solar hydrodynamics. If, however, the rotation is so small that the Coriolis number, $\Omega^* = 2\tau_c\Omega$ (which is also the reciprocal of the Rossby number) is much smaller than unity, then only one parameter $V^{(0)}$ (the "fundamental mode" of the Λ-effect) survives. It is, however, impossible to reproduce the observed profile of solar meanflow using this mode alone. It is essential to include the other four coefficients also.

Some technical remarks should be made. Unlike the coming developments of Part III, we have in this section chosen to use the non-rotating, or "inertial", frame of reference. The fractional depth of the SCZ is assumed to be 0.3. The radiative angular momentum transport, as well as the radiative transport of heat, is ignored. This assumption is, of course, readily defended for the angular momentum, but it is not so obviously justified in the case of heat. Future calculations will have to include a layer, 500—1000 km thick, above the top of the SCZ through which energy is mainly transported by radiation. A brief estimate in Chapter 8 suggests that such a layer does indeed possess screening properties.

Chapter 5. Meridional Circulation and Differential Rotation

Since we accepted, in the previous Chapter, meridional flow as one of the main transporters of angular momentum, it is appropriate that a detailed investigation of its influence should now follow. The basic equation is simply (4.17), in which we may express the circulation by a stream function A:

$$\bar{u}_r = (r^2 \varrho \sin\theta)^{-1}\,\partial A/\partial\theta\,, \qquad \bar{u}_\theta = -(r\varrho \sin\theta)^{-1}\,\partial A/\partial r\,. \tag{1}$$

It can be seen immediately that the streamlines are A = constant, since $\bar{\boldsymbol{u}} \cdot \nabla A = 0$. If the stream function is given by the series

$$A = \sum_{n=2,4,\ldots} A_n P_n^{\,1} \sin\theta\,, \tag{2}$$

we obtain

$$\begin{aligned} \bar{u}_r &= \sum_{n=2,4,\ldots} (\varrho r^2)^{-1}\, n(n+1)\, A_n P_n\,, \\ \bar{u}_\theta &= -\sum_{n=2,4,\ldots} (\varrho r)^{-1}\,(\mathrm{d}A_n/\mathrm{d}r)\, P_n^{\,1}\,. \end{aligned} \tag{3}$$

The angular velocity will be expressed by the series

$$\Omega = \Omega_\odot\,\Big(1 + \sum_{n=1,3,\ldots} \omega_{n-1} P_n^{\,1}/\sin\theta\Big)\,, \tag{4}$$

where $\Omega_\odot$ is the basic rotation.

From now on we shall use a summation convention with respect to the index n and, later on, the index l, that is if n and/or l appear more than once as an index in any product then summation is implied and we shall drop the summation symbol $\sum$. [Note the slight difference from the usual summation convention: a dummy (summed) suffix may appear more than twice in a product, as for instance in the first term on the right-hand side of (5) below. This does not create ambiguities in our context.]

5.1. Slow Flow

If we ignore all the non-uniform parts of the angular velocity on the right-hand side of our basic equation and put $\nu = \nu_T$, we find

$$[r^2\omega''_{n-1} + (4 + r\varrho^{-1}\varrho')\, r\omega'_{n-1} - (n-1)(n+2)\,\omega_{n-1}]\, P_n^1$$
$$= 2(\varrho\nu_T)^{-1}\,[r^{-1}n(n+1)\,A_nP_n \sin\theta - A_n'P_n^1 \cos\theta]. \tag{5}$$

Of course, this approximation is only valid for a nearly rigid rotation, i.e. for slow flow. At the end of the Chapter we shall abandon this assumption.

Using the elementary relations (A. 12) and (A.13) in the appendix, we may rewrite equation (5) as

$$x^2\omega_n'' + (4 - G)\, x\omega_n' - n(n+3)\,\omega_n$$
$$= \frac{2}{R\varrho(x)\,\nu_T}\left[\frac{n(n+1)\,A_n}{(2n+1)\,x} - \frac{(n+2)(n+3)\,A_{n+2}}{(2n+5)\,x}\right.$$
$$\left. - \frac{nA_n'}{2n+1} - \frac{(n+3)\,A'_{n+2}}{2n+5}\right], \tag{6}$$

in which both the normalized radius $x = r/R$ and the density gradient

$$G = -\mathrm{d}\ln\varrho/\mathrm{d}\ln x \tag{7}$$

have been introduced.

Let us apply this equation first to the simplest non-trivial case, $n = 2$: we find

$$\omega_2'' + (4 - G)\, x^{-1}\omega_2' - 10x^{-2}\omega_2 = F_2, \tag{8}$$

where

$$F_2 = 4[5R\nu_Tx^2\varrho(x)]^{-1}\,(3x^{-1}A_2 - A_2'). \tag{9}$$

(We have set $A_4 \equiv 0$ for simplicity.) Only the simplest flow pattern, i.e.

$$\overline{u}_r = \frac{3A_2(2 - 3\sin^2\theta)}{\varrho x^2R^2}, \qquad \overline{u}_\theta = -\frac{3A_2'\sin\theta\cos\theta}{\varrho xR^2}, \tag{10}$$

has been used. Since the radial drift does not cross the boundaries of the SCZ at $x = 1$ and $x = x_i$ we have

$$A_2(x_i) = A_2(1) = 0. \tag{11}$$

A stream that at the surface flows equatorwards must have negative $A'(1)$ and therefore positive A beneath the surface. Fig. 5.1 shows the general picture. The number of nodes of A' in $(x_i, 1)$ corresponds to the number of cells in the flow.

Let us now formally solve equation (8) with the density profile $G =$ constant, i.e.

$$\varrho = \varrho_{\odot} x^{-G}. \tag{12}$$

The characteristic equation is then

$$\lambda^2 + (3 - G)\,\lambda - 10 = 0, \tag{13}$$

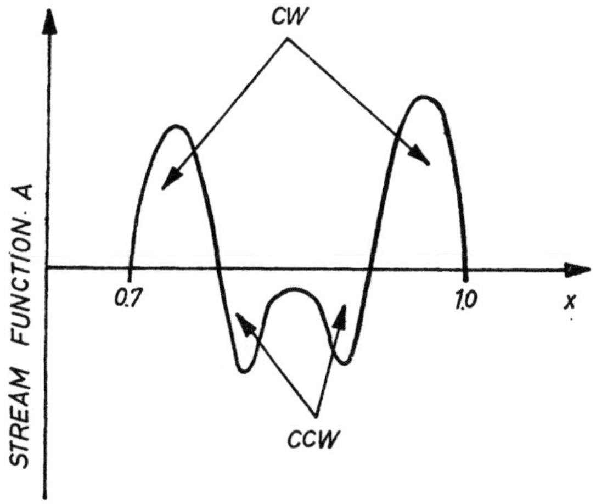

Fig. 5.1. The general shape of the stream function A for a circulation pattern having 3 cells. In both the top and bottom cells the flow is clockwise (CW).

which possesses two roots, λ_1 and λ_2. Hence, it follows that

$$\omega_2 = c_1 x^{\lambda_1} + c_2 x^{\lambda_2} + \int\limits_1^x (\lambda_2 - \lambda_1)^{-1} F_2(\xi)\,(x^{\lambda_2}\xi^{1-\lambda_2} - x^{\lambda_1}\xi^{1-\lambda_1})\,\mathrm{d}\xi. \tag{14}$$

Because stress-free conditions apply at the boundaries, the gradient of the angular velocity must vanish there:

$$\omega_2'(x_i) = \omega_2'(1) = 0. \tag{15}$$

From these conditions we deduce

$$\begin{aligned} c_1 &= \int\limits_{x_i}^1 \frac{F_2(\xi)}{\lambda_1(\lambda_2 - \lambda_1)} \frac{(\lambda_1 x_i^{\lambda_1}\xi^{1-\lambda_1} - \lambda_2 x_i^{\lambda_2}\xi^{1-\lambda_2})}{x_i^{\lambda_2} - x_i^{\lambda_1}}\,\mathrm{d}\xi, \\ c_2 &= \int\limits_{x_i}^1 \frac{F_2(\xi)}{\lambda_2(\lambda_2 - \lambda_1)} \frac{(\lambda_2 x_i^{\lambda_2}\xi^{1-\lambda_2} - \lambda_1 x_i^{\lambda_1}\xi^{1-\lambda_1})}{x_i^{\lambda_2} - x_i^{\lambda_1}}\,\mathrm{d}\xi, \end{aligned} \tag{16}$$

so that at the surface

$$\omega_2(1) = \frac{1}{10} \int_{x_i}^{1} F_2(\xi) \frac{\lambda_2 x_i^{\lambda_2} \xi^{1-\lambda_2} - \lambda_1 x_i^{\lambda_1} \xi^{1-\lambda_1}}{x_i^{\lambda_2} - x_i^{\lambda_1}} \, d\xi . \tag{17}$$

Substituting for F_2 and then performing one partial integration using conditions (11), we finally obtain the result

$$\omega_2(1) = -\frac{2}{25 R \nu_T \varrho_\odot} \int_{x_i}^{1} A_2(\xi) \times \frac{\lambda_1(2 + G - \lambda_1)\, x_i^{\lambda_1} \xi^{G-2-\lambda_1} - \lambda_2(2 + G - \lambda_2)\, x_i^{\lambda_2} \xi^{G-2-\lambda_2}}{x_i^{\lambda_2} - x_i^{\lambda_1}} \, d\xi . \tag{18}$$

An explicit representation of both solutions of the characteristic equation (13) is

$$\lambda_1 = \frac{1}{2}(G-3) + \left\{\left[\frac{1}{2}(G-3)\right]^2 + 10\right\}^{1/2} > 0,$$
$$\lambda_2 = \frac{1}{2}(G-3) - \left\{\left[\frac{1}{2}(G-3)\right]^2 + 10\right\}^{1/2} < 0. \tag{19}$$

This choice of notation brings out the positivity of the denominator in the integrand of (18), but it also implies the negativity of the last term in its numerator. Since a simple calculation with λ_1 proves that the first term of the numerator is positive,

$$\lambda_1(2 + G - \lambda_1) = (5/2)\,\{[8 + (G-7)^2]^{1/2} + G - 7\} > 0, \tag{20}$$

we can conclude that the kernel in (18) is quite generally positive. Only the form of A_2 can influence the sign of ω_2. For example, for non-stratified media we find simply

$$\omega_2(1) = -14 (5 R \nu_T \varrho_\odot)^{-1} \int_{x_i}^{1} (1 - x_i^7)^{-1} \, \xi^3 A_2(\xi) \, d\xi . \tag{21}$$

If A_2 is positive-semidefinite, the meridional circulation consists of one cell, with equatorward surface flow, and $\omega_2(1)$ is negative, which implies *equatorial acceleration*. Considered in the first quadrant of a meridional plane (cf. Fig. 2.2) the equatorward surface flow will seem to be "clockwise", and we shall use this adjective, or its antithesis "counter-clockwise", frequently below in describing motions viewed in this way. If the function A_2 has one or more zeros (two or more cells) only a more detailed analysis of the integral can determine the rotation law. It is surprising but true that the flow structure deep within the convection zone can strongly influence surface features. Nevertheless (21) also shows that the outermost layers have the greater effect.

The usual conclusion drawn from (21) is that a flow which is directed equatorward at the surface creates equatorial acceleration. But the smallness of the observed North-South motion makes this claim unconvincing in the case of the Sun, and a closer inspection of (21) reveals that the real situation is more complex. Our expression shows that a certain integral over $(x_i, 1)$ is solely responsible for determining the form of the rotation law. Even if $\mathrm{d}A/\mathrm{d}x = 0$ at $x = 1$, so that the meridional surface flow is zero, the equatorial region will be accelerated if $A_2 \gtreqless 0$, and more generally this conclusion is quite independent of the derivative of A at $x = 1$. From this point of view the apparent absence of $\overline{u}_\theta$ at the solar surface has essentially no significance for this mechanism of equatorial acceleration.

Let us illustrate this point by an example. Suppose that

$$A_2 = \left(1 + \sum_{k=1} a_k x^k\right) A\,. \tag{22}$$

Because the bounding surfaces are stress-free, four boundary conditions have to be satisfied:

$$A_2 = (\mathrm{d}/\mathrm{d}x)\,[(\varrho x^2)^{-1}\,\mathrm{d}A_2/\mathrm{d}x] = 0 \quad \text{at} \quad x = x_i,\, 1\,. \tag{23}$$

Using again the approximate density profile (12), we find that these require

$$\sum a_k = -1\,, \qquad \sum a_k x_i^k = -1\,, \tag{24}$$

$$\sum k(k-3+G)\,a_k = 0\,, \qquad \sum k(k-3+G)\,x_i^k a_k = 0\,, \tag{25}$$

and these can be satisfied if just four parameters, a_k, are included in (22). An additional coefficient is needed only if the solution is required to satisfy a further constraint. Let us make $\overline{u}_\theta$ vanish at the surface. Then the further

Table V.i. Parameters defining various stream functions

x_0	1	1	0.9
G	0	20(IV)	20(III)
a_1	−6.061224	−5.380952	−5.26363
a_2	14.516034	11.503401	10.96451
a_3	−17.180758	−12.224498	−11.30472
a_4	10.058309	6.462585	5.77098
a_5	−2.332362	−1.360544	−1.16713
$a'(1)$	0	0	$5.40216 \cdot 10^{-4}$

condition is

$$\sum_{k=1}^{5} k a_k = 0. \tag{26}$$

The five parameters are given in Table V.i for a non-stratified zone ($G = 0$) and for one with a density stratification ($G = 20$, giving $\varrho_i/\varrho_\odot = 1253$ for $x_i = 0.7$). Fig. 5.2 shows the stream function for $G = 20$. These circulation patterns generate an equatorial acceleration even though no observable N-S motion exists on the surface. (The Roman numerals "IV" and "III" on the last two columns of the table are labels used to distinguish particular models: models I and II are introduced later.)

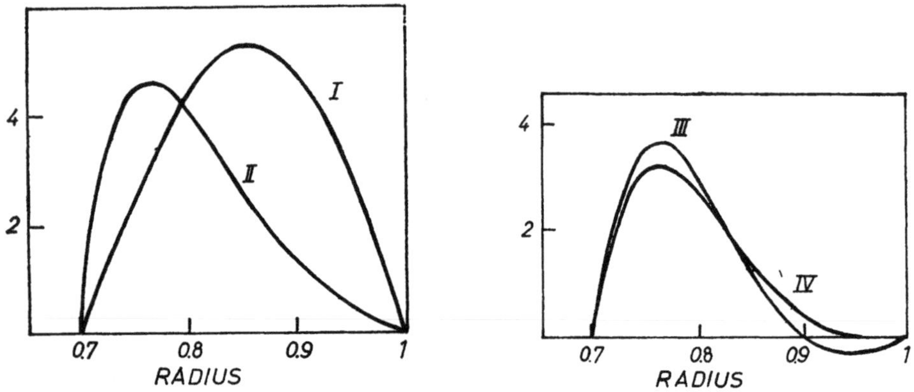

Fig. 5.2. Graphs of the stream functions used in our computations:
I: one clockwise cell in an unstratified medium,
II: one clockwise cell in a stratified medium [$G = 20$: see (40)],
III: two cells in a stratified medium (Table V.i),
IV: one clockwise cell in a stratified medium, without surface drift (Table V.i).

A further step is possible. As relation (21) is non-local, the function A need be positive only for some x, and may be negative for others. Why not in the outer part? This would describe a flow which is polewards at the surface. Table V.i. and Fig. 5.2 show a stream function which obeys the four boundary conditions and which vanishes at $x_0 = 0.9$. At the top and bottom of the convection zone the meridional flow is polewards but an equatorial acceleration is, nevertheless, present. In this two-cell model, the bottom cell has clockwise motion and the top cell counterclockwise motion (in the northern hemisphere). Let us emphasize once more that no simple conclusion can be drawn from the non-local condition (21). A very rough rule of thumb is that to fit the observations upward streaming should "dominate" near the rotation axis, and downward streaming should dominate near the equator. Also, the clockwise circulation ($A > 0$) should dominate the counterclockwise circulation.

5.2. Fast Flow

Although the arguments of § 5.1 are the truth, they are not the whole truth. The fact that the interaction between flow and rotation is in principle a nonlinear one has so far been ignored. The approximation we have developed above holds only when the rotation is nearly-uniform. The full equation governing the angular velocity is

$$\nabla \cdot (\varrho r^2 \sin^2 \theta \cdot \nu_T \nabla \Omega) - \varrho \bar{u} \cdot \nabla (r^2 \sin^2 \theta \Omega) = 0, \qquad (27)$$

from which it follows that, for small viscosity or strong circulation, the solution $\Omega r^2 \sin^2 \theta =$ constant plays a special role. Clearly, such a solution has nothing to do with equatorial acceleration. Consequently the latter can only be extracted from a special solution of (27) with appropriately chosen parameters (Köhler, 1966).

Let us work from equation (27) with the meridional flow

$$\bar{u}_r = (U/dx^2)\,(3 \cos^2 \theta - 1)\, a(x), \qquad \bar{u}_\theta = -(U/dx) \sin \theta \cos \theta\, a'(x). \qquad (28)$$

The normalized density $\varrho/\varrho_\odot$ is denoted by d. Then, with the general ansatz

$$\Omega = \sum \Omega_{n-1} P_n{}^1/\sin \theta, \qquad (29)$$

we obtain, using equation (A.17) of the appendix,

$$\begin{aligned}&[x^2 \Omega''_{n-1} + (4 + \varrho^{-1}\varrho' x)\, x\Omega'_{n-1} - (n-1)(n+2)\, \Omega_{n-1}]\, P_n{}^1 \\ &= (RU/d\nu_T)\, [(\Omega'_{n-1} + 2x^{-1}\Omega_{n-1})\,(3 \cos^2 \theta - 1)\, a P_n{}^1 \\ &\quad - n(n+1)\,(2n+1)^{-1}\,(P^1_{n+1} - P^1_{n-1}) \cos \theta \cdot (\mathrm{d}a/\mathrm{d}x)\, \Omega_{n-1}],\end{aligned} \qquad (30)$$

which, after some manipulation helped by (A.13), leads to

$$\begin{aligned}&x^2 \Omega''_{n-1} + [4 + x(\ln \varrho)']\, x\Omega'_{n-1} - (n-1)(n+2)\, \Omega_{n-1} \\ &- \frac{Re}{d(x)} \left\{ \left[\frac{(3n+6)(n+3)}{(2n+3)(2n+5)} \left(\Omega'_{n+1} + \frac{2}{x} \Omega_{n+1} \right) + \frac{2n^2 + 2n - 6}{(2n-1)(2n+3)} \right. \right. \\ &\times \left. \left(\Omega'_{n-1} + \frac{2}{x} \Omega_{n-1} \right) + \frac{(n-2)(3n-3)}{(2n-3)(2n-1)} \left(\Omega'_{n-3} + \frac{2}{x} \Omega_{n-3} \right) \right] a \\ &- \left[\frac{(n-1)^2 (n-2)}{(2n-3)(2n-1)} \Omega_{n-3} + \frac{n(n+1)}{(2n-1)(2n+3)} \Omega_{n-1} \right. \\ &- \left. \left. \frac{(n+2)^2 (n+3)}{(2n+3)(2n+5)} \Omega_{n+1} \right] a' \right\} = 0, \quad n = 1, 3, \ldots.\end{aligned} \qquad (31)$$

In (31) we have introduced the global Reynolds number of the flow:

$$Re = UR/\nu_T \simeq U[\mathrm{m/s}]. \tag{32}$$

The boundary conditions for these equations are different for $n = 1$ and $n > 1$. It is a well-known fact that the latitude-independent solution, Ω_0, is exceptional: if one of the boundaries is assumed to be stress-free, the other one automatically transmits no stress either. This can easily be shown. Since the first equation ($n = 1$) can be written as

$$(\mathrm{d}/\mathrm{d}x)\,[x^4 d\Omega_0' - (2/5)\,Re\;x^2(18\Omega_2/7 - \Omega_0)\,a(x)] = 0, \tag{33}$$

we find

$$\Omega_0' = (\text{constant}/dx^4) + (2/5)\,Re(a/dx^2)\,(18\Omega_2/7 - \Omega_0). \tag{33a}$$

This gradient has to vanish at a slippery surface, and since $a(1) = 0$, the constant must be zero, giving

$$\Omega_0' = 2Re(18\Omega_2/7 - \Omega_0)\,a/5dx^2. \tag{34}$$

As $a(x)$ is zero at the lower boundary $x = x_1$, the Ω-gradient automatically vanishes there as does, in addition, the stress. The corresponding boundary condition — that the inner surface be slippery — can no longer be thought of as an independent condition. It is redundant and, both to close the problem and to ensure the rotation of the shell, it must be replaced by some other demand, e.g. that the angular velocity at some intermediate level takes some prescribed value.

Summing up, the boundary conditions for the full equation (31) must be:

$$\begin{aligned} &\Omega_0(1) = 1(\text{say}), \qquad \Omega_0'(1) = 0, \\ &\Omega_n'(1) = \Omega_n'(x_i) = 0 \qquad n \geqq 2. \end{aligned} \tag{35}$$

We shall first use the very simplest stream function that obeys (24) and (25), with in both cases $G = 0$: the modifications required for non-vanishing G will be discussed later. Then the four coefficients in the ansatz (22) are

$$a_1 = -5.0653, \qquad a_2 = 9.3219, \qquad a_3 = -7.3848, \qquad a_4 = 2.1282. \tag{36}$$

The corresponding stream function (defining "model I") is positive taking values between 0.7 and 1 (Fig. 5.2); at the surface its gradient is $a'(1) = -0.06$. It is a single-celled, clockwise stream. At mid-latitudes in the northern hemisphere we find

$$\bar{u}_\theta = -0.5a'(1)\,U \simeq 0.03\nu_T Re/R \approx 2Re\;[\mathrm{cm/s}], \tag{37}$$

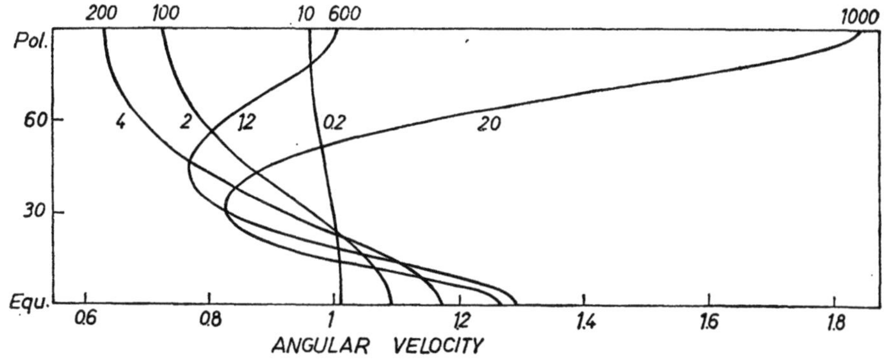

Fig. 5.3. Surface rotation laws generated by the clockwise flow I, dependent on the Reynolds number of the mean flow. The curves are labelled with the surface drift velocity in m/s. The ordinate gives the latitude, i.e. equatorial distance.

where we have taken $\nu_T \approx 5 \cdot 10^{12}$ cm²/s. In Fig. 5.3 we present further results of these calculations for a uniform medium. Small positive values of *Re* produce a differential surface rotation of solar type, with both equatorial acceleration and polar deceleration. With increasing *Re*, the latter changes into an acceleration, giving a form of polar vortex. At about $Re \approx 800$, the poles rotate with the same angular velocity as the equator with deceleration confined to mid-latitudes. Near the equator all profiles have $\cos\theta\partial\Omega/\partial\theta > 0$, i.e. equatorial acceleration. This is a basic result, as we shall now show.

The Reynolds equation (4.11) may be solved by writing $\boldsymbol{t} = \boldsymbol{\nabla} \times \boldsymbol{V}$, or

$$t_r = (x \sin\theta)^{-1}\, \partial(V \sin\theta)/\partial\theta\,, \qquad t_\theta = -x^{-1}\, \partial(xV)/\partial x\,. \tag{38}$$

where the surface traction, $\boldsymbol{t}$, is related to the stress tensor T_{ij} by (4.12). Due to the stress-free state of the surface we have $t_r(1, \theta) = 0$, and equatorial symmetry requires that $t_\theta(x, \pi/2) = 0$. The former gives $\partial V/\partial\theta = 0$ at the

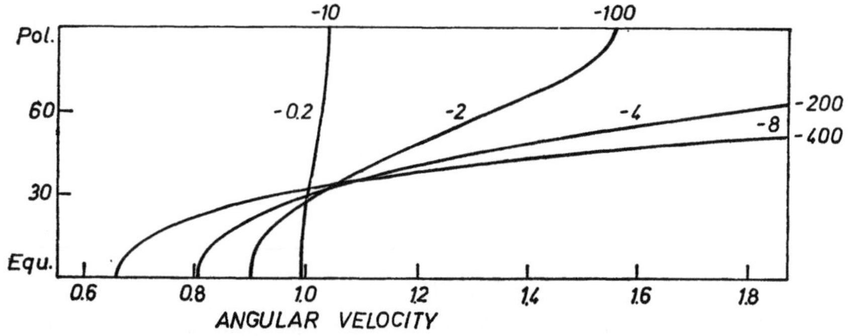

Fig. 5.4. The same situation as in Fig. 5.3 but for a counterclockwise flow (I → —I).

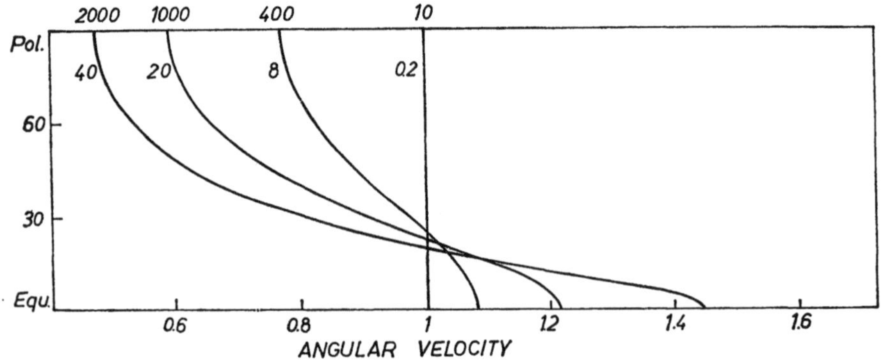

Fig. 5.5. The same situation as in Fig. 5.3 but for the clockwise flow II.

equator. Hence, by $(38)_2$, we have $\partial t_\theta/\partial\theta = 0$ on $\theta = \pi/2$. We conclude that not only does t_θ vanish at the equator, but it also remains small at low latitudes. Thus $t_\theta \approx 0$ there, i.e.

$$R\Omega\bar{u}_\theta - \nu_T \partial\Omega/\partial\theta \approx 0, \tag{39}$$

so that $\partial\Omega/\partial\theta > 0$ near the equator whenever the meridional surface flow is equatorward. The opposite is true for counter-clockwise flows: $\partial\Omega/\partial\theta < 0$. Examples of the latter ($Re < 0$) are depicted in Fig. 5.4. The poles are accelerated and the equator is decelerated in all cases. The differences in the rotation profiles for slow and fast flows are only of a quantitative nature. Obviously, there is no hope in either case of producing a solar-type rotation with a counter-clockwise flow and a single cell.

Density stratification can be included with ease. It suffices to multiply the constants given in (36) by

$$A = x^{-G/2} A_2|_{G=0} \tag{40}$$

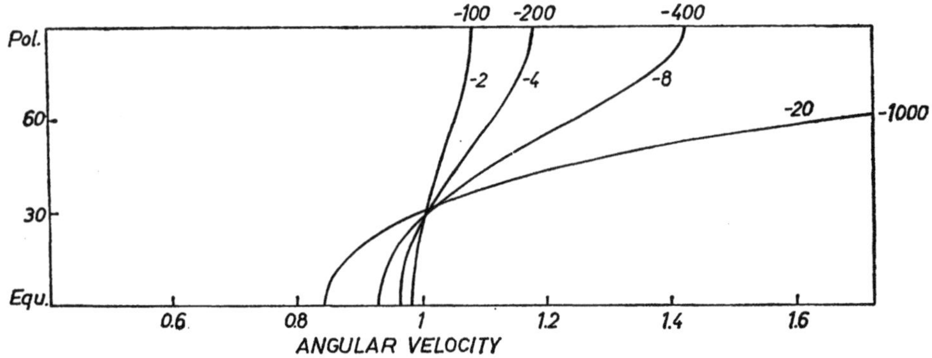

Fig. 5.6. The same situation as in Fig. 5.5 but for the opposite stream direction (II → —II).

Table V.ii. Stream functions which generate the observed rotation law.

model	G	cells	Re	$\bar{u}_\theta(1)$	$\bar{u}_\theta(x_i)$
I	0	1	100	2 m/s	−3.5 m/s
II	20	1	400	8 m/s	−0.5 m/s

(Fig. 5.2). The resulting stream-function (defining "model II") obeys the four required conditions (23). It even possesses the same gradient at the surface, so that (37) remains true. The velocities at the bottom of the zone are proportional to $A'(x_i)/\varrho(x_i)$, i.e. they are reduced to a fraction $x_i^{G/2} \approx 0.03$ of the corresponding velocities when no stratification is present. Thus we expect that a rapidly rotating regime ("polar vortex") will appear for Reynolds numbers greater than 800. Our computer results (Fig. 5.5) reveal that no polar vortex arises unless the Reynolds number exceeds 2,000, corresponding to a surface drift velocity of about 40 m/s. Because of convergence problems our numerical results became unreliable for larger values of Re.

It can be seen from Fig. 5.3 and 5.5 that the surface profiles of the angular velocity for

$$\text{model I}: Re = 100 \quad \text{and} \quad \text{model II}: Re = 400$$

resemble one another. They both approach the observed rotation law for the Sun and are characterized by the parameters in Table V.ii. In order to obtain the observed Ω-profile when the SCZ is stratified, the upper surface drift velocity must be larger than when stratification is absent; also, the bottom drift velocity must be smaller. The required velocity of 8 m/s is close to the lower limit for observational detection.

These conclusions must not, however, be given too much weight. If we assume the two-cell profile ("model III") of Table V.i, which is positive for $0.7 < x < 0.9$ and negative for $0.9 < x < 1$, then the meridional motions

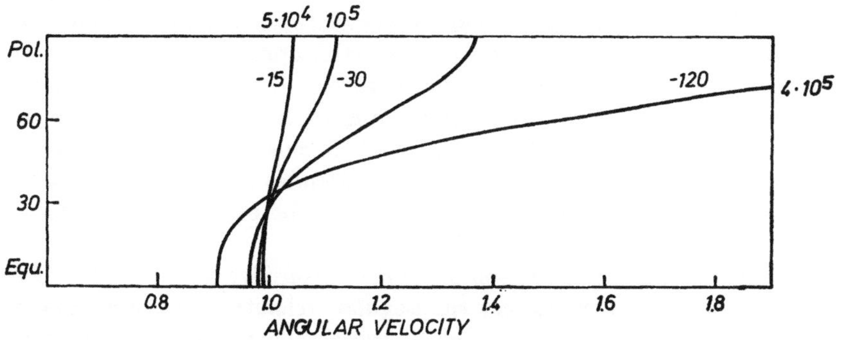

Fig. 5.7. Surface rotation laws for the two-cell stream III.

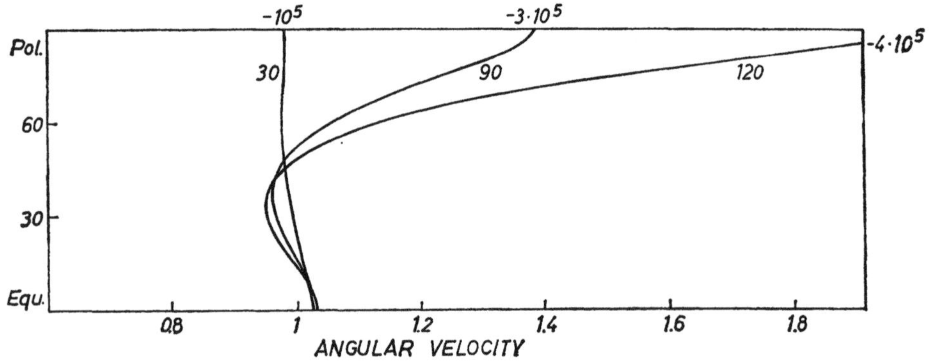

Fig. 5.8 The same situation as in Fig. 5.7 but for the opposite stream direction (III → −III).

are directed polewards both at the bottom and the top of the SCZ. For $Re = 10^4$ (which corresponds at the surface to a poleward flow of 3 m/s), a solar-type differential rotation is generated (Fig. 5.7). Incorporating the actual values with the basic $\Omega = 1$ shows us that the rotation is better thought of as polar deceleration than equatorial acceleration. We have to accept that statements made about the surface drift velocities required in order to obtain a solar-type rotation law are valueless when the number of cells is unknown. From the literature we know that two or three cells are *not* exceptional predictions of mean-field hydrodynamics of the SCZ. Our calculations demonstrate convincingly that it is the overall profile of the stream function, and not merely its surface values, that determines the form of the rotation law.

We may also examine stream functions that are negative throughout the region under consideration or are, at least, mainly negative. In practice, we use the profiles I, II and III with negative *Re*. The bottom drift is then, in all cases, equatorwards whereas the surface drift is polewards in models I and II. In the two-cell model III the surface drift is equatorwards. In other words, the bottom cell is always counter-clockwise while a top cell, if present, is clockwise. One can immediately translate this apparently small difference into a statement about the corresponding surface rotation laws. The results for models I and II are not surprising: we have an equatorial deceleration and/or a polar acceleration (Fig. 5.4, 5.6). The small top cell of model III produces, however, a deceleration in mid-latitudes and this is, in fact, a characteristic of fast clockwise cells. As was to be expected, this effect is small but consistent with (39), cf. Fig. 5.8.

To study the radial distribution of rotation, consider equation (34). If Ω_0 is the dominant term we expect the angular velocity to have a negative gradient for positive $a(x)$. Thus a clockwise circulation normally produces sub-rotation. Fig. 5.9 shows the depth-dependence of the angular velocity

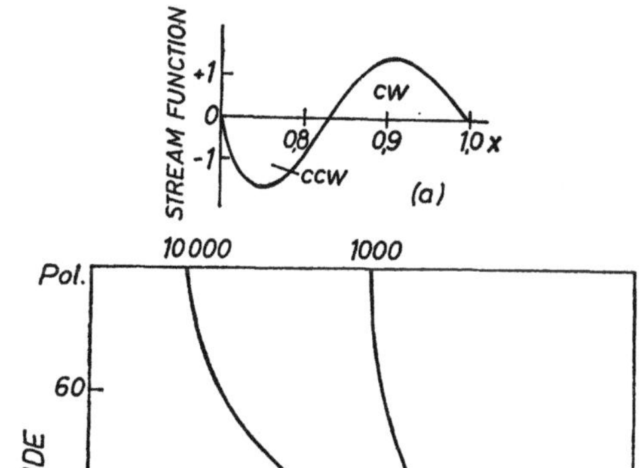

Fig. 5.9. The effect on the differential rotation of a meridional flow with two nearly equal cells. (b) The surface rotation laws show the top cell dominating. The curves are marked with the Reynolds numbers of the originating flows. For one of them ($Re = 10^4$) the inner rotation laws are shown in figure (c), the fractional radius x being shown on the curves. The outer layers are subrotating but below the maximum at $x = 0.83$ super-rotation occurs.

in the equatorial region for a model with $x_0 = 0.82$, i.e. with two nearly equal cells. The top-cell is clockwise, the bottom cell counter-clockwise. The angular velocity at first increases inwards (sub-rotation) but, at a greater depth, it starts to decrease.

5.3. Sources of Meridional Circulation

In Chapter 4 it was demonstrated that two strong generators of differential rotation exist: meridional circulation and Reynolds stress. The latter becomes important only if the basic rotation influences the turbulence in a non-trivial manner. The question arises whether a global rotation can, by itself, generate a meridional circulation. As the answer is "Yes", we will later have to compare the strength of the Λ-effect with that of the meridional flow and consider their different contributions to the observed equatorial acceleration of the Sun. As a first step we here deal with the meridional components of the Reynolds equation. We shall often find it useful to regard the terms involving Ω as sources that must, when nonzero, be balanced by meridional flows, such as the well-known Biermann-Kippenhahn circulation, or by Reynolds stresses, such as the Λ-effect. In this section, we concentrate on the former. Apart from the Biermann-Kippenhahn circulation, we shall find that baroclinicity is an additional cause for meridional flow. (Baroclinic flow appears whenever the mean density is not a function of the pressure; see below.) This complication will force us (Chapter 10) to incorporate pole-equator differences of density and temperature into the theory, but this step must be postponed until we have acquired an understanding of how the energy is transported in a rotating turbulent medium.

Let us turn the Reynolds equation in the form

$$\varrho(\boldsymbol{u} \cdot \nabla)\, \boldsymbol{u} + \nabla \cdot (\varrho Q) = -\nabla p + \varrho \boldsymbol{g} . \tag{41}$$

We are now interested in the meridional components of this quite traditional equation, expressed relative to an inertial frame of reference. They are

$$\begin{aligned}
&\varrho\left(\bar{u}_r \frac{\partial \bar{u}_r}{\partial r} + \frac{\bar{u}_\theta}{r}\frac{\partial \bar{u}_r}{\partial \theta} - \frac{\bar{u}_\theta^2 + \bar{u}_\phi^2}{r}\right) + \frac{1}{r^2}\frac{\partial}{\partial r}(r^2 \varrho Q_{rr}) + \frac{1}{r\sin\theta}\frac{\partial}{\partial\theta}(\sin\theta\, \varrho Q_{r\theta}) \\
&\qquad - \frac{\varrho}{r}(Q_{\theta\theta} + Q_{\phi\phi}) = -\frac{\partial p}{\partial r} - \varrho g , \\
&\varrho\left(\bar{u}_r \frac{\partial \bar{u}_\theta}{\partial r} + \frac{\bar{u}_\theta}{r}\frac{\partial \bar{u}_\theta}{\partial \theta} + \frac{\bar{u}_r \bar{u}_\theta}{r} - \frac{\cot\theta}{r} u_\phi^2\right) + \frac{1}{r^3}\frac{\partial}{\partial r}(r^3 \varrho Q_{r\theta}) + \frac{1}{r}\frac{\partial}{\partial\theta}(\varrho Q_{\theta\theta}) \\
&\qquad + \frac{\varrho\cot\theta}{r}(Q_{\theta\theta} - Q_{\phi\phi}) = -\frac{1}{r}\frac{\partial p}{\partial\theta} .
\end{aligned} \tag{42}$$

As is customary for relatively slow flows, we shall neglect the terms nonlinear in $\bar{u}_\theta$ and/or $\bar{u}_r$. Of the surviving nonlinear advection terms, only the rotational force then remains

$$\nabla p = \varrho \boldsymbol{g} - \nabla \cdot (\varrho Q) + \varrho \Omega^2 \boldsymbol{s} , \tag{43a}$$

which in component form becomes

$$\begin{aligned}
\partial p/\partial r &= -\varrho g + \varrho r \sin^2\theta \cdot \Omega^2 - r^{-2}\, \partial(r^2 \varrho Q_{rr})/\partial r \\
&\quad - (r\sin\theta)^{-1}\, \partial(\varrho \sin\theta \cdot Q_{r\theta})/\partial\theta + r^{-1}\varrho(Q_{\theta\theta} + Q_{\phi\phi}) , \\
\partial p/\partial\theta &= \varrho r^2 \sin\theta \cdot \cos\theta \cdot \Omega^2 - r^{-2}\, \partial(r^3 \varrho Q_{r\theta})/\partial r \\
&\quad - \partial(\varrho Q_{\theta\theta})/\partial\theta - \varrho \cot\theta (Q_{\theta\theta} - Q_{\phi\phi}) .
\end{aligned} \tag{43b}$$

The term on the right-hand side of equation (43a), which is quadratic in Ω, can be written in a number of equivalent ways:

$$Z = \varrho\Omega^2\mathbf{s} = -\varrho\mathbf{\Omega} \times (\mathbf{\Omega} \times \boldsymbol{r}) = \varrho[\Omega^2\boldsymbol{r} - (\boldsymbol{r} \cdot \mathbf{\Omega})\, \mathbf{\Omega}]$$

$$= \frac{1}{2}\, \varrho\nabla(|\mathbf{\Omega} \times \boldsymbol{r}|^2) = \frac{1}{2}\, \varrho\nabla(s^2\Omega^2). \tag{44}$$

The vector distance from the rotation axis is $\mathbf{s}$, and $|\mathbf{s}| = s = r \sin\theta$; its representation in spherical coordinates is given in (A4) in the appendix. The last two expressions in (44) hold only when the rotation is uniform, in which case (43) agrees with the well-known formulation that uses the co-rotating frame of reference. In that formulation, a Coriolis acceleration $2\mathbf{\Omega} \times \bar{\boldsymbol{u}}$ and a centrifugal force $\varrho\Omega^2\mathbf{s}$ arise, where the velocity $\bar{\boldsymbol{u}}$ is then measured relative to the rotating frame. Note that in none of the correlations that occur in (43a) does the suffix ϕ occur an odd number of times; it appears twice or not at all.

The traditional approach, via the Boussinesq relation (2.2), would exclude the effects of a purely rigid rotation on the turbulence, and would therefore omit from Q_{ij} tensors such as $\Omega_i\Omega_j$ and $(g_i\Omega_j + \Omega_i g_j)$. Let us discuss briefly what effects such additional terms might be expected to have. For example, $\Omega_i\Omega_j$ wuold lead to an off-diagonal component in the correlation tensor:

$$Q_{r\theta} = -\nu_T\, \partial\bar{u}_\theta/\partial r + \cdots - \mu_T\Omega^2 \sin\theta\cos\theta\,, \tag{45}$$

where we may make the simple estimates: $\nu_T = \overline{u'^2}/L$ and $\mu_T = L^2$. The relative order of magnitude of the extra term is

$$L^2\Omega^2R/Lu'\bar{u}_\theta \simeq (\bar{u}_\phi/\bar{u}_\theta)\, \Omega L/u'\,, \tag{46}$$

and, if $L\Omega \gg \bar{u}_\theta$, this greatly exceeds unity. The quantity $L\Omega$, as seen on the solar surface, is about 10 m/s, and it seems *a priori* really dangerous to ignore such a term. But as its existence is as yet not convincingly established, we will, for the present, consider neither the $\Omega_i\Omega_j$ nor the $(g_i\Omega_j + \Omega_i g_j)$ tensors further.

Equations (43) determine both components of the meridional flow. This is more transparent after the pressure has been eliminated:

$$\frac{1}{r^2}\frac{\partial^2}{\partial r\,\partial\theta}(r^2\varrho Q_{rr}) - \frac{1}{\sin\theta}\frac{\partial^2}{\partial r\,\partial\theta}(\varrho\sin\theta Q_{\theta\theta}) + \cot\theta\frac{\partial}{\partial r}(\varrho Q_{\phi\phi})$$

$$-\frac{1}{r}\frac{\partial}{\partial\theta}\left[\varrho(Q_{\theta\theta} + Q_{\phi\phi}) - \frac{1}{\sin\theta}\frac{\partial}{\partial\theta}(\varrho\sin\theta\, Q_{r\theta})\right] - \frac{\partial}{\partial r}\left[\frac{1}{r^2}\frac{\partial}{\partial r}(r^3\varrho Q_{r\theta})\right]$$

$$= r\sin^2\theta\frac{\partial}{\partial\theta}(\varrho\Omega^2) - r^2\sin\theta\cos\theta\frac{\partial}{\partial r}(\varrho\Omega^2) - g\frac{\partial\varrho}{\partial\theta}. \tag{47}$$

Those terms on the right-hand side of this equation, that involve the differential rotation through $\partial\Omega/\partial r$ and/or $\partial\Omega/\partial\theta$, are the cause of the Biermann-Kippenhahn circulation. Note however that, for a rigid rotation, the right-hand side of (47) becomes $\partial(p, \varrho)/\partial(r, \theta)$, which vanishes for polytropic media $[p = p(\varrho)]$ or, more generally, for $p = p(\Psi)$ and $\varrho = \varrho(\Psi)$. It is only when the isoplanes of pressure and density do not coincide, that we generate a nonzero "baroclinic" flow.

Equation (47), whose right-hand side is identical to the ϕ-component of

$$\nabla \times [-\varrho \boldsymbol{g} + \varrho \boldsymbol{\Omega} \times (\boldsymbol{\Omega} \times \boldsymbol{r})], \tag{48}$$

reveals that there is more than one generator of meridional flow: the non-uniform centrifugal force density $\varrho\Omega^2$, and the non-uniform density, ϱ. One exception, however, must be borne in mind: if $\varrho\Omega^2$ depends only on the distance from the axis, $s = r \sin\theta$, then $\sin\theta\, \partial(\varrho\Omega^2)/\partial\theta = \cos\theta\, \partial(\varrho\Omega^2)/\partial r$ and hence the Biermann-Kippenhahn circulation disappears. This may be traced to the fact that the centrifugal force is then conservative, i.e. it can be expressed as a potential

$$\boldsymbol{Z} = \varrho \boldsymbol{s}\Omega^2 = \nabla \int^{s} \xi f(\xi)\, \mathrm{d}\xi . \tag{49}$$

We have established here more sources of meridional drift than Kippenhahn did in his original paper. There, the depth-dependent centrifugal force, $r^2\varrho \sin\theta \cos\theta\, \partial\Omega^2/\partial r$, was considered [cf. his equation (51)]. This stems from his neglect of the term $\nabla \times (\varrho^{-1}\nabla p)$, based on the assumption that the gradients of ϱ and p are parallel. In this way it was possible to ignore totally the thermodynamics of the convection zone.

It may be worth writing the right-hand side of equation (47) more explicitly as

$$\begin{aligned} &-r^2 \sin\theta \cos\theta\varrho\, \partial\Omega^2/\partial r + r \sin^2\theta\, \varrho\, \partial\Omega^2/\partial\theta \\ &+ (-g + r \sin^2\theta\Omega^2)\, \partial\varrho/\partial\theta - r^2 \sin\theta \cos\theta\Omega^2\, \partial\varrho/\partial r . \end{aligned} \tag{50}$$

The first two terms correspond to the Biermann-Kippenhahn circulation, whose existence can be explained quite simply. Fig. 2.2 demonstrates the general situation. Due to frictional losses the flow needs energy which only the centrifugal force can provide. The gain of energy is positive at AB and negative at BA, its magnitude being proportional to Ω^2. For a rigid rotation the two energies cancel each other out, and no flow can be maintained. The situation is different, however, if the angular velocity is larger at AB than at BA. Then energy gains exceed losses and circulation takes place. Hence, a necessary condition for the existence of the circulation is

$$\partial\Omega^2/\partial r > 0 . \tag{51}$$

The rotation resulting from this law creates a clockwise flow which is directed towards the equator at the surface and towards the poles at the bottom of the SCZ (assuming only one cell exists). As was demonstrated above, it is just this circulation which creates an equatorial acceleration. Indeed, it was at one time believed that to solve the solar rotation problem it was only necessary to find the reason for such an over-all super-rotation! Of course, in seeking the reason for equatorial acceleration we must investigate the effects of turbulence too. The theoretical ideas about differential rotation were developed by Lebedinsky, Wasiutynski, Biermann, Kippenhahn, Köhler and others with that idea in mind, but our equation (47) reveals that the circulation may also have other causes. If the polar and equatorial densities are not the same, the difference in buoyancy also drives a global flow. Any possible pole-equator difference in density is always connected with a pole-equator difference in temperature. From the observational, as well as the theoretical, point of view the latter was often described in terms of the mean-field thermodynamics of the convection zone.

These suggestions were based on the idea that the rotation not only influences the turbulent transport of angular momentum, but also that of energy. Therefore, not only the angular velocity but also the mean temperature might be latitude-dependent. A brief discussion of (50) may clarify the situation. Only terms involving $\sin\theta\cos\theta$ need be included and we set

$$\varrho = \varrho_0(1 - \delta\varrho \sin^2\theta), \qquad \Omega = \Omega_0(1 - \delta\Omega\sin^2\theta), \tag{52}$$

with Ω_0 and ϱ_0 having gradients

$$\mathrm{d}\Omega_0/\mathrm{d}r = V\Omega_0/r, \qquad \mathrm{d}\varrho_0/\mathrm{d}r = -G\varrho_0/r: \tag{53}$$

Taking the ratio $\alpha = \Omega^2 L/g$ of the centrifugal and gravitational force to be 10^{-5}, we obtain (50)

$$\cdots = \varrho_0 g(-2\alpha Vx + \alpha Gx + 2\delta\varrho). \tag{54}$$

Three distinct effects, therefore, contribute and interact. Due to the smallness of α the dominant feature must be the "latitude-dependent density". The remaining terms are associated with the "depth-dependent density", and the "Kippenhahn-Biermann effect" as represented by V. They are both of order α and involve different powers of x. For positive V (super-rotation) they even have different signs. It was precisely this super-rotation that in Kippenhahn's model led to the correct rotation law. This model must encounter difficulties if a real density stratification is included.

As we know, the Biermann-Kippenhahn circulation for positive V is clockwise in the first quadrant. Hence, we may, in general, expect this will also be true if the combination of terms explicitly shown in (54) is negative. In particular we expect it will be true when $\delta\varrho$ is negative, i.e. if the density decreases towards the poles. On the basis of (50) we further suggest that

Table V.iii. Sources of meridional flow

effect	"cause"		"action" (in northern hemisphere) surface flow	circulation
Biermann-Kippenhahn circulation	sub-rotation	$\partial\Omega/\partial r < 0$	SN	ccw
	equatorial acceleration	$\cos\theta\ \partial\Omega/\partial\theta \gtreqless 0$	SN	ccw
baroclinic flow	radial density stratification	$\partial\varrho/\partial r < 0$	SN	ccw
	polar decrease in density	$\cos\theta\ \partial\varrho/\partial\theta \gtreqless 0$	NS	cw

clockwise circulation will arise when $\cos\theta\ \partial\Omega/\partial\theta < 0$. Table V.iii. summarizes the situation. The four "causes" of meridional flow, viz. $\partial\Omega/\partial r$, $\partial\Omega/\partial\theta$, $\partial\varrho/\partial r$ and $\partial\varrho/\partial\theta$, are specified in terms of their character, and the related "actions" are given, being either clockwise or counter-clockwise circulation. 'SN' means that the meridional motion is from the South towards the North, as seen in the northern hemisphere, i.e. from equator to poles.

As we have seen in the previous section, a clockwise flow generates sub-rotation and equatorial acceleration, but these in turn, as Table V.iii. shows, create the *opposite* circulation: counter-clockwise. We cannot expect a substantial effect of the right sign to survive this cancellation. We note that the conclusion depended on choosing a flow pattern having only one cell, and in § 11.4 it is suggested that self-excitation of differential rotation and meridional circulation can occur but only when there is more than one cell.

In order to discuss equation (47) in more detail we need to write the required correlations in the following form:

$$
\begin{aligned}
Q_{rr} &= -2\nu_T\, \partial\bar{u}_r/\partial r\,, \\
Q_{\theta\theta} &= -2\nu_T r^{-1}(\bar{u}_r + \partial\bar{u}_\theta/\partial\theta)\,, \\
Q_{\phi\phi} &= -2\nu_T r^{-1}(\bar{u}_r + \cot\theta\, \bar{u}_\theta)\,, \\
Q_{r\theta} &= -\nu_T r^{-1}(\partial\bar{u}_r/\partial\theta + r\, \partial\bar{u}_\theta/\partial r - \bar{u}_\theta)\,.
\end{aligned}
\tag{55}
$$

These expressions are the traditional ones. We could also have introduced an "anisotropic viscosity", as have other authors (see Chapter 7), but as the value of such a procedure is strictly limited, we shall pass over this complication. Since the main contributor to the eddy viscosity is small-scale turbulence, which is relatively isotropic, we expect that the eddy viscosity will also be nearly isotropic. Using our stream function (2) we find that the components of flow are as in equation (3), so we obtain

$$\begin{aligned} Q_{rr} &= -2\nu_T n(n+1)\,(A_n/\varrho r^2)'\,P_n, \\ Q_{\theta\theta} + Q_{\phi\phi} &= 2\nu_T(\varrho r^3)^{-1}\,n(n+1)\,(-2A_n + rA_n')\,P_n, \\ Q_{r\theta} &= \nu_T(\varrho r^3)^{-1}\,[n(n+1)\,A_n + \varrho r^4(A'/\varrho r^2)']\,P_n^1. \end{aligned} \tag{56}$$

These expressions can now be substituted into the left-hand side of equation (47). All that remains is to consider the right-hand side. We must, of course, linearize the centrifugal force. Using a series expansion, we can write [cf. equation (29)]

$$\Omega^2 = \Omega_0^2 + 2\Omega_0 \sum_{n=3,5\ldots} \Omega_{n-1} P_n^1/\sin\theta. \tag{57}$$

When considering the density we make two assumptions. First, we take the density, wherever it occurs in the Reynolds stress term of (47), to have its undisturbed value, i.e. be latitude-independent. Second, and this is easy to accept, we also assume that the density involved in the centrifugal force is also latitude-independent. The second assumption simply means that gravity is supposed to be large compared with the centrifugal force, $g \gg \Omega^2 R$ (i.e. $\alpha \ll 1$). Then, with

$$\varrho = \sum_{n=0,2,\ldots} \varrho_n P_n, \tag{58}$$

the right-hand side of (47) becomes

$$\left\{-\frac{r^2}{3}\,(\varrho_0\Omega_0^2)'\,\delta_{2n} + 2r\Omega_0\varrho_0\left[\frac{n(n-1)\,(1-\delta_{2n})}{2n-1}\,\Omega_{n-2} - \frac{(n+1)\,(n+2)}{2n+3}\,\Omega_n\right]\right.$$
$$\left.-\frac{(1-\delta_{2n})\,(2n-2)}{2n-1}\,(r^2\varrho_0\Omega_0\Omega_{n-2})' - \frac{2n+4}{2n+3}\,(r^2\varrho_0\Omega_0\Omega_n)' + \varrho_n g\right\}P_n^1. \tag{59}$$

Remembering that $P_n^1 = -\mathrm{d}P_n/\mathrm{d}\theta$, we arrive at the final result

$$\begin{aligned} \mathcal{D}A_n = {} & \frac{r^2}{3\nu_T}\,(\varrho_0\Omega_0^2)'\,\delta_{2n} + \frac{2n+4}{(2n+3)\,\nu_T}\,[(r^2\varrho_0\Omega_0\Omega_n)' + (n+1)\,r\varrho_0\Omega_0\Omega_n] \\ & + \frac{2n-2}{2n-1}\cdot\frac{1-\delta_{2n}}{\nu_T}\,[(r^2\varrho_0\Omega_0\Omega_{n-2})' - nr\Omega_0\Omega_{n-2}] - \frac{g}{\nu_T}\,\varrho_n, \end{aligned} \tag{60}$$

$n = 2, 4, \ldots$

where the operator $\mathcal{D}$ is given by

$$\mathcal{D}A_n = A_n^{\mathrm{IV}} - (\ln \varrho)' A_n''' - 2\left[\frac{n(n+1)}{r^2} + (\ln \varrho)'' + \frac{(\ln \varrho)'}{r}\right] A_n''$$
$$+ \left[\frac{4n(n+1)}{r^3} + (n^2+n+2)\frac{(\ln \varrho)'}{r^2} - \frac{2}{r}(\ln \varrho)'' - (\ln \varrho)'''\right] A_n'$$
$$+ n(n+1)\left[\frac{n^2+n-6}{r^4} + \frac{2(\ln \varrho)''}{r^2}\right] A_n. \tag{61}$$

Let us add a brief numerical discussion of this equation. As usual the density profile is taken to be x^{-G}. Then for $n = 2$, equation (60) becomes

$$x^4 A_2^{\mathrm{IV}} + x^3 G A_2''' - 12x^2 A_2'' + (24 - 8G)\, x A_2' + 12 G A_2 = F_2 x^f, \tag{62}$$

where

$$F_2 = (g\alpha R^4 \varrho_0/\nu_T)\left[\frac{1}{3}(2V - G) - (\varrho_2/\alpha\varrho_0 x)\right], \tag{63}$$

and $f = 5 - G$. In order to simplify matters, F_2 will be assumed to be constant. For a surface flow at $\theta = 45°$, we have

$$\overline{u}_\theta(1, 45°) \equiv -A_2'(1)/4\varrho_0(1)\, R^2, \tag{64}$$

which can be written in the alternative form

$$\overline{u}_\theta(1, 45°) = -\frac{1}{4}\alpha g R^2 \nu_T^{-1}\left[\frac{1}{3}(2V - G) - \varrho_2(\varrho_0\alpha)^{-1}\right] A_2'(1)/F_2. \tag{65}$$

If the solar values for the parameters are adopted, this is equivalent, in c.g.s. units, to

$$\overline{u}_\theta(1, 45°) \simeq 10^8 \cdot \left[\frac{1}{3}(2V - G) - \varrho_2(\alpha\varrho_0)^{-1}\right][-A_2'(1)/F_2]. \tag{66}$$

In order to obtain the relation between $A_2'(1)$ and F_2 we solved equation (62) with the standard boundary conditions (23), and with the results given in Table V.iv.

Obviously, the ratio $A_2'(1)/F_2$ is negative and of order 10^{-3}. $A_2'(1)$ has the opposite sign to F_2. From (64) it can be seen that the surface flow is directed polewards for negative F_2 and equatorwards for positive F_2. Positive F_2 results from a positive V and/or from a negative density difference ϱ_2, the latter implying that the equatorial region is denser than the polar zones. Such a density distribution (polar deficiency) causes a meridional flow that at the surface is directed towards the equator. This is a rather general result.

Table V.iv. Surface values of A_2'/F_2 found by solving equation (62)

f ↓ G →	0	10	20	30
30	$-9.3 \cdot 10^{-5}$	$-5.3 \cdot 10^{-5}$	$-2.8 \cdot 10^{-5}$	$-1.6 \cdot 10^{-5}$
25	$-1.3 \cdot 10^{-4}$	$-7.1 \cdot 10^{-5}$	$-3.6 \cdot 10^{-5}$	$-2.1 \cdot 10^{-5}$
20	$-1.8 \cdot 10^{-4}$	$-1.0 \cdot 10^{-4}$	$-5.0 \cdot 10^{-5}$	$-2.8 \cdot 10^{-5}$
15	$-2.9 \cdot 10^{-4}$	$-1.5 \cdot 10^{-4}$	$-7.2 \cdot 10^{-5}$	$-4.0 \cdot 10^{-5}$
10	$-4.8 \cdot 10^{-4}$	$-2.5 \cdot 10^{-4}$	$-1.1 \cdot 10^{-4}$	$-6.0 \cdot 10^{-5}$
5	$-9.3 \cdot 10^{-4}$	$-8.5 \cdot 10^{-4}$	$-2.0 \cdot 10^{-4}$	$-1.0 \cdot 10^{-4}$
0	$-2.0 \cdot 10^{-3}$	$-9.5 \cdot 10^{-4}$	$-3.9 \cdot 10^{-4}$	$-1.9 \cdot 10^{-4}$
−5	$-5.3 \cdot 10^{-3}$	$-2.4 \cdot 10^{-3}$	$-9.2 \cdot 10^{-4}$	$-4.4 \cdot 10^{-4}$
−10	$-1.6 \cdot 10^{-2}$	$-6.8 \cdot 10^{-3}$	$-2.5 \cdot 10^{-3}$	$-1.2 \cdot 10^{-3}$
−15	$-5.4 \cdot 10^{-2}$	$-2.2 \cdot 10^{-2}$	$-8.1 \cdot 10^{-3}$	$-3.7 \cdot 10^{-3}$
−20	$-2.0 \cdot 10^{-1}$	$-8.3 \cdot 10^{-2}$	$-2.9 \cdot 10^{-2}$	$-1.3 \cdot 10^{-2}$
−25	$-8.2 \cdot 10^{-1}$	$-3.3 \cdot 10^{-1}$	$-1.2 \cdot 10^{-1}$	$-5.2 \cdot 10^{-2}$
−30	-3.5	-1.4	$-4.8 \cdot 10^{-1}$	$-2.2 \cdot 10^{-1}$

Substituting in numerical values we find that, for the Sun,

$$|\overline{u}_\theta(1, 45°)| \simeq 10^5 \, [2V - G - 3\varrho_2(\alpha\varrho_0)^{-1}], \tag{67}$$

so that a normalized density difference of 10^{-7} yields a flow of about 10 m/s. Pole-equator differences in density or temperature are, therefore, strong generators of meridional circulation. Furthermore, we can isolate in (65) the role of the overall radial density gradient. Positive G produces negative $\overline{u}_\theta$, in contrast to positive V (super-rotation). Hence, density stratification leads to counter-clockwise flow which at the surface is directed polewards. The result is equatorial deceleration. If this is true, Kippenhahn's model ($V \approx 1$) cannot be valid, because of the strong density stratification of the SCZ.

Our simple computation does not, however, reveal the number of cells which develop. This is still an open question: just how many cells appear as a consequence of a given inhomogeneity F remains to be determined. As we have seen, the rotation problem cannot be solved completely using the Reynolds equation alone. In particular, the latitude-dependence of the density must be determined by solving a further equation, namely the heat transfer equation. This determines, roughly speaking, the temperature profile for any latitude which, together with a constitutive relation and the latitudinal component of the Reynolds equation, provides the density profile.

Since an equation for $\partial p/\partial\theta$ is needed before information can be extracted from the Reynolds equation, it may be more convenient to consider the two meridional components (43) of the Reynolds equation, separately, rather than their fourth-order version (47). In this case the pressure must also be expanded:

$$p = \sum_{0,2,\dots} p_n P_n . \tag{68}$$

Using the relations (A.14), together with (57), we find after a short calculation that

$$\begin{aligned} p_n' = & -\varrho_n g + \frac{2}{3} r\varrho\Omega_0^2(\delta_{n0} - \delta_{n2}) \\ & + 2r\varrho(1 - \delta_{n0}) \left[\frac{(n+1)(n+2)}{2n+3}\,\Omega_n - \frac{(n-1)\,n(1-\delta_{2n})}{2n-1}\,\Omega_{n-2}\right]\Omega_0 \\ & + \nu_T \frac{n(n+1)}{r^2}\left\{A_n'' - (\ln\varrho)'\,A_n' - \left[\frac{n(n+1)}{r^2} + 2(\ln\varrho)''\right]A_n\right\}, \\ n = & \; 0, 2, \dots \end{aligned} \tag{69}$$

The latitudinal component of (43) can be manipulated in the same way. Since this equation is not applicable to a latitude-independent pressure and density, it can only be used for $n > 0$. We obtain

$$\begin{aligned} -p_n = & \frac{1}{3} r^2\varrho\Omega_0^2\,\delta_{n2} + 2r^2\varrho\Omega_0\left[\frac{n+2}{2n+3}\,\Omega_n + \frac{(n-1)(1-\delta_{n2})}{2n-1}\,\Omega_{n-2}\right] \\ & + \nu_T\left\{-A_n''' + (\ln\varrho)'\,A_n'' + \left[\frac{n(n+1)}{r^2} + \frac{2(\ln\varrho)'}{r} + (\ln\varrho)''\right]A_n' \right. \\ & \left. - \frac{2n(n+1)}{r^3}\,A_n\right\}, \qquad n = 2, 4, \dots \end{aligned} \tag{70}$$

Its differentiation and addition to (69) leads directly to (60), the fourth-order expression given earlier for the stream function A.

Chapter 6. The Lambda — Effect and Differential Rotation

Having discussed the angular momentum transport by meridional circulation, we shall now study the direct contribution made by turbulence to the maintenance of the non-uniform rotation. We shall assume the medium to be statistically steady and non-magnetic. The Reynolds equation (4.36) then reduces to

$$x^{-2}\, \partial(\varrho x^3 Q_{r\phi})/\partial x + (\varrho/\sin^2\theta)\, \partial(\sin^2\theta\, Q_{\theta\phi})/\partial\theta = 0\,. \tag{1}$$

This equation can be analyzed if the structure of the tensor Q_{ij} is known. We shall neglect the effect of the meridional circulation on the turbulence because it does not directly affect the $Q_{i\phi}$-correlations. We shall therefore discuss only the reduced stress relation

$$Q = Q^{(0)} + Q^{(d)} + Q^{(r)}\,. \tag{2}$$

The second tensor on the right-hand side of (2) vanishes for uniform rotation and can depend only on combinations of $\partial\Omega/\partial r$ and $\partial\Omega/\partial\theta$, as was indeed implied by Boussinesq's original ansatz. Although no proof will be given here, we must appeal to the well-known fact that two different eddy viscosities, ν_{vv} and ν_{hh}, appear in the TMT when the turbulence is anisotropic:

$$\begin{aligned} Q_{r\phi} &= -\nu_{vv} r \sin\theta\, \partial\Omega/\partial r + Q^{(r)}_{r\phi}\,, \\ Q_{\theta\phi} &= -\nu_{hh} \sin\theta\, \partial\Omega/\partial\theta + Q^{(r)}_{\theta\phi}\,. \end{aligned} \tag{3}$$

For simplicity, we shall ignore such details here (but see the next Chapter) and focus our attention on the last tensor in (2) which has been separately defined as

$$Q^{(r)}_{ij} = \Lambda_{ijk}\Omega_k\,. \tag{4}$$

This represents the influence of the global rotation on turbulence in uniform rotation for which, in contrast to the earlier Boussinesq formulation, nonvanishing TMT may be expected. It prevents rigid rotation from being a solution of the Reynolds equation, and so plays an essential role in all theories of differential rotation.

6.1. The Fundamental Λ-Mode

We cannot construct an acceptable correlation tensor from $\boldsymbol{\Omega}$ by itself, and so rotation alone can never lead to the non-diffusive expressions, $Q_{ij}^{(r)}$. This tensor must be both symmetric and polar. With the addition of one further field, $\mathring{\boldsymbol{g}}$, we can construct such a tensor,

$$Q_{ij}^{(r)} = \Lambda_V(\varepsilon_{ipk}\mathring{g}_j + \varepsilon_{jpk}\mathring{g}_i)\,\mathring{g}_p\Omega_k, \tag{5}$$

in which $\boldsymbol{\Omega}$ and $\mathring{\boldsymbol{g}}$, the radial direction, are preferred. It is perhaps worth mentioning here that (5) remains unaltered after the transformation $\mathring{\boldsymbol{g}} \to -\mathring{\boldsymbol{g}}$, in contrast to the α-tensor arising in mean field electrodynamics (Krause and Rädler, 1980). In spherical coordinates the tensor (5) possesses only one non-zero component, $Q_{r\phi}$, and so the modified Boussinesq expressions (3) become

$$\begin{aligned} Q_{r\phi} &= -\nu_{vv}x \sin\theta \cdot \partial\Omega/\partial x + \Lambda_V \sin\theta \cdot \Omega, \\ Q_{\theta\phi} &= -\nu_{hh} \sin\theta \cdot \partial\Omega/\partial\theta. \end{aligned} \tag{6}$$

The presence of the non-diffusive Λ-effect in $(6)_1$ can readily be shown to be a direct consequence of the Coriolis force acting on anisotropic eddies.

In view of the above remarks, we shall consider the influence of slow rotation, in which case Λ_V is independent of Ω, i.e.

$$\Lambda_V = \nu_{vv} V^{(0)}(x), \tag{7}$$

with

$$\partial V^{(0)}/\partial\theta = \partial\nu_{vv}/\partial\theta = \partial\nu_{hh}/\partial\theta = 0. \tag{8}$$

Because of its central role, we shall often call $V^{(0)}$ the "fundamental" mode of TMT. In what follows we will show that the effect just outlined can never produce a latitudinal differential rotation but will instead give rise to a characteristic radial rotation law. This result agrees well with the numerical work of Gilman and Foukal (1979).

Before discussing the equations we shall comment on the boundary conditions. Since very little angular momentum leaves the Sun, we may assume that the solar surface is stress-free, so that the vertical angular momentum flux vanishes there:

$$\partial\Omega/\partial x = \nu_{vv}(\nu + \nu_{vv})^{-1}\, V^{(0)}\Omega \quad \text{at} \quad x = 1. \tag{9}$$

Our knowledge of conditions at the base of the convection zone is limited, so we will distinguish between two possible extreme cases. At one extreme it is assumed that there is rigid rotation at $x = x_i$ because of the large viscosity there, while at the other extreme the stress-free boundary condition is applied (see § 4.7).

The first of these cases is the more amenable to analytic treatment. The choice of boundary conditions leads to no significant differences when the rotation rate is small or when the convection zone is thick.

According to our definitions, the Reynolds equation (1) now takes the form

$$\frac{1}{\varrho x^2}\frac{\partial}{\partial x}\left\{\varrho x^3\left[(\nu+\nu_{vv})\,x\frac{\partial\Omega}{\partial x}-\nu_{vv}V^{(0)}\Omega\right]\right\}=-\frac{\nu+\nu_{hh}}{\sin^3\theta}\frac{\partial}{\partial\theta}\left(\sin^3\theta\frac{\partial\Omega}{\partial\theta}\right). \quad (10)$$

Kippenhahn (1963) has shown by direct calculation that, when $\nu+\nu_{vv}$, $\nu+\nu_{hh}$ and $V^{(0)}$ are constants, (10) has no solution in which the differential rotation varies with latitude. Consequently, he was able to obtain differential rotation (in latitude) only by adding a meridional circulation. We can confirm his result and extend it to the case when both the density and the mean properties of the turbulence depend on radius alone.

To this end, let us seek a separable solution of (10) of the form $\Omega=X(x)\times\Theta(\theta)$, for which (10) reduces to

$$\frac{(\mathrm{d}/\mathrm{d}x)\,\{\varrho x^3[(\nu+\nu_{vv})\,x\,\mathrm{d}X/\mathrm{d}x-\nu_{vv}V^{(0)}X]\}}{\varrho(\nu+\nu_{hh})\,x^2X}=-\frac{1}{\sin^3\theta}\frac{\partial}{\partial\theta}\left(\sin^3\theta\frac{\partial\Theta}{\partial\theta}\right). \quad (11)$$

This separation ensures that each side of (11) equals the same constant k (say) and hence leads to the two distinct equations

$$(\mathrm{d}/\mathrm{d}x)\,\{\varrho x^3[(\nu+\nu_{vv})\,x\,\mathrm{d}X/\mathrm{d}x-\nu_{vv}V^{(0)}X]\}=k(\nu+\nu_{hh})\,\varrho x^2X, \quad (12)$$

$$(\mathrm{d}/\mathrm{d}\theta)\,(\sin^3\theta\,\mathrm{d}\Theta/\mathrm{d}\theta)=-k\sin^3\theta\,\Theta. \quad (13)$$

Since the rotation law must be symmetric with respect to both the rotation axis and the equatorial plane, the function Θ satisfies

$$\mathrm{d}\Theta/\mathrm{d}\theta=0 \quad \text{at} \quad \theta=0,\ \ \pi/2. \quad (14)$$

As there is no tangential stress on the outer boundary, the function

$$S=(\nu+\nu_{vv})\,x\,\mathrm{d}X/\mathrm{d}x-\nu_{vv}V^{(0)}X \quad (15)$$

vanishes at the surface, i.e. $S(1)=0$. We denote the particular solution, for which S vanishes everywhere, by X_0 and note that it is the eigensolution having the eigenvalue $k=0$. Furthermore, if X_0 vanishes at a particular point, then all its derivatives must vanish at that point also, and it must therefore vanish everywhere. Since this trivial case is (at this stage of the argument) of no interest, it may be assumed without loss of generality that $X_0>0$.

On making the substitution $X=X_0\xi$ in (12) and the boundary condition (9), we now obtain

$$(\mathrm{d}/\mathrm{d}x)\,[\varrho(\nu+\nu_{vv})\,x^4X_0\,\mathrm{d}\xi/\mathrm{d}x]=\varrho k(\nu+\nu_{hh})\,x^2X_0\xi, \qquad \xi'(1)=0. \quad (16)$$

Multiplying (16) by ξ and integrating, we find

$$k \int_{x_i}^{1} \varrho(\nu + \nu_{hh})\, x^2 X_0 \xi^2 \,\mathrm{d}x = -\int_{x_i}^{1} \varrho(\nu + \nu_{vv})\, x^4 X_0 \xi'^2 \,\mathrm{d}x - \varrho(\nu + \nu_{vv})\, x_i^4 X_0(x_i)\, \xi'(x_i)\, \xi(x_i). \tag{17}$$

The last term on the right-hand side vanishes if the inner boundary is either slippery or rotating rigidly. For positive ν_{vv} and ν_{hh} it is clear that $k \leqq 0$. Similar manipulations of (13) and (14) yield the relation

$$k \int_{0}^{\pi/2} \sin^3\theta \cdot \Theta^2 \,\mathrm{d}\theta = \int_{0}^{\pi/2} \sin^3\theta (\mathrm{d}\Theta/\mathrm{d}\theta)^2 \,\mathrm{d}\theta, \tag{18}$$

so that the same eigenvalue k must be non-negative: $k \geqq 0$. Together these two results imply that the eigenvalue k vanishes. Consequently, the eigen-solutions of (13) and (16) must be constants, which means that the only possible solutions for media that are weakly influenced by rotation are latitude-independent. Thus the rotation law is necessarily identical with the solution, X_0, for which the stress vanishes everywhere i.e. for which $S \equiv 0$. We then have by (15)

$$\mathrm{d} \ln \Omega/\mathrm{d} \ln x = \nu_{vv}(\nu + \nu_{vv})^{-1} V^{(0)}, \tag{19}$$

or

$$\Omega = \Omega_{\odot} \exp \int_{1}^{x} \nu_{vv}(\nu + \nu_{vv})^{-1} V^{(0)}(x')\, dx'/x'. \tag{20}$$

Solution (20) satisfies the stress-free condition (9) not only at the outer boundary but everywhere in the SCZ. The angular momentum transport due to the turbulence vanishes everywhere but the sign of its non-diffusive part determines the sign of the radial gradient of Ω. Positive $V^{(0)}$ generates an angular velocity which increases outward (super-rotation) and negative $V^{(0)}$ generates one that increases inward (sub-rotation). Indeed, it seems that the latter situation is the one pertinent to the upper layers of the SCZ since observational evidence has indicated that at, and just beneath, the surface the radial gradient is

$$\mathrm{d} \ln \Omega/\mathrm{d} \ln x \simeq -1.2. \tag{21}$$

Clearly, the observations can be explained by the existence of a negative fundamental mode of TMT of magnitude 2, i.e. $V^{(0)} = -2$. The opposite case, i.e. super-rotation, appears to arise in the Earth's atmosphere which, on average, rotates faster by about 10 m/s than the solid Earth (see Hide, 1984). The equatorial velocity is 436 m/s, hence the effect is somewhat smaller than that of (21):

$$\mathrm{d} \ln \Omega/\mathrm{d} \ln x \simeq 1 \cdot 3, \tag{22}$$

where we have used $\Delta r \approx 100$ km in our estimate. A positive TMT mode $V^{(0)}$ of order unity is thus required to explain the Earth's super-rotation (Rüdiger, 1975). Venus also super-rotates but in an extreme way: the atmosphere moves about ten times faster than the solid surface. We doubt, however, whether such steep profiles can be explained by our simple eddy viscosity concept, which is based not on linearization with respect to Ω but with respect to its gradients, both radial and latitudinal. So we must also be careful when trying to apply the theory to the equatorial jet streams of Jupiter and Saturn. Observations of the latter reveal an equatorial eastward flow of 400 m/s, four times the speed of the Jovian jets, the width of the streams being only about 10^0.

6.2. The Higher Λ-Modes

When the rotation is not slow there is, in addition to the Λ_V in $Q_{r\phi}$, also a Λ_H in $Q_{\theta\phi}$ (see Section 4.6)

$$Q_{\theta\phi} = -\nu_{hh} \sin\theta \, \partial\Omega/\partial\theta + \Lambda_H \cos\theta \, \Omega . \tag{23}$$

Again a normalization will be used: $\Lambda_H = \nu_{vv} H$. In general, the Λ-quantities depend on the latitude θ and may therefore be expressed in powers of $\sin^2\theta$, as in (2.21), rather than as the more usual sum of Legendre polynomials. This ansatz is a matter of mathematical convenience rather than physics. The eddy viscosity ν_{vv} is included for dimensional reasons and, based on the Sun's small-scale eddies, its value is taken to be $5 \cdot 10^{12}$ cm^2/s. Denoting the flux mode number by the integer l, the mode $l = 0$ is the fundamental flux mode which only exists when there is vertical TMT. The latitudinal profiles of the functions V and H depend on the forms assumed for the fluctuating velocity. At this stage of our discussion two apparently contradictory formalisms have been employed, with consequences reflected in the modal expansions (2.21). Whereas one formalism is useful in studying the influence of global rotation on nearly homogeneous turbulence, it is natural to adopt the other when dealing with the same rotational effect but on strongly inhomogeneous turbulence.

For mathematical reasons we will work with the polynomial expansion (5.4) taking $\omega_0(1) = 0$, a condition which determines $\Omega_\odot$. Often, however, the modal representation (3.6) is used. The transformation between the two forms can be effected quite simply by using

$$\sin^{2l}\theta = q_n^{(l)} P_n^1/[n(n+1)\sin\theta], \tag{24}$$

where

$$q_n^{(l)} = 2(2n+1)\,(l+1) \int\limits_0^1 x(1+x^2)^l \, P_n(x)\,\mathrm{d}x . \tag{25}$$

Table VI.i. The transformation coefficient $q_n^{(l)}$ given by equation (25)

l ↓ n →	1	3	5	7
0	2	0	0	0
1	1.6	−1.6	0	0
2	1.371	−2.133	0.762	0
3	1.218	−2.328	1.407	−0.297

(We should here remind readers of the summation convention introduced immediately before § 5.1.) The first few coefficients are given in Table VI.i. The relation between the two sets of rotation coefficients is

$$\omega_n = q_{n+1}^{(l)} \omega^{(l)}/(n+1)\,(n+2). \tag{26}$$

If we further postulate

$$\nu_{rv} = \nu_T, \qquad \nu_{hh} = s\nu_T, \qquad \mathrm{d}\ln \varrho/\mathrm{d}\ln x = -G, \tag{27}$$

the Reynolds equation (1) takes the form

$$\frac{\partial}{\partial x}\left(x^4 \frac{\partial \Omega}{\partial x}\right) - Gx^3 \frac{\partial \Omega}{\partial x} + \frac{sx^2}{\sin^3\theta}\frac{\partial}{\partial \theta}\left(\sin^3\theta \frac{\partial \Omega}{\partial \theta}\right)$$
$$= \frac{\partial}{\partial x}(x^3 V\Omega) - Gx^2 V\Omega + \frac{x^2}{\sin^3\theta}\frac{\partial}{\partial\theta}(H\cos\theta\sin^2\theta\Omega). \tag{28}$$

Due to the dependence of the functions V and H on the angular velocity itself, (28) is highly non-linear in Ω. There is one exceptional case, though: for very slow rotation, the functions V and H lose their Ω-dependence. Only then does a general analytical theory seem to be possible. The question arises whether we can linearize (28) without too great a loss of physical content. Such a step appears permissible in the "approximation of weak differential rotation". The assumption is that the angular velocity can be adequately represented everywhere by the one basic constant value, $\Omega_\odot$

$$\Omega = \Omega_\odot + \tilde{\Omega}(r, \theta), \qquad \tilde{\Omega} \ll \Omega_\odot. \tag{29}$$

In this case the leading term on the right-hand side of (28) contains combinations such as $V(\Omega_\odot)\,\Omega_\odot$ ..., which provide the inhomogeneous part of the differential equation for $\tilde{\Omega}$. Indeed, because the angular momentum is given, the determination of the differential rotation poses, quite generally, an inhomogeneous problem, in sharp contradistinction to the dynamo theory of cosmic magnetic fields.

Physically, the mathematical linearization is tantamount to assuming that the angular velocity is fixed (namely $\Omega_\odot$) insofar as the Λ-effect is concerned. It is not necessarily a question of how strong the rotational influence is on the turbulence although, the weaker this influence, the better the approximation.

Substituting the expressions (2.21) and (5.4) into (28), we find

$$x^2\omega_n'' + (4 - G)\,x\omega_n' - sn(n+3)\,\omega_n$$

$$= [(3 - G)\,V^{(l)} + xV^{(l)\prime} - n(n+3)\,H^{(l)}/2l]\,q_{n+1}^{(l)}/(n+1)\,(n+2), \tag{30}$$

where we have used the relation

$$(\partial/\partial\theta)\,(\cos\theta\,\sin^{2l+2}\theta) = -[(n-1)\,(n+2)/2ln(n+1)]\,q_n^{(l)}\sin^2\theta\cdot P_n^1, \tag{31}$$

which can be derived from (24).

Assume that the depth-dependence of the Λ-functions is given by

$$V^{(l)} = (x_i/x)^\lambda\,V_i^{(l)}, \qquad H^{(l)} = (x_i/x)^\lambda\,H_i^{(l)}, \tag{32}$$

and introduce the abbreviations

$$\tilde V_n = q_{n+1}^{(l)}x_i^\lambda V_i^{(l)}/(n+1)\,(n+2), \quad \tilde H_n = q_{n+1}^{(l)}x_i^\lambda H_i^{(l)}/2l(n+1)\,(n+2), \tag{33}$$

$V_i^{(l)}$ and $H_i^{(l)}$ being the values at the base of the convection zone. In this case some manipulations transform the linearized Reynolds equation into

$$x^2\omega_n'' + (4 - G)\,x\omega_n' - sn(n+3)\,\omega_n = x^{-\lambda}[\lambda(\lambda + G - 3) - sn(n+3)]\,\tilde S_n, \tag{34}$$

where

$$\tilde S_n = [(3 - G - \lambda)\,\tilde V_n - n(n+3)\,\tilde H_n]/[\lambda(\lambda + G - 3) - sn(n+3)]. \tag{35}$$

The solution of (34) is

$$\omega_n = C\delta_{n0} + C_1x^{\lambda_1} + C_2x^{\lambda_2} + x^{-\lambda}\tilde S_n, \tag{36}$$

where λ_1 and λ_2 are the roots of the quadratic equation

$$\lambda^2 + (3 - G)\,\lambda - sn(n+3) = 0. \tag{37}$$

Hence $\lambda_1 + \lambda_2 = G - 3$ and $\lambda_1\lambda_2 = -sn(n+3)$. The coefficients C_1 and C_2 can be determined from the boundary conditions

$$x\omega_n' - x^{-\lambda}\tilde V_n = 0 \quad \text{at} \quad x = x_i, 1. \tag{38}$$

Note that the differential equation and boundary conditions determine the solution to within the arbitrary additive constant C, which appears on the right-hand side of (36) and which represents the given angular momentum of the shell.

In the present context it is sufficient to present only the solution at the surface:

$$\omega_n(1) = C\delta_{n0} + \frac{(\lambda_1\lambda_2\psi_n - \lambda_1 - \lambda_2 - \lambda)\,\tilde{V}_n + (1 + \psi_n\lambda)\,\lambda_1\lambda_2 s^{-1}\tilde{H}_n}{\lambda_1\lambda_2 + \lambda(\lambda + \lambda_1 + \lambda_2)}. \tag{39}$$

This depends basically on the "geometry functions"

$$\psi_n = [(\lambda_1 - \lambda_2)\,x_i^{-\lambda} + \lambda_2 x_i^{\lambda_2} - \lambda_1 x_i^{\lambda_1}]/\lambda_1\lambda_2(x_i^{\lambda_2} - x_i^{\lambda_1}), \tag{40}$$

from which we can form

$$\Psi = \frac{(\lambda_1\lambda_2\psi_n - \lambda_1 - \lambda_2 - \lambda)\,x_i^{\lambda}}{\lambda_1\lambda_2 + \lambda(\lambda + \lambda_1 + \lambda_2)} = \frac{\lambda_1 - \lambda_2 - (\lambda + \lambda_1)\,x_i^{\lambda+\lambda_2} + (\lambda + \lambda_2)\,x_i^{\lambda+\lambda_1}}{[\lambda_1\lambda_2 + \lambda(\lambda + \lambda_1 + \lambda_2)]\,(x_i^{\lambda_2} - x_i^{\lambda_1})}. \tag{41}$$

This quantity will play a fundamental role in what follows. For example, (39) can be rewritten as

$$\omega_n(1) = C\delta_{n0} + [\Psi V_i^{(l)} + (x_i^{\lambda} + \lambda\Psi)\,H_i^{(l)}/2ls]\,q_{n+1}^{(l)}/(n + 1)\,(n + 2). \tag{42}$$

The following analytical discussion of Ψ will lead to relatively simple expressions for the rotation law.

Let us, for the moment, assume that Ψ does not depend on the integer n. Then the summation procedure (26) would yield

$$\omega^{(l)} = \Psi V_i^{(l)} + (x_i^{\lambda} + \lambda\Psi)\,H_i^{(l)}/2ls, \tag{43}$$

and the rotation law at the surface would become

$$\Omega/\Omega_\odot = \text{constant} + \sum_{l=1} [\Psi V_i^{(l)} + (x_i^{\lambda} + \lambda\Psi)\,H_i^{(l)}/2ls] \sin^{2l}\theta. \tag{44}$$

This result becomes more transparent if we re-express it in terms of the overall differential rotation, i.e. the pole-equator difference (3.7) which is about -0.3 for the Sun:

$$\delta\Omega = -\sum_{l=1} [\Psi V_i^{(l)} + (x_i^{\lambda} + \lambda\Psi)\,H_i^{(l)}/2ls]. \tag{45}$$

These general results are indeed useful since the dependence of Ψ on n is actually rather weak in some important cases.

Next let us consider the horizontal Reynolds stress $Q_{\theta\phi}$, which can be calculated most easily through its special derivative $\partial(\sin^2\theta Q_{\theta\phi})/\partial\theta$. On the one hand, relations (3.16) and (3.17) give

$$\partial(\sin^2\theta \cdot Q_{\theta\phi})/\partial\theta|_1 = -\nu_{vv}\Omega_\odot x_i^{-\lambda}(n - 1)\,(n + 2)\,\tilde{W}_n P_n^1 \sin^2\theta, \tag{46}$$

where

$$\tilde{W}_n = x_i^{\lambda} q_n^{(l)} w^{(l)}/2n(n + 1)\,l, \tag{47}$$

while, on the other hand, equation (23) together with our rotation law (42) lead to

$$\partial(\sin^2\theta \cdot Q_{\theta\phi})/\partial\theta = \nu_{vv}\Omega_{\odot} x_i^{-\lambda}(n-1)(n+2)\,\Psi(s\tilde{V}_{n-1} + \lambda\tilde{H}_{n-1})\,P_n^1 \sin^2\theta\,. \tag{48}$$

A comparison of these two expressions yields

$$\tilde{W}_n + \Psi(s\tilde{V}_{n-1} + \lambda\tilde{H}_{n-1}) = 0\,, \tag{49}$$

which, due to the structure of $q_n^{(l)}$, implies

$$w^{(l)} = -\Psi(2lsV_i^{(l)} + \lambda H_i^{(l)})\,, \tag{50}$$

if Ψ is again assumed to be independent of n. Note that the role Ψ plays in (50) for $w^{(l)}$ is very similar to the role it played in (43) for $\omega^{(l)}$. We should always expect non-uniformly rotating convection zones to give non-trivial w-profiles, and for nonconstant w-profiles to be linked with non-uniform rotation. The only exception to this rule is the case of extremely small Ψ, for which the horizontal Reynolds stress vanishes but not the differential rotation.

Before trying to make estimates for the function Ψ, which will be the principal topic of the remainder of this chapter, we should consider yet another aspect of the problem. With the expressions (43) and (50) we possess relations which connect the observed profiles Ω and W to the unobservable profiles $V(\theta)$ and $H(\theta)$. It should be possible, therefore, to determine the Λ-effect inside the convection zone merely from surface observations. There is a complication, however, for the parameters s and λ are unknown, and plausible arguments have to be devised for estimating them. Relations (43) and (50) yield, in general,

$$\begin{aligned} V_i^{(l)} &= -\lambda x_i^{-\lambda}\omega^{(l)} - (\lambda x_i^{-\lambda} + \Psi^{-1})\,w^{(l)}/2ls\,, \\ H_i^{(l)} &= x_i^{-\lambda}(2ls\omega^{(l)} + w^{(l)})\,. \end{aligned} \tag{51}$$

One sees at a glance that if λ is large then so must be $V_i^{(l)}$ and $H_i^{(l)}$, in order to establish the observed $\omega^{(l)}$ and $w^{(l)}$. This is quite believable since large λ confines the Λ-effect to the base of the convection zone. It will be important to find out whether the differential rotation suffers a "screening" due to the density stratification and/or the outward decrease of Λ. Clearly, a Λ-effect (strictly speaking, the part that excludes fundamental mode) exists in the bulk of the convection zone. The question is whether the corresponding differential rotation can survive the smoothing action of viscosity in the upper layers of the convection zone where (with the exception of $V^{(0)}$) the Λ-effects is almost zero. To make further progress, we must investigate the geometry function Ψ in greater detail and, to this end, the following approximation proves to be useful.

6.3. The Thin-Shell Approximation

For sufficiently shallow convection zones ($\mathsf{d} = 1 - x_i \ll 1$) an expansion of Ψ_n in d is possible and leads to

$$\psi_n = \psi_0 + \frac{1}{2}(1 + \lambda\psi_0)\,\mathsf{d}\,, \tag{52}$$

where

$$\psi_0 = (\lambda_1 + \lambda_2 + \lambda)/\lambda_1\lambda_2 = (3 - \lambda - G)/sn(n+3)\,. \tag{53}$$

Due to their simplicity, expressions (52) and (53) yield relatively simple rotation laws. Substituting (52) and (53) into (41), we find

$$\Psi = \frac{1}{2}\,\mathsf{d}\,, \tag{54}$$

which, when applied to (43), yields

$$\omega^{(l)} = \frac{1}{2}\left[\mathsf{d}\,V_i^{(l)} + \left(1 - \frac{1}{2}\,\mathsf{d}\,\lambda\right)H_i^{(l)}/ls\right], \tag{55}$$

a result very appealing in its simplicity. After substituting (54) into (44) and expanding x_i, we arrive at the rotation law

$$\Omega(1,\theta)/\Omega_\odot = C + \frac{1}{2}\sum_{l=1}\left[\mathsf{d}\,V_i^{(l)} + \left(1 - \frac{1}{2}\,\mathsf{d}\,\lambda\right)H_i^{(l)}/ls\right]\sin^{2l}\theta\,. \tag{56}$$

The absence of the density gradient here suggests that, even more generally for larger d, it does not have a strong influence on the latitudinal rotation profile. If $\Omega_\odot$ denotes the true value of the angular velocity on the polar axis, C is unity. The overall differential rotation (45) is given by

$$\delta\Omega = -\frac{1}{2}\sum_{l=1}\left[\mathsf{d}\,V_i^{(l)} + \left(1 - \frac{1}{2}\,\mathsf{d}\,\lambda\right)H_i^{(l)}/ls\right], \tag{57}$$

to which all the non-fundamental TMT-modes contribute. Hence, one can formally produce the observed equatorial acceleration ($\delta\Omega < 0$) from just a single positive mode, provided only that $l \geqq 1$.

Our result suggests the following conclusions:

(i) Equatorial acceleration requires the sum (57) to be negative. Thus, some positive ("outward") $V^{(l)}$ and/or positive ("equatorward") $H^{(l)}$ are always necessary for equatorial acceleration. No conclusion can be drawn, however, concerning the sign of the TMT-functions Λ_V and Λ_H. The quantitative analysis actually involves negative modes, as well as the positive ones;

(ii) The non-fundamental Λ-modes can be portrayed very directly as modes of the surface rotation law. Each $\sin^{2l}\theta$-mode of the Λ-effect appears as a $\sin^{2l}\theta$-mode of the angular velocity. Free-free models prove to be simpler in this respect than rigid-free models for which the rotation law is

$$\frac{\Omega(1,\theta)}{\Omega_\odot} = 1 + \sum_{l=1}\left[\mathsf{d}\, V_i^{(l)} \sin^{2l}\theta - \frac{1}{2}\,\frac{\mathsf{d}^2}{\sin^3\theta}\, H_i^{(l)}\,\frac{\mathrm{d}}{\mathrm{d}\theta}(\cos\theta\,\sin^{2l+2}\theta)\right]; \qquad (58)$$

(iii) In both these models the vertical Λ-effect increases with d, but the horizontal effect behaves differently for each; it depends on d^2 for rigid-free models and on $\mathsf{d}^0 = 1$ for free-free models. For models with merely a vertical Λ-effect, an astonishing relation emerges:

$$\delta\Omega = -\frac{1}{2}\,\mathsf{d}\sum_{l=1} V_i^{(l)}, \qquad (59)$$

suggesting that even for thicker layers, the dependence of the pole-equator difference on the thickness of the layer is principally linear;

(iv) Density stratification and the depth-dependence of Λ are of minor significance. Only for $\lambda \gtrsim 10$ does the rotation law become modified.

Now let us apply the thin-shell approximation (54) to the calculation of the horizontal Reynolds stress $Q_{\theta\phi}$. Substituting this expression into (50) we obtain

$$w^{(l)} = -\mathsf{d}\left(lsV^{(l)} + \frac{1}{2}\,\lambda H_i^{(l)}\right), \qquad (60)$$

where the λ appears to play a more prominent role than it does in the rotation profile. For small λ, at least, the almost paradoxical consequence is that the horizontal Reynolds stress is due to the vertical Λ-effect. Hence an analysis of observations of the horizontal Reynolds stress would lead to statements about the vertical Λ-modes alone. No general findings about the differential rotation can be expected from such investigations, however, since this also depends on the horizontal modes. For example, (55) and (60) give, for $l = 1$ and $s = 1$,

$$2\omega^{(1)}/w^{(1)} = -1 - 2(1 - \mathsf{d}\,\lambda)\, H_i^{(1)}/\mathsf{d}(2V_i^{(1)} + \lambda H_i^{(1)}), \qquad (61)$$

so that the observed positivity of the left-hand side is inconsistent with vanishing $H^{(1)}$.

Obviously, the right-hand side of (61) contains more than trivial geometric quantities. No simple causal relation betwen the observable profiles $\Omega(\theta)$ and $w(\theta)$ exists; they are independent of each other insofar as Λ-functions can always be found that precisely reproduce a given $\Omega(\theta)$ and $w(\theta)$. The observable $w^{(l)}$ are proportional to d. Thus, if a convection zone possesses significant finite values for one or more $w^{(l)}$, its thickness cannot be too small. Measurements of the horizontal Reynolds stress $\overline{u_\theta' u_\phi'}$ therefore provide evidence that the SCZ cannot be extremely thin. This conclusion is inconsistent with the calculations of Christensen-Dalsgaard *et al.* (1979) who argue that the solar convection zone may be much shallower than those of standard solar models, and might even be only a few percent of the solar radius. Observe that in the limit ($\mathsf{d} \to 0$) of a very thin convection zone, the induced $w^{(1)}$ vanishes, in sharp contrast to the rotation coefficients, $\omega^{(l)}$, which tend to $H_i^{(l)}/2$. According to (55) and (60) one can express the unobservable $V^{(l)}$ and $H^{(l)}$ in terms of the observable $\omega^{(l)}$ and $w^{(l)}$:

$$V_i^{(l)} = -\lambda(1 + \lambda\,\mathsf{d})\,\omega^{(l)} - (2\,\mathsf{d}^{-1} + \lambda + \lambda^2\,\mathsf{d})\,w^{(l)}/2ls\,, \tag{62}$$

$$H_i^{(l)} = (1 + \lambda\,\mathsf{d})\,(2ls\omega^{(l)} + w^{(l)})\,. \tag{63}$$

These formulae also illustrate the importance of determining the $w(\theta)$ profile better than we know it at the present time. Let us tentatively insert into (62) and (63) the observational values (3.2) and (3.23), $\mathsf{d} \approx 0.2$, and the simplest values of the parameters, i.e. $s = 1$ and $\lambda \approx 0$, we find

$$V_i^{(1)} \approx -10\,, \quad H_i^{(1)} \approx 3\,, \quad V_i^{(2)} \approx 0\,, \quad H_i^{(2)} \approx -1\,. \tag{64}$$

Within the framework of our assumptions these Λ-modes create simultaneously both the observed rotation law and the observed w-profile. Any Reynolds theory of solar differential rotation has to explain the existence of

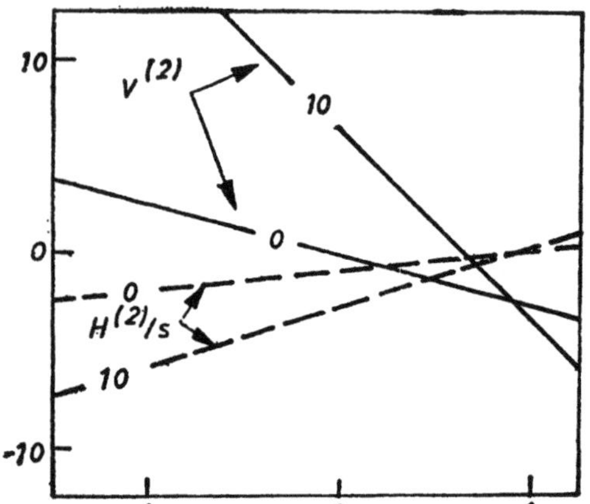

Fig. 6.1. The dependence of the required Λ-modes $V^{(2)}$ and $H^{(2)}$ on the, as yet, unknown but observable parameter $w^{(2)}$. The (straight) lines are labelled by their λ values, representing the inward increase of Λ.

values such as these. The turbulence model that reproduces (64) should reflect the real solar situation.

There is a remarkable difference between these and the corresponding results for rigid-free shells, for which both $V_i^{(1)}$ and $H_i^{(1)}$ turn out to be positive (Rüdiger, 1983a). Strikingly, the inner boundary condition strongly influences the sign of $V_i^{(1)}$. Yet this is only true of thin-shell models: the thicker the shell, the smaller the difference between the two models. The results of the following thick-shell approximation are therefore valid for either a stress-free or a no-slip inner surface. For future reference, the relationship between the as yet unknown $w^{(2)}$, and the required $V_i^{(2)}$ and $H_i^{(2)}$ is shown in Fig. 6.1. It can be seen at once that $H_i^{(2)}$ is hardly influenced by $w^{(2)}$ and must always be negative. In contrast, negative $w^{(2)}$ would imply negative $V_i^{(2)}$. All these results are completely independent of the details of the mechanism which generates the Reynolds stress. They hold for all situations that can be modelled by the Λ-effect.

6.4. The Thick-Shell Approximation

This approximation assumes that d is of order unity. To avoid lengthy calculations we restrict ourselves to the case $s = 1$. Clearly, this is to some extent at variance with the real situation. The non-diffusive TMT is due exclusively to the anisotropic large-scale turbulence modes. The diffusive TMT, represented by ν_{vv} and ν_{hh}, is created both by isotropic small-scale motions and by anistropic large-scale modes. The latter contribute to both viscosities, but with differing weight, so that in reality $s \neq 1$. Nevertheless, we believe that the dominant contributions are made by the small-scale turbulence.

It remains to consider models with large G and with values of λ which are probably the most realistic ones. The actual density profile of the SCZ requires $G \approx 50$. If we accept this, the roots of the characteristic equation (37) are approximately $\lambda_1 \approx G$ and $\lambda_2 \approx 0$. With such values, the function Ψ becomes, for sufficiently deep convection zones,

$$\Psi \simeq (1 - x_i^{\lambda})\, G/\lambda(\lambda + G). \tag{65}$$

Provided λ is small, we find

$$\Psi \simeq -\ln x_i \approx \mathrm{d}\,, \tag{66}$$

which is twice as large as the result given in (54), the value obtained in the thin-shell approximation. Approximation (66) gives

$$\begin{aligned} \delta\Omega &\simeq -\sum_{l=1} 0.3 V_i^{(l)} + H_i^{(l)}/2l\,, \\ V_i^{(l)} &\simeq -1.5 w^{(l)}/l\,, \qquad H_i^{(l)} \simeq w^{(l)} + 2l\omega^{(l)}\,, \end{aligned} \tag{67}$$

so that Λ-modes with

$$V_i^{(1)} \approx -3, \qquad H_i^{(1)} \approx 3, \tag{68}$$

reproduce the observations.

The case of non-vanishing λ is, however, more realistic because the lifetime of an eddy decreases with its distance from the solar center, so that x_i^λ is very small and equation (65) yields

$$\Psi \simeq G/\lambda(\lambda + G) \simeq 1/\lambda. \tag{69}$$

This shows that the density gradient has little effect in either case. We find

$$\delta\Omega \simeq -\sum_{l=1} [(V_i^{(l)}/\lambda) + (H_i^{(l)}/2ls)], \tag{70}$$

and

$$V_i^{(l)} \simeq -(2ls\omega^{(l)} + w^{(l)})\,\lambda/2lsx_i^\lambda, \qquad H_i^{(l)} \simeq (2ls\omega^{(l)} + w^{(l)})/x_i^\lambda. \tag{71}$$

From (71) we have the general result

$$V_i^{(l)}/H_i^{(l)} \simeq -\lambda/2ls, \tag{72}$$

which shows that $V^{(l)}/H^{(l)}$ is a large (negative) number. Now, at last, it becomes obvious that the modes $V^{(l)}$ and $H^{(l)}$ always have opposite signs. The observational values give positive $H^{(1)}$ and (probably) negative $H^{(2)}$. For the first mode with $\lambda \approx 10$ we find

$$V_i^{(1)}/H_i^{(1)} \approx -5, \qquad H_i^{(1)} \approx 30. \tag{73}$$

6.5. Conclusions

Let us summarize the main results of the formal theory presented above. The latitudinal surface profiles of the angular velocity and the horizontal Reynolds stress can be interpreted as consequences of the action of the modes $V^{(l)}$ and $H^{(l)}$ for $l > 0$. It has been demonstrated that the characteristic function Ψ controls the efficiency of these modes, and this is also shown in Fig. 6.2, which compares the estimates derived above with numerical results based on (41). In some important cases the dependence of Ψ on n actually disappears. If the natural outward decrease of Λ is ignored for a moment, the value of Ψ merely varies from d/2 (weak density gradient) to d (strong density gradient). Hence, the addition of a really strong density stratification only slightly increases the differential rotation. There is smoothing for neither the angular velocity profiles nor the horizontal Reynolds stress. This finding agrees with that of Stix (1981) who found that "a shear flow pattern with such" (low) "values of n would be transmitted to the surface even from the bottom of the convection zone".

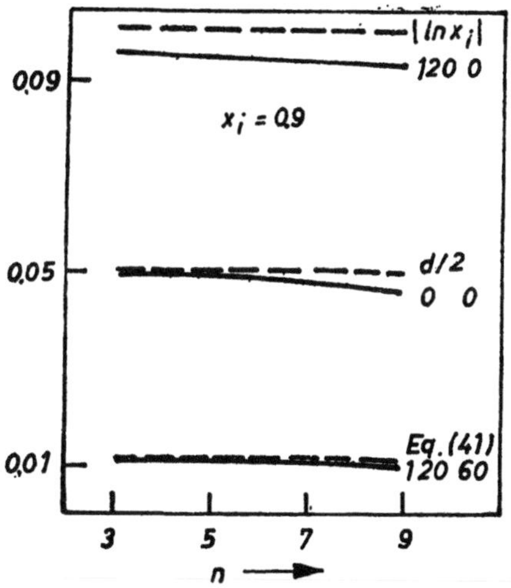

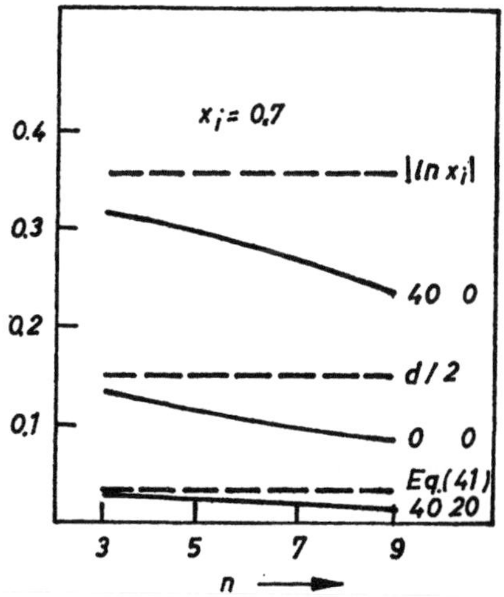

Fig. 6.2. Comparison of our approximate Ψ (dashed lines) with the exact numerical values calculated from (41). The latter (solid) lines are labelled with the values of G and λ. For deep convection zones ($x_i \lesssim 0.7$) the approximation is only good for "large G, large λ".

We have to take into account other functions that decrease rapidly with radius, namely the non-fundamental ($l > 0$) modes of Λ, which exist mainly in the deeper layers of the convection zone. The radial profiles of, for example, $V^{(1)}$ and $H^{(1)}$, depend essentially on the ratio of the eddy turnover time to the rotation period, and this ratio diminishes with the radius. The decreases used in Fig. 6.2 are very rapid so that the concomitant Λ-effect would exist only at the base of the SCZ. As a consequence, the required values of $V^{(1)}$ and $H^{(1)}$ increase in accordance with the relations

$$V_i^{(1)}/H_i^{(1)} \simeq -\lambda/2s\,, \qquad H_i^{(1)} \simeq x_i^{-\lambda}\,, \tag{74}$$

provided the density stratification is strong. These conclusions are confirmed by Table VI.ii which gives the exact values for a zone with $\mathsf{d} = 0.2$ and with $w^{(1)} = 4\omega^{(1)} = 2.5$. The results in the Table were obtained using the simpli-

Table VI.ii. The required Λ-modes for various density profiles and correlation time models.

$\varrho_i/\varrho_\odot$	$\Lambda_i/\Lambda_\odot$	Ψ	$V_i^{(1)}$	$H_i^{(1)}$	$V_\odot^{(1)}$	$H_\odot^{(1)}$
1	1	0.096	−12	3.7	−12	3.7
10^6	1	0.203	−6	3.7	−6	3.7
10^6	10	0.071	−130	37	−13	3.7
10^6	100	0.034	−4100	370	−41	3.7

fying assumption $s = 1$, and surface values are labelled with "$\odot$". Observe in particular that density stratification has practically no influence on $V^{(1)}$ and $H^{(1)}$. In contrast, the inclusion of the outward decrease in the Λ-effect requires rather large values of $V^{(1)}$ and $H^{(1)}$ at the base of the SCZ.

Next, let us discuss the (artificial) situation in which the horizontal Reynolds stress vanishes at the surface, $w(1) = 0$. Then (51) become

$$V_i^{(l)} = -\lambda x_i^{-\lambda} \omega^{(l)}, \qquad H_i^{(l)} = 2ls x_i^{-\lambda} \omega^{(l)}, \tag{75}$$

and again we have

$$V_i^{(l)}/H_i^{(l)} = -\lambda/2ls, \tag{76}$$

which is negative for a positive eddy viscosity ($s = 1$). Equation $(32)_2$ implies

$$H^{(1)}(x) = 2\omega^{(1)} x^{-\lambda} \simeq x^{-\lambda}, \tag{77}$$

which is unity at the surface and increases with depth. Hence, for $w^{(1)} = 0$, the function $H^{(1)}$ is required to attain the value "1" even at the surface and not just deep within the convection zone. This is a consequence of trying to explain non-negative surface cross-correlations $\cos \theta \overline{u_\theta' u_\phi'}$ using a positive eddy viscosity.

The required Λ would become very small only if the observations were to imply relations such as $2ls\omega^{(l)} + w^{(l)} \approx 0$ for $l = 1, 2 \ldots$, but this does not seem to be the case on the Sun. Were it true, the required Λ-modes would be extremely simple [cf. (62)]:

$$V_i^{(l)} = 2\omega^{(l)}/\mathrm{d}, \qquad H_i^{(l)} \simeq 0, \tag{78}$$

so that the functions $V^{(l)}(x) = 2x_i^{\lambda} x^{-\lambda} \omega^{(l)}/\mathrm{d}$ would take very small values at the surface. There is, therefore, reason enough to state again that the observation of equatorward (or vanishing) eddy momentum flux is a fundamental and delicate finding. It would be very tempting to explain it by means of a negative eddy viscosity ($s < 0$) but we shall ignore here such a possibility.

Equations (55) and (60) should be regarded as the main results of this Chapter. They demonstrate some important consequences of a latitude-dependent Λ. As we did not adopt any special turbulence model, our findings are quite general, and are valid for wide choices of the defining parameters. If solar values are used, we find that the first modes, i.e. $V^{(1)}$ and $H^{(1)}$, are definitely dominant and have opposite signs. In the spirit of an order of magnitude estimate, the values

$$V_i^{(1)} \simeq -150, \qquad H_i^{(1)} \simeq 30, \tag{79}$$

should not be too wide of the mark. These values are those inferred from the direct observations using mean-field hydrodynamics. We shall keep them in mind when we consider the observed solar situation and seek a rotating turbulence model that is compatible with the structure of the SCZ. As a consequence we will require the Λ-modes to be of order unity even at the surface. That is why $H^{(1)}$ must overcompensate for the negative $Q_{\theta\phi}$-correlation which was due to $\partial\Omega/\partial\theta$ alone, and this requirement must hold at the surface. That part of $Q_{\theta\phi}$ which arises from the diffusion part of (23) alone is given by

$$w^{(1)}|_D \approx -0.9\,, \qquad w^{(2)}|_D \approx 0.6\,. \tag{80}$$

On the one hand, if the observed $w^{(1)}$ is indeed positive then the surface value of the mode $H^{(1)}$ *must* exceed 0.9. On the other hand, since $H^{(1)}$ ought to be small there, it seems unlikely that w will be positive. If the horizontal TMT were observed to be poleward (i.e. $w < 0$), the constraints on the theory would be very much weaker.

Do we really know the true w-profile of the Sun? If the answer is "Yes" then

$$H_{\odot}{}^{(1)} > 1 \tag{81}$$

is an immediate and fundamental result which has a strong theoretical implications. It does not disappear were $w \approx 0$ closer to the truth than the values given by Ward.

Our second finding, negative $V^{(1)}$, is confirmed in a more complete theory involving turbulent heat transport, but only for small λ. For the more realistic case of a marked outward decrease of the Λ-modes $V^{(1)}$ and $H^{(1)}$, $V^{(1)}$ is positive. These results are derived in Chapter 11. In the last analysis, however, such findings are only re-formulations of the observations. A set of observed data, as input, is transformed into a set of turbulence parameters, as output. In Chapter 15 these output parameters are considered from the viewpoint of turbulence theory.

Chapter 7. Anisotropic Viscosity and Differential Rotation

7.1. On the Wasiutyński-Kippenhahn Theory

Even when magnetic effects are ignored (as they are here), there are in a stratified rotating turbulence always two preferred directions. Both rotation ($\boldsymbol{\Omega}$) and gravity ($\boldsymbol{g}$) will modify the turbulence and, consequently, they will both appear in the correlation tensor. This is the reason why the first workers in this field felt that the old Boussinesq relation (2.2) was too simple to be applicable to astrophysics. The statistics are quite different from those of an ordinary gas. The mean free path and mean free time between collisions are much too short for molecular motions to be influenced noticeably by rotation or gravity, but the relatively long lifetime of the turbulent eddies invites the introduction of an anisotropic "Austausch" (exchange). Just such a course was followed by Wasiutyński (1946). His reasoning is, however, hard to follow. Since his results have often been used, we shall record them here:

$$\begin{aligned}
Q_{rr} &= -2\nu_1\, \partial\bar{u}_r/\partial r\,, \qquad Q_{\theta\theta} = -2\nu_2 r^{-1}(\bar{u}_r + \partial\bar{u}_\theta/\partial\theta)\,,\\
Q_{\phi\phi} &= -2\nu_3 r^{-1}(\bar{u}_r + \cot\theta\cdot\bar{u}_\theta)\,,\\
Q_{r\theta} &= -r^{-1}[\nu_1\, \partial(r\bar{u}_\theta)/\partial r + \nu_2(\partial\bar{u}_r/\partial\theta - 2\bar{u}_\theta)]\,,\\
Q_{r\phi} &= -r^{-1}[\nu_1\, \partial(r\bar{u}_\phi)/\partial r - 2\nu_3\bar{u}_\phi]\,,\\
Q_{\theta\phi} &= -(r\sin\theta)^{-1}\,[\nu_2\, \partial(\sin\theta\cdot\bar{u}_\phi)/\partial\theta - 2\nu_3\cos\theta\cdot\bar{u}_\phi]\,.
\end{aligned} \tag{1}$$

The three free parameters ν_1, ν_2, ν_3 reflect the "Austausch". On making some assumptions, Elsässer (1966) was able to reconstruct expressions (1) convincingly. Meanwhile, Kippenhahn (1963) had simplified them further by removing the distinction between ν_2 and ν_3: he allowed only the radial direction to play an exceptional role. Although in this way he had ignored the preferred status of the rotation axis for the turbulence, this does *not* imply a complete neglect to rotational effects. The mean velocity gradient still contains the angular velocity, and rotation therefore still appears in the correlation tensor. But as Gilman (1974) pointed out, this approach "will be true only if the motions parametrized have very fast turnover times compared with the rotation time scale Ω^{-1}". Let us call such an approximation "slow rotation".

Kippenhahn assumed the turbulent viscosities to be constant and wrote

$$\nu_1 = \nu_T, \quad \nu_2 = \nu_3 = s\nu_T.$$

In this way he obtained from (1) the expressions

$$\begin{aligned}
Q_{rr} &= -2\nu_T\, \partial\bar{u}_r/\partial r, \quad Q_{\theta\theta} = -2s\nu_T r^{-1}(\bar{u}_r + \partial\bar{u}_\theta/\partial\theta),\\
Q_{\phi\phi} &= -2s\nu_T r^{-1}(\bar{u}_r + \cot\theta\, \bar{u}_\theta),\\
Q_{r\theta} &= \nu_T r^{-1}[-r\, \partial\bar{u}_\theta/\partial r - s\, \partial\bar{u}_r/\partial\theta + (2s-1)\,\bar{u}_\theta], \qquad (2)\\
Q_{r\phi} &= -\nu_T r(\partial\Omega/\partial r)\sin\theta + 2(s-1)\,\nu_T\Omega\cdot\sin\theta,\\
Q_{\theta\phi} &= -\nu_T s(\partial\Omega/\partial\theta)\sin\theta.
\end{aligned}$$

The last two equations are of particular interest. They demonstrate a way in which the concept of an anisotropic viscosity can produce formally the fundamental mode of the Λ-effect. When $s \neq 1$, the correlation $Q_{r\phi}$ does not vanish even for a rigid-body rotation, and non-uniform rotation is therefore a consequence of the Reynolds equation. Note that no Λ occurs in $Q_{\theta\phi}$, a consequence of assuming that ν_2 and ν_3 are equal. Wasiutyński's original expression was

$$Q_{\theta\phi} = -\nu_2 \sin\theta \cdot \partial\Omega/\partial\theta + 2(\nu_3 - \nu_2)\,\Omega\cos\theta, \qquad (3)$$

which implies a horizontal Λ-effect for $\nu_2 \neq \nu_3$. It is perhaps tempting to appeal to such a term in the theory of differential rotation, too. Yet this would be feasible only if the difference, $\nu_3 - \nu_2$, were to vanish at the poles, i.e. $\nu_3 - \nu_2 \sim \sin^2\theta$. Otherwise (i) the correlation $Q_{\theta\phi}$ would not vanish at the poles [cf. (3.16)] and (ii) the Reynolds equation (6.1) would be regular along the axis only if H vanishes there. Both consequences are unacceptable, so excluding the possibility of constant Λ_H.

In what follows we present an independent description of anisotropic viscosity which, surprisingly, does not completely coincide with those given above. In a completely general turbulence theory, the single scalar "viscosity" is replaced by a fourth rank tensor which links the correlation tensor to the rate of mean strain tensor:

$$Q_{ij} = \cdots - \mathcal{N}_{ijkl}\bar{u}_{k,l}. \qquad (4)$$

This tensor $\mathcal{N}_{ijkl}$ is symmetric in i and j by definition of Q_{ij}, and in general it depends on rotation, $\boldsymbol{\Omega}$, and gravity, $\boldsymbol{g}$:

$$\mathcal{N} = \mathcal{N}(\boldsymbol{g}, \boldsymbol{\Omega}). \qquad (5)$$

If the dependence on $\boldsymbol{\Omega}$ is neglected, only the *radial* unit vector $\hat{\boldsymbol{g}}$ can be used in the construction of the viscosity tensor. In this case we expect as a result of our calculations the same expressions as Kippenhahn. The following form for $\mathcal{N}_{ijkl}$ can be readily verified:

$$\mathcal{N}_{ijkl} = \nu_1(\delta_{ik}\delta_{jl} + \delta_{jk}\delta_{il}) + \nu_2(\mathring{g}_i\delta_{jl} + \mathring{g}_j\delta_{il})\,\mathring{g}_k + \nu_3(\mathring{g}_i\delta_{jk} + \mathring{g}_j\delta_{ik})\,\mathring{g}_l + \nu_4\delta_{ij}\mathring{g}_k\mathring{g}_l + \nu_5\mathring{g}_i\mathring{g}_j\mathring{g}_k\mathring{g}_l + \delta_{kl}\cdots. \tag{6}$$

All the terms involving the Kronecker delta δ_{lk} can be ignored since, for an incompressible fluid, $\delta_{lk}u_{l,k} \equiv 0$. Let us now introduce the two vector quantities

$$A_l = \mathring{g}_k\overline{u}_{l,k}, \qquad B_l = \mathring{g}_k\overline{u}_{k,l}, \tag{7}$$

in terms of which (6) gives

$$\mathcal{N}_{ijkl}\overline{u}_{k,l} = \nu_1(\overline{u}_{i,j} + \overline{u}_{j,i}) + \nu_2(\mathring{g}_iB_j + \mathring{g}_jB_i) + (\mathring{\boldsymbol{g}}\cdot\boldsymbol{B})\,(\nu_4\delta_{ij} + \nu_5\mathring{g}_i\mathring{g}_j) + \nu_3(\mathring{g}_iA_j + \mathring{g}_jA_i). \tag{8}$$

Hence, the anisotropic part of the correlation tensor is

$$Q_{ij} = \cdots - \nu_2(\mathring{g}_iB_j + \mathring{g}_jB_i) - (\mathring{\boldsymbol{g}}\cdot\boldsymbol{B})\,(\nu_4\delta_{ij} + \nu_5\mathring{g}_i\mathring{g}_j) - \nu_3(\mathring{g}_iA_j + \mathring{g}_jA_i), \tag{9}$$

and its components are

$$\begin{aligned} Q_{rr} &= -(2\nu_2 + \nu_4 + \nu_5)\,B_r - 2\nu_3A_r, \\ Q_{\theta\theta} &= Q_{\phi\phi} = -\nu_4B_r, \quad Q_{r\theta} = -\nu_2B_\theta - \nu_3A_\theta, \\ Q_{r\phi} &= -\nu_2B_\phi - \nu_3A_\phi, \quad Q_{\theta\phi} = 0. \end{aligned} \tag{10}$$

It should be noticed that there are five free parameters in (8): $\nu_1 \ldots \nu_5$. Therefore our result must differ from that of Kippenhahn who isolated only two free parameters, ν_T and s. Can the expressions (2) be reproduced by a special choice of our ν's? As we shall see the answer is "No". Since $\nabla \times \boldsymbol{g} = 0$,

$$\boldsymbol{B} - \boldsymbol{A} = \mathring{\boldsymbol{g}} \times (\nabla \times \overline{\boldsymbol{u}}) = r^{-1}[0, \partial\overline{u}_r/\partial\theta - \partial(r\overline{u}_\theta)/\partial r, -\partial(r\overline{u}_\phi)/\partial r]. \tag{11}$$

Thus, the radial components of $\boldsymbol{A}$ and $\boldsymbol{B}$ are identical. Further it can easily be demonstrated that

$$\boldsymbol{A} = \partial\overline{\boldsymbol{u}}/\partial r, \tag{12}$$

and hence

$$\boldsymbol{B} = r^{-1}(r\,\partial\overline{u}_r/\partial r, \quad \partial\overline{u}_r/\partial\theta - \overline{u}_\theta, -\overline{u}_\phi). \tag{13}$$

The final expressions for the correlations are therefore

$$\left.\begin{aligned} Q_{rr} &= -(2\nu_1 + 2\nu_2 + 2\nu_3 + \nu_4 + \nu_5)\,\partial\overline{u}_r/\partial r, \\ Q_{\theta\theta} &= -2\nu_1r^{-1}(\overline{u}_r + \partial\overline{u}_\theta/\partial\theta) - \nu_4\,\partial\overline{u}_r/\partial r, \\ Q_{\phi\phi} &= -2\nu_1r^{-1}(\overline{u}_r + \cot\theta\cdot\overline{u}_\theta) - \nu_4\,\partial\overline{u}_r/\partial r, \end{aligned}\right\} \tag{14}$$

and

$$Q_{r\theta} = -\nu_1 r^{-1}(\partial \bar{u}_r/\partial\theta + r\,\partial\bar{u}_\theta/\partial r - \bar{u}_\theta) - \nu_2 r^{-1}(\partial\bar{u}_r/\partial\theta - \bar{u}_\theta)$$
$$- \nu_3\,\partial\bar{u}_\theta/\partial r, \tag{15}$$
$$Q_{r\phi} = -[(\nu_1 + \nu_3)\,r\,\partial\Omega/\partial r - (\nu_2 - \nu_3)\,\Omega]\sin\theta, \tag{16}$$
$$Q_{\theta\phi} = -\nu_1 \sin\theta\,\partial\Omega/\partial\theta. \tag{17}$$

With the exception of $Q_{r\theta}$, we can make our correlations coincide with those of Kippenhahn if we set

$$\nu_1 = s\nu_T, \quad \nu_2 = -\nu_3 = (s-1)\,\nu_T, \quad \nu_4 = 0, \quad \nu_5 = 2(1-s)\,\nu_T. \tag{18}$$

When this choice of parameters is substituted in (1), however, we find that the correlation $Q_{r\theta}$ is

$$Q_{r\theta} = \nu_T r^{-1}[-r\,\partial\bar{u}_\theta/\partial r + (1-2s)\,\partial\bar{u}_r/\partial\theta + (2s-1)\,\bar{u}_\theta], \tag{19}$$

which differs significantly from Kippenhahn's expression $(2)_3$. The difference is of order $(s-1)\,\partial\bar{u}_r/\partial\theta$.

It is rather surprising that we are *not* able to reproduce with our analysis the formulae which are so often used in hydrodynamical calculations. The reasons for this discrepancy are still unclear. As the form (6) is the most general one possible of its type, we are forced to doubt the formulation (1) of Wasiutyński.

7.2. Negative Viscosity?

Chapter 6 dealt with mean-field aspects of the phenomenon of differential rotation. Assuming that the Λ-effect is the most prominent generator of non-uniform rotation, we showed how the observed rotation law could be reproduced by suitably choosing a latitudinal profile, $\Lambda(\theta)$, but to justify that choice we required the observations to yield a further profile. We established that the horizontal cross-correlation, $\overline{u_\theta' u_\phi'}$, of the surface velocities could provide what we needed. We found that this profile [normalized by the introduction of the function $w(\theta)$] is generated by precisely the same double-set of non-diffusive TMT modes as is the rotation law. We could infer the unobservable parameters $V^{(1)}$, $H^{(1)}$, $V^{(2)}$, $H^{(2)}$ which occur in the expansions (2.21) from direct observations of the cross-correlation. At that stage we regretted that, up to the present time, these observations determined the function $w(\theta)$ so poorly. In accord with the work of Ward, we used only a rough representation of his results, taking $w^{(2)} \approx 0$. No difficulties with the rotation law arose as a consequence. Both profiles could be produced simultaneously by the Λ-effect and, moreover, with an eddy viscosity that was both isotropic and positive.

In his analysis of the motion of sunspot groups with respect to the horizontal cross-correlations, Tuominen (1982) obtained relatively small numerical values for the cross-correlation. The position of recurrent groups near the central meridian were measured after one rotation in order to derive the fluctuating velocities. Only a small coefficient, $w \approx 0.04$, was found, giving a TMT about two orders of magnitude smaller than Ward's. At the end of Chapter 6 we discussed the effects on the required Λ-modes of assuming w to be everywhere non-negative, and we found that the function $H^{(1)}$ must attain the value "1", not merely deep in the SCZ but also on its upper surface. This was a consequence of our general belief that it should be possible to reconcile non-negative surface cross-correlations, $\cos\theta\overline{u_\theta' u_\phi'}$, with a *positive* eddy viscosity.

We should not completely ignore one possibility of explaining the observations: "negative viscosity". Starr (1968) suggested this was a solar phenomenon when he compared Ward's observations with the old Boussinesq ansatz (2.2). Observations of the Sun indicated that it rotates with $\cos\theta\, \partial\Omega/\partial\theta \geqq 0$, and that could be immediately explained by invoking a negative ν_T. Even were it possible to deduce a negative ν_T from theoretical reasoning, and even were the Boussinesq relation true, the observation of positive $w(\theta)$ could never, as has sometimes been claimed, obviate the need for a theory of the equatorial acceleration. A proper theory must be able to explain simultaneously both observations: equatorial acceleration *and* equatorward TMT.

Let us return to negative eddy viscosity. We will use this term in a restricted sense to describe all cases of negative s. We must emphasize that, as we demonstrated earlier, the observations do *not* force us to accept such values. Even the equatorward TMT found by Ward is not inconsistent with a positive s. It is very striking however that, according to (6.62), the values

$$s = -1, \quad V^{(1)} \simeq \mathsf{d}^{-1}, \quad H^{(1)} \approx 0, \tag{20}$$

reproduce the observations quite clearly. More information is contained in (6.62). With the observed values its first equation can be written as

$$sV^{(1)} = -\lambda(1 + 0.62\, s) - 10 < 0, \tag{21}$$

so that s and $V^{(1)}$ always have opposite signs if s is not too large. Hence, an appeal to negative eddy viscosity will always lead to positive $V^{(1)}$. Furthermore, setting $s = -1$ in (6.63), we find also the other modes to be positive.

Table VII.i. The required Λ's for a model with $w^{(1)} = 2\omega^{(1)}$ and $\mathsf{d} = 0.2$.

	$\lambda = 0$	$\lambda = 10$
$V_i^{(1)}/\omega^{(1)}$	$-10/s$	$-30 - 40/s$
$H_i^{(1)}/\omega^{(1)}$	$2(s + 1)$	$6(s + 1)$

For a model with (say) $w^{(1)} = 2\omega^{(1)}$ and $\mathsf{d} = 0.2$, Table VII.i. gives values of the Λ-fluxes at the base of the SCZ which are required in order to reproduce the observations. Again we find the case $s = -1$ mildly exotic. No $H^{(1)}$ is necessary in order to reproduce the observations. Although a negative eddy viscosity may seem to be rather bizarre, one special turbulence model does exist that reproduces precisely the required numbers (20). This is strictly a two-dimensional turbulence with motion lying wholly in the horizontal plane. This subject will be dealt with in greater detail in Chapter 16.

Table VII.ii. The same as Table VI.ii. but for negative viscosity ($s = -1$)

$\varrho_i/\varrho_\odot$	$\Lambda_i/\Lambda_\odot$	Ψ	$V_i^{(1)}$	$H_i^{(1)}$
10^6	1	0.210	6	1.2
10^6	10	0.074	80	12.4
10^6	100	0.036	1300	120

When the geometry function Ψ defined in (6.41) is calculated using $s = -1$, the results shown in Table VII.ii are obtained. Here, the relative thickness, d, of the zone was taken to be 0.2 and again we used $w^{(1)} = 4\omega^{(1)} \approx 2.5$. The required $V_i^{(1)}$ and $H_i^{(1)}$ are only slightly smaller than those for $s = 1$, the largest discrepancy being the opposite sign of $V^{(1)}$. We always find that the product $sV_i^{(1)}$ is negative. The quantitative demands on Λ are similar for $s = 1$ and $s = -1$. It is unlikely, therefore, that we can make any progress by postulating a negative eddy viscosity. That the geometry function Ψ increases (slightly) for negative s is connected with the existence of eigensolutions for negative s.

Surprisingly enough, the Reynolds equation (6.1) possesses eigensolutions if negative s is allowed and $V^{(1)}, H^{(1)}, \ldots$ are ignored. Physically, this would mean that differential rotation can be excited even when the higher, non-fundamental modes of Λ are absent, i.e. for slowly rotating turbulence. This conclusion seems, however, to be dubious for the following reasons. Let us introduce the positive quantity $\bar{s} = -s$. Then the Reynolds equation, which is linear in Ω where

$$\Omega = e^{\sigma t}\Omega_{n-1}(x)P_n^1/\sin\theta, \qquad (22)$$

is

$$x^2\Omega_n'' + (4 - V^{(0)})\,x\Omega_n' + [\bar{s}n(n+3) - 3V^{(0)} - \sigma x^2]\,\Omega_n = 0, \qquad (23)$$

for $n = 2, 4, 6\ldots$ Positive σ denotes growing modes and negative σ decaying modes. Equation (23) ignores, of course, meridional flow and density stratification. The fundamental mode $V^{(0)}$ of the Λ-effect is involved, yet it can be set zero without losing the eigenfunctions. The marginal state ($\sigma = 0$) can easily be solved to give

$$\Omega_n^{(0)} = x^{r_1}[c_1 \sin(\ln x^{r_2}) + c_2 \cos(\ln x^{r_2})], \tag{24}$$

where the exponents are

$$r_1 = \frac{1}{2}(V^{(0)} - 3), \quad r_2 = \frac{1}{2}[4\bar{s}n(n+3) - (V^{(0)} + 3)^2]^{1/2}. \tag{25}$$

The outer stress-free condition, $\partial\Omega/\partial x = V^{(0)}\Omega$, requires

$$r_2 c_1 + (r_1 - V^{(0)})\, c_2 = 0, \tag{26}$$

and, if we assume for simplicity a rigidly rotating base to the convection zone, i.e. $\Omega_n = 0$ for $n > 0$, then

$$c_1 \sin(\ln x_i^{r_2}) + c_2 \cos(\ln x_i^{r_2}) = 0. \tag{27}$$

Hence, the eigencondition is

$$\tan(\ln x_i^{r_2}) = r_2/(r_1 - V^{(0)}). \tag{28}$$

The smallest solution of (28) is approximately

$$r_2 = \pi/2 \ln x_i \quad \text{or} \quad \bar{s}^{(0)} = [(\pi/\ln x_i)^2 + (3 + V^{(0)})^2]/4n(n+3). \tag{29}$$

For $x_i = 0.7$, $n = 2$ and $V^{(0)} = 0$, (29) gives $s = -2.2$. For such a "negative eddy viscosity" an eigensolution, i.e. a differential rotation, exists in the absence of any generator to maintain it! It obeys the correct boundary conditions, in contrast to the stress-free solutions of $\Delta^2 \boldsymbol{u} = 0$ which Emden (1936) criticised in detail. He was quite correct in suggesting that "they do not say anything about the solar rotation law but can only serve as a description of the surface distribution of Ω". Our eigensolution is dubious, however, for another reason.

We observe that the eigenvalues decrease with increasing n. Even the smallest, but still negative, s are able to generate marginal solutions for the differential surface rotation. Yet, the larger the n is, the more easily the modes are excited. If they grow in the vicinity of an eigenvalue with $d\sigma/ds \neq 0$ we have an "illposed problem": the higher the mode number, the more rapid the growth. This is obviously nonsense. It is physically an impossibility for s to be both negative and *constant*. The negativity of s must disappear for all sufficiently large n, i.e. the turbulence parameter s must be n-dependent. Indeed, the Reynolds equation is valid only for large-scale flows; it is meaningless for scales of the order of the turbulence. Thus, the theory is not as absurd as it at first seems. Indeed, it has been shown (Krause and Rüdiger, 1974b) that the negative eddy viscosity of a two-dimensional turbulence disappears as the scales of the mean flow and turbulence become comparable. Hence, at a certain (large) n, the parameter s must become positive, whereupon the eigensolutions no longer exist.

It may be of some interest to compute the s-dependence of the growth rate near a marginal state, i.e. the derivative $\sigma' = \mathrm{d}\sigma/\mathrm{d}s$ at $s = -\bar{s}^{(0)}$. This can be accomplished by a perturbation method. We write

$$s = s^{(0)} + \delta s, \quad \sigma = \sigma' \delta s, \quad \Omega = \Omega^{(0)} + \Omega^{(1)} \delta s, \tag{30}$$

where the superscript $^{(0)}$ denotes an eigensolution, i.e.

$$s^{(0)} = -\bar{s}^{(0)}, \qquad \Omega^{(0)} = (V^{(0)} - r_1)\, y_1 + r_2 y_2, \tag{31}$$

with

$$y_1 = x^{r_1} \sin (\ln x^{r_2}), \qquad y_2 = x^{r_1} \cos (\ln x^{r_2}). \tag{32}$$

The differential equation is clearly

$$\Omega_n^{(1)''} + x^{-1}(4 - V^{(0)})\, \Omega_n^{(1)'} + x^{-2}[\bar{s}^{(0)} n(n+3) - 3V^{(0)}]\, \Omega_n^{(1)} = F, \tag{33}$$

with

$$F = [\sigma' + n(n+3)\, x^{-2}]\, \Omega_n^{(0)},$$

which can be solved to yield the expression

$$\Omega_n^{(1)} = c_1 y_1 + c_2 y_2 + y, \tag{34}$$

with the particular solution

$$y = r_2^{-1} \int_1^x F(\xi)\, \xi^{4-V^{(0)}} [y_1(x)\, y_2(\xi) - y_1(\xi)\, y_2(x)]\, \mathrm{d}\xi, \tag{35}$$

which obeys $y(1) = y'(1) = 0$. Since the homogeneous version of (33) possesses a nontrivial solution [namely $\Omega_n^{(0)}$], the inhomogeneous equation has no solution unless it satisfies a certain "solubility condition". Using the lower boundary condition $y(x_i) = 0$, we may easily prove that this constraint is simply

$$\int_{x_i}^1 [\sigma' + \xi^{-2} n(n+3)]\, [(V^{(0)} - r_1)\, y_1(\xi) + r_2 y_2(\xi)]\, \xi^{4-V^{(0)}}$$
$$\times\ [y_1(x_i)\, y_2(\xi) - y_2(x_i)\, y_1(\xi)]\, \mathrm{d}\xi = 0. \tag{36}$$

For the purposes of numerical work, condition (36) may be written as $\alpha\sigma' + n(n+3)\,\beta = 0$, where α and β are independent of n. It is immediately clear from this relation that $|\sigma'|$ increases with n. Hence, the larger the n, the more rapidly the corresponding mode grows. In particular, we find that for $x_i = 0.7$ the relation (36) generates the values given in Table VII.iii.

Table VII.iii. Numerical solutions of the dispersion relation (36).

	$V^{(0)} = -1$			$V^{(0)} = 1$		
	$n = 2$		$n = 4$	$n = 2$		$n = 4$
$s^{(0)}$	−2.0		−0.7	−2.7		−0.8
α		−1.15			−0.69	
β		−1.37			−0.80	
β/α		1.19			1.16	
$d\sigma/ds$	−11.9		−33.3	−11.6		−32.5

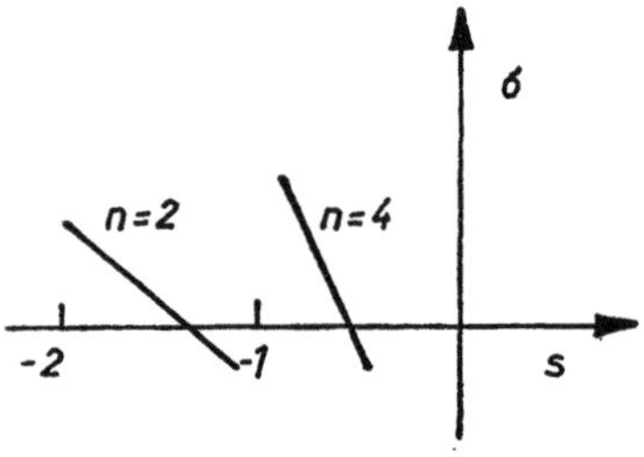

Fig. 7.1. The general dependence of the growth rate σ of the eigensolutions on the anisotropy parameter s. The smaller the latitudinal scale, the faster the growth.

Obviously, the differences between the various $V^{(0)}$ are small, and we may set $V^{(0)} = 0$ without serious error. With the values shown, we can determine the characteristic shape of the σs-diagram and this is illustrated in Fig. 7.1

7.3. Diffusive Contributions to the Turbulent Momentum Transport

The basic rotation influences the coefficients that multiply the spatial derivatives of mean flow in the expressions for the turbulent momentum transport (TMT), i.e. the viscosities. Again, this influence vanishes at the poles and does not differ from hemisphere to hemisphere. The viscosities may be thought of as functions of $\sin^2 \theta$ and may be written as

$$\begin{Bmatrix} \nu_{vv} \\ \nu_{hh} \end{Bmatrix} = \nu_T \sum_{l=0} \begin{Bmatrix} \nu_{vv}^{(l)} \\ \nu_{hh}^{(l)} \end{Bmatrix} \sin^{2l} \theta . \tag{37}$$

The normalization has been chosen to make $\nu_{vv}^{(0)} = 1$ and $\nu_{hh}^{(0)} = s$. This, however, is only half the story. Equation (6.6) lacks some spatial derivatives, whose coefficients vanish as $\Omega \to 0$. The complete expressions for the eddy momentum fluxes ought to be

$$\begin{aligned} Q_{r\phi} &= -\nu_{vv} r \sin\theta \cdot \partial\Omega/\partial r - \nu_{vh} \cos\theta \cdot \partial\Omega/\partial\theta + \Lambda_V \sin\theta \cdot \Omega , \\ Q_{\theta\phi} &= -\nu_{hv} r \cos\theta \cdot \partial\Omega/\partial r - \nu_{hh} \sin\theta \cdot \partial\Omega/\partial\theta + \Lambda_H \cos\theta \cdot \Omega , \end{aligned} \tag{38}$$

(cf. Kichatinov, 1987) with, in analogy to (37),

$$\begin{Bmatrix} \nu_{vh} \\ \nu_{hv} \end{Bmatrix} = \nu_T \sum_{l=1} \begin{Bmatrix} \nu_{vh}^{(l)} \\ \nu_{hv}^{(l)} \end{Bmatrix} \sin^{2l} \theta . \tag{39}$$

Observe that the $\sin \theta$ and $\cos \theta$ in (38) preserve the desired symmetry with respect to the equatorial plane. It should be stressed that all the correlations involving an odd number of u_ϕ'-factors must change sign under the transformation $\Omega \to -\Omega$, while those with an even number of u_ϕ'-factors must not. This is the reason why the eddy momentum fluxes are generally odd in Ω in contrast to, for example, the meridional heat flux components.

Let us briefly describe the role of the newly established viscosity terms. For this purpose, we may ignore both $\nu_{hh}^{(l)}$ and $\nu_{vh}^{(l)}$ for non-vanishing l: these coefficients are small because of their rotational origin, and they appear only in products with $\partial\Omega/\partial\theta$, which is also small for the same reason. The correlations under consideration therefore take the form

$$\begin{aligned} Q_{r\phi} &= \nu_T \sin \theta \left[-r\, \partial\Omega/\partial r + V^{(0)}\Omega + \sum_l \left(\Omega V^{(l)} - \nu_{vv}^{(l)} r\, \partial\Omega/\partial r \right) \sin^{2l} \theta \right], \\ Q_{\theta\phi} &= \nu_T \left[-s \sin \theta \cdot \partial\Omega/\partial\theta + \sum_l \left(\Omega H^{(l)} - \nu_{hv}^{(l)} r\, \partial\Omega/\partial r \right) \sin^{2l} \theta \cos \theta \right], \end{aligned} \tag{40}$$

and they differ from those used in previous sections only by giving a more complex structure to the non-fundamental modes: the $V^{(l)}$ and $H^{(l)}$ used earlier must be replaced by

$$\hat{V}^{(l)} = V^{(l)} - \nu_{vv}^{(l)}\, \partial \ln \Omega/\partial \ln r , \quad \hat{H}^{(l)} = H^{(l)} - \nu_{hv}^{(l)}\, \partial \ln \Omega/\partial \ln r . \tag{41}$$

Also, the rotationally-created viscosity terms influence the latitudinal profile of the angular velocity, provided Ω possesses a radial gradient. This can easily be produced by the fundamental mode ($l = 0$) of the vertical TMT which is linear in Ω. If this is the most important effect, the equation

$$x\, \partial\Omega/\partial x = V^{(0)}\Omega \tag{42}$$

provides an excellent approximation to the radial profile of Ω. Thus we arrive at

$$\hat{V}^{(l)} = V^{(l)} - \nu_{vv}^{(l)} V^{(0)}, \quad \hat{H}^{(l)} = H^{(l)} - \nu_{hv}^{(l)} V^{(0)} . \tag{43}$$

When these effects are included, the rotation law (6.56) at the surface of a shallow convection zone becomes

$$\Omega/\Omega_\odot = C + \frac{1}{2} \sum_{l=1} \left[\mathrm{d}\left(V_i^{(l)} - \nu_{vv}^{(l)} V_i^{(0)} \right) + l^{-1} \left(H_i^{(l)} - \nu_{hv}^{(l)} V_i^{(0)} \right) \right] \sin^{2l} \theta . \tag{44}$$

Unfortunately, little is known about rotationally-created viscosity coefficients. We suspect, however, that the corresponding terms in (44) are of minor interest in the theory of differential rotation as they are of second order in the Strouhal number defined by the turbulence.

Chapter 8. Energy Transport in Rotating Turbulence

8.1. Screening Effects

As we have seen, one effect of rotation on large-scale eddies is to create a non-uniformity in that rotation. Analysing this mechanism in greater detail, we discerned the underlying cause: through the rotation, physical quantities such as the Λ-fluxes become strongly dependent on the angle θ between the direction of the rotation axis and the gravitational field. The effect is, of course, appreciable only when the influence of rotation is not too weak, i.e. when the Coriolis frequency 2Ω is not too small compared with the eddy decay rate $2\pi/\tau_c$. If the basicr otation can cause a θ-dependent momentum flux, it is only to be expected that it will also create a θ-dependent turbulent heat flux. All attempts to detect latitude dependence in the mean temperature and mean heat flux on the solar surface have, however, met with failure. We must ask, "Why do we see a uniform temperature and energy flux on the Sun, and yet a non-uniform rotation?"

Spruit (1977) tried to answer this question by analysing the result of applying an arbitrary disturbance in the heat flux at a general point within the convection zone. He discovered that there was a strong screening between the source of disturbance and its manifestation on the solar surface. "This is due to the high efficiency of turbulent diffusion of heat in the deeper layers of the convection zone, and to the behaviour of the solar surface, which acts as an isolating layer with respect to flux disturbances" (Spruit, 1977). There is no doubt that this is a valid conclusion, and the discussion below will confirm it. The discovery of Stix (1981) that a like screening does not exist for temperature disturbances is, however, surprising.

In order to analyze the approaches of both Spruit and Stix, we shall suppose that a disturbance, $\tilde{\boldsymbol{F}}$ in heat flux obeys the extremely simplified equation

$$\nabla \cdot \tilde{\boldsymbol{F}} \equiv r^{-2}\, \partial(r^2 \tilde{F}_r)/\partial r + (r \sin\theta)^{-1}\, \partial(\sin\theta \cdot \tilde{F}_\theta)/\partial\theta = 0\,. \qquad (1)$$

The relation between the flux and the temperature will, however, be the traditional one, viz:

$$\tilde{\boldsymbol{F}} = -\varrho C_p \chi_T \nabla \tilde{T}\,, \qquad (2)$$

where χ_T is the thermal conductivity (supposed constant) due to the small-scale turbulence. According to mixing length theory, its magnitude should be of order 10^{13} cm²/s. As usual C_p is the specific heat at constant pressure. If we use the series expansion

$$\tilde{T} = \sum_{n=0,2,\ldots} \tilde{T}_n P_n, \tag{3}$$

then we find that (1) becomes

$$\frac{1}{\varrho} \frac{\mathrm{d}}{\mathrm{d}x} \left(x^2 \varrho \frac{\mathrm{d}\tilde{T}_n}{\mathrm{d}x} \right) - n(n+1)\, \tilde{T}_n = 0, \tag{4}$$

and finally, assuming the density profile (5.12) as usual, we obtain

$$x^2 \tilde{T}_n'' + (2 - G)\, x \tilde{T}_n' - n(n+1)\, \tilde{T}_n = 0. \tag{5}$$

In what follows we shall prescribe a certain disturbance at depth x_i and then determine its visible manifestation at the surface. The question to be answered is whether or not the prescribed disturbance is discernable to observation.

The general solution of (5) may be written as

$$\tilde{T}_n = C_1 x^{\lambda_1} + C_2 x^{\lambda_2}, \tag{6}$$

where λ_1 and λ_2 are the roots of the characteristic equation

$$\lambda^2 + (1 - G)\, \lambda = n(n+1). \tag{7}$$

The arbitrary constant C_1 and C_2 follow from the boundary conditions. There is a considerable heat flux through the top and bottom of the convective region. The energy leaves the surface in the form of electromagnetic radiation, hence

$$\tilde{F}_r = 4\sigma T_\odot{}^3 \tilde{T} \quad \text{at} \quad x = 1, \tag{8}$$

where $T_\odot$ is the surface temperature. Thus, at $x = 1$, we have

$$4f^* \tilde{T}_n + \mathrm{d}\tilde{T}_n/\mathrm{d}x = 0, \tag{9}$$

with

$$f^* = \sigma R T_\odot{}^3 / \varrho_\odot \chi_T C_p. \tag{10}$$

Substituting the solar values (in c.g.s. units)

$$\begin{aligned} &R = 7 \cdot 10^{10}, \quad T_\odot = 6.2 \cdot 10^3, \quad \varrho_\odot = 10^{-6}, \\ &C_p = 4 \cdot 10^8, \quad \chi_T = 10^{13}, \end{aligned} \tag{11}$$

into (10), we find

$$f^* = 100 - 200. \tag{12}$$

The papers of Spruit and Stix give even larger values for this quantity, namely $f^* = 2000$.

As a first relation between the coefficients C_1 and C_2, we obtain from (9)

$$C_1(\lambda_1 + 4f^*) + C_2(\lambda_2 + 4f^*) = 0, \tag{13}$$

which must be complemented by another equation arising from the lower boundary condition, in the choice of which we have some flexibility. First we shall follow Stix and postulate that a certain *temperature* disturbance exists at the depth x_i:

$$T_n(x_i) = \delta T. \tag{14}$$

This leads to a second relation between C_1 and C_2,

$$C_1 x_i^{\lambda_1} + C_2 x_i^{\lambda_2} = \delta T, \tag{15}$$

and we obtain

$$C_{1,2} = \mp(\lambda_{2,1} + 4f^*)\, \delta T/[(\lambda_1 + 4f^*)\, x_i^{\lambda_2} - (\lambda_2 + 4f^*)\, x_i^{\lambda_1}]. \tag{16}$$

The visible temperature disturbance at $x = 1$ is the sum of C_1 and C_2, i.e.

$$\tilde{T}_n(1)/\delta T = (\lambda_1 - \lambda_2)/[(\lambda_1 + 4f^*)\, x_i^{\lambda_2} - (\lambda_2 + 4f^*)\, x_i^{\lambda_1}] = R^T, \quad \text{say}. \tag{17}$$

For x_i close to unity, the temperature response function, R^T, proves to be

$$R^T \simeq 1/(1 + 4f^* \mathsf{d}), \tag{18}$$

which is small for large f^* and approximately 1 for small f^*. In other words we have strong screening for large f^* and no screening at all for small f^*. No extraordinary dependence on n is apparent. The numerical values of R^T confirm these statements (see Table VIII.i, where $f^* = 100$). We note the original disturbance is screened if R^T is significantly smaller than unity. Surprisingly, the screening is reduced if G is large. Large G is, however, the most realistic case for the Sun. Note in particular that the response is only 0.01 for $G = 20$, $x_i = 0.7$ and $n = 10$.

Next let us apply a different lower boundary condition. A *flux* disturbance δF is assumed to exist at a depth x_i. This is directly connected to the tempera-

Table VIII.i. The temperature response function, 100 R^T, for $f^* = 100$ and for different values of n, G and x_i.

x_i	$n = 2$		$n = 10$	
	$G = 0$	$G = 20$	$G = 0$	$G = 20$
0.7	0.5	4	0.1	1
0.9	2	5	0.8	4

ture gradient so that

$$\partial \tilde{T}_n/\partial x = -R x_i^G \delta F/\varrho_\odot \chi_T C_p \tag{19}$$

is prescribed at $x = x_i$. The relation corresponding to (15) reads

$$\lambda_1 C_1 x_i^{\lambda_1 - 1} + \lambda_2 C_2 x_i^{\lambda_2 - 1} = -R x_i^G \delta F/\varrho_\odot \chi_T C_p, \tag{20}$$

and the coefficients C_1 and C_2 become

$$C_{1,2} = \pm \frac{(\lambda_{2,1} + 4f^*) R x_i^{G+1} \delta F/\varrho_\odot \chi_T C_p}{\lambda_2(\lambda_1 + 4f^*) x_i^{\lambda_2} - \lambda_1(\lambda_2 + 4f^*) x_i^{\lambda_1}}. \tag{21}$$

At the solar surface, therefore, we have a visible temperature disturbance of

$$\tilde{T}_n(1) = \frac{(\lambda_2 - \lambda_1) R x_i^{G+1} \delta F/\varrho_\odot \chi_T C_p}{\lambda_2(\lambda_1 + 4f^*) x_i^{\lambda_2} - \lambda_1(\lambda_2 + 4f^*) x_i^{\lambda_1}}, \tag{22}$$

which, due to (8), is identical to the flux disturbance

$$\frac{\tilde{F}_r(1)}{\delta F} = \frac{4(\lambda_2 - \lambda_1) f^* x_i^{G+1}}{\lambda_2(\lambda_1 + 4f^*) x_i^{\lambda_2} - \lambda_1(\lambda_2 + 4f^*) x_i^{\lambda_1}} = R^F, \text{ say.} \tag{23}$$

If we expand the response function, R^F, for small d, we obtain

$$R^F = 4f^*[1 - (G + 1)\,\mathsf{d}]/\{4f^* + \mathsf{d}[n(n + 1) - 4f^*(G - 1)]\}, \tag{24}$$

so that for large f^*

$$R^F = [1 - (G + 1)\,\mathsf{d}]/[1 - (G - 1)\,\mathsf{d}], \tag{25}$$

which is near unity. Thus, we can no longer speak of screening. For deep disturbances, large G and large n, expansion (25) is, however, not valid. Then the response has to be calculated numerically. Table VIII.ii. gives R^F for two characteristic values of f^*.

Table VIII.ii. Flux response R^F for $f^* = 0.01$ and for $f^* = 100$, and for different values of n, G and x_i.

f^*	x_i	$n = 2$		$n = 10$	
		$G = 0$	$G = 20$	$G = 0$	$G = 20$
100	0.7	$3.6 \cdot 10^{-1}$	$2.9 \cdot 10^{-2}$	$2.6 \cdot 10^{-2}$	$6.1 \cdot 10^{-4}$
	0.9	$7.8 \cdot 10^{-1}$	$7.5 \cdot 10^{-1}$	$4.8 \cdot 10^{-1}$	$3.1 \cdot 10^{-1}$
0.01	0.7	$9.3 \cdot 10^{-3}$	$6.5 \cdot 10^{-5}$	$1.1 \cdot 10^{-4}$	$1.1 \cdot 10^{-6}$
	0.9	$5.0 \cdot 10^{-2}$	$1.6 \cdot 10^{-2}$	$2.4 \cdot 10^{-3}$	$7.3 \cdot 10^{-4}$

For small f^* there is always screening, but what about the more realistic case of large f^*? For $x_i = 0.7$ we do have screening, and this increases as it should with increasing G and n. The most relevant values of R^F (those for

$G = 20$, $x_i = 0.7$) are small, even that for $n = 10$ being less than 0.001. To sum up, we see that only for deep-seated, small-scale disturbances in a strongly stratified SCZ, Spruit's model suggests a far more effective screening than does that of Stix. The same is true for small f^*. Then (18) gives $R^T \approx 1$ (i.e. almost no screening) while (24) gives small R^F (i.e. strong screening). Table VIII.ii also contains values of R^F for this case.

Finally, in order to analyze the role of the outer boundary condition (8), solutions satisfying other surface conditions will be considered. The disturbance in surface temperature may, for example, be set to zero and then the flux disturbance calculated. This procedure has already been used by Gilman in his numerical surveys. We expect to find a correspondence with the results already obtained for large f^* since the new ansatz is equivalent to taking the limit $f^* \to \infty$ in (9). The upper boundary condition $\tilde{T}_n(1) = 0$ yields $C_1 + C_2 = 0$, which must be used in conjunction with (20). This leads to

$$C_{1,2} = \mp R x_i^{G+1} \delta F / \varrho_\odot \chi_T C_p (\lambda_1 x_i^{\lambda_1} - \lambda_2 x_i^{\lambda_2}) . \tag{26}$$

The desired surface heat flux,

$$\tilde{F}_r = -(\varrho_\odot \chi_T C_p / R)\,(C_1 \lambda_1 + C_2 \lambda_2), \tag{27}$$

then becomes

$$\tilde{F}_r / \delta F = (\lambda_2 - \lambda_1)\, x_i^{G+1} / (\lambda_2 x_i^{\lambda_2} - \lambda_1 x_i^{\lambda_1}) . \tag{28}$$

This response function is consistent with (23) if we let f^* become infinite. Equation (28) thus predicts a large screening for deep, small-scale disturbances. We have to stress that the application of the Stefan-Boltzmann law with a large value of f^* is tantamount to requiring that the temperature disturbance vanishes at the surface. This result is almost trivial. Far less trivial, however, is the result that under such conditions the flux disturbances are small. This conclusion is due solely to the large density gradient. The observed uniformities in temperature and in heat flux at the solar surface require two separate explanations: the Stefan-Boltzmann law keeps the temperature independent of latitude and the density gradient likewise maintains the flux independent of latitude.

At the other extreme, the application of the Stefan-Boltzmann law with small f^* forces the flux disturbances to remain small:

$$\tilde{F}_r(1) = 0 . \tag{29}$$

Let us consider whether an effective screening may also exist in such an circumstances. In this case the temperature response function is given by (17) with $f^* = 0$, so that

$$R^T = (\lambda_1 - \lambda_2) / (\lambda_1 x_i^{\lambda_2} - \lambda_2 x_i^{\lambda_1}), \tag{30}$$

which approaches unity as $x_i \to 1$. Values for the response function are given in Table VIII.iii. They do not display any appreciable screening of temperature disturbances, consequently, the observations cannot be explained with $f^* \ll 1$.

Table VIII.iii. Temperature response function for boundary condition (29)

x_i	$n = 2$		$n = 10$	
	$G = 0$	$G = 20$	$G = 0$	$G = 20$
0.7	0.7	0.9	0.04	0.2
0.9	0.97	0.98	0.6	0.7

It seems clear that boundary conditions do play a significant role. Either the temperature fluctuations at the surface are artificially set to zero or the Stefan-Boltzmann law applies in conditions of large f^*. In order to understand the observational situation we must consider the disturbances in heat flux rather than those in the temperature. Nor can we be sure "whether it is of any value to consider the effects of the turbulent transport of certain disturbances without simultaneously considering the mechanisms which produce such disturbances" (Stix).

The formulation below, which attempts to answer such objections, is based on a mean-field concept and does not rely on the postulated existence of arbitrary disturbances. Our approach will lead to valid results only for lateral profiles of a mean-field temperature. Nevertheless one should not forget the temperature screening of the thin layer above the top of the SCZ though we shall still ignore this effect in the calculations based on our model below. Possibly, the embarrassment of a small but unwanted residual pole-equator temperature difference, which we shall encounter in one case below, could be explained and removed by the effect.

Rotation, as such, strongly modifies the structure of the eddy conductivity tensor; latitudinally varying temperature and/or heat flux profiles are natural consequences. Adopting the Boltzmann law as the realistic outer boundary condition, we can finally determine the pole-equator difference of the surface temperature.

Stix also looked for a possible screening of velocity disturbances. In the simplest approach an angular velocity disturbance, $\delta\Omega$, is prescribed at any depth, x_i, and the surface is assumed to be stress-free. The equation governing the longitudinal velocity is (5.6) with its right-hand side set to zero. So we have

$$\omega_n = C_1 x^{\lambda_1} + C_2 x^{\lambda_2},$$

with

$$\lambda_1 C_1 + \lambda_2 C_2 = 0, \quad x_i^{\lambda_1} C_1 + x_i^{\lambda_2} C_2 = \delta\Omega,$$

which is obvious from our earlier considerations. The response function is given by (30), where now λ_1 and λ_2 are the roots of

$$\lambda^2 + (3 - G)\,\lambda = n(n + 3).$$

When both G and n are large, these hardly differ from the roots of (7) so that the values of Table VIII.iii for the response (30) apply. The conclusion may be drawn that disturbances in the mean angular velocity are *not* effectively screened. This statement also holds for small x_i. The differential rotation of a convective solar core should even detectably influence the rotation profile of the solar surface!

8.2. Energy Transport

The proper determination of the energy transport through a convective/turbulent medium is certainly the key problem of stellar hydrodynamics. Its complexity has been described in reviews by several authors (e.g. Gough, 1976; Zahn, 1979). Pioneering concepts such as nonlocality (Spiegel, 1963; Ulrich, 1976; Maeder, 1975), the anelastic approximation (Latour *et al.*, starting in 1976), or that of Roxburgh (1978) has led to some enlightening results.

The basic points can be taken from the fundamental entropy equation:

$$\varrho T \,\mathrm{d}S/\mathrm{d}t = \varepsilon^V - \nabla \cdot \boldsymbol{F}^D, \tag{31}$$

where the right-hand side represents the heat gained, due to friction

$$\varepsilon^V = \frac{1}{2}\, \varrho \nu (u_{i,k} + u_{k,i})^2 \tag{32}$$

and diffusion

$$\boldsymbol{F}^D = -\varrho C_p \chi_D \nabla T. \tag{33}$$

Here χ_D is the sum of the thermal conductivities of molecular and radiative origins. From the thermodynamic relation

$$\varrho T \,\mathrm{d}S = \varrho \,\mathrm{d}H - \mathrm{d}p \tag{34}$$

(where H is the enthalpy) we readily obtain

$$\varrho \,\mathrm{d}H/\mathrm{d}t - \mathrm{d}p/\mathrm{d}t = \varepsilon^V - \nabla \cdot \boldsymbol{F}^D, \tag{35}$$

and, in addition,

$$\partial(\varrho H - p)/\partial t + \nabla \cdot (\varrho H \boldsymbol{u} + \boldsymbol{F}^D) = \boldsymbol{u} \cdot \nabla p + \varepsilon^V, \tag{36}$$

which, on substituting for the internal energy $U = H - p/\varrho$, may be written

$$\partial(\varrho U)/\partial t + \nabla \cdot (\varrho U \boldsymbol{u} + \boldsymbol{F}^D) = -p\, \nabla \cdot \boldsymbol{u} + \varepsilon^V. \tag{37}$$

The well-known temperature equation for an ideal gas with $U = C_v T$ follows directly from (37) for constant C_v

$$C_v \varrho (\partial T/\partial t + \boldsymbol{u} \cdot \nabla T) = \nabla \cdot (\varrho C_p \chi_D \nabla T) - p \nabla \cdot \boldsymbol{u} + \varepsilon^V . \tag{38}$$

From the Navier-Stokes equation one may deduce that

$$\boldsymbol{u} \cdot \nabla p + \varepsilon^V = \varrho \boldsymbol{u} \cdot \boldsymbol{g}^{\text{eff}} - \frac{1}{2} \frac{\partial}{\partial t} (\varrho \boldsymbol{u}^2) - \frac{1}{2} \nabla \cdot (\varrho \boldsymbol{u}^2 \boldsymbol{u})$$
$$+ \frac{\partial}{\partial x_j} [\varrho \nu u_i (u_{i,j} + u_{j,i})] , \tag{39}$$

which, together with (36), yields

$$\partial \left(\varrho U + \frac{1}{2} \varrho \boldsymbol{u}^2 \right) \Big/ \partial t + \nabla \cdot \left(\varrho \boldsymbol{u} H + \boldsymbol{F}^D + \boldsymbol{F}^V + \frac{1}{2} \varrho \boldsymbol{u}^2 \boldsymbol{u} \right) = \varrho \boldsymbol{u} \cdot \boldsymbol{g}^{\text{eff}} , \tag{40}$$

with the friction term $\boldsymbol{F}^V$ being given by

$$F_j^V = -\varrho \nu u_i (u_{i,j} + u_{j,i}) , \tag{41}$$

and where $\boldsymbol{g}^{\text{eff}}$ is the effective gravity which differs from $\boldsymbol{g}$ by the centrifugal acceleration. If we apply this equation to the steady flow of an ideal gas, using $H = C_p T$ with a constant C_p, and neglect molecular viscosity and conductivity, we obtain

$$\nabla \cdot \left(\varrho C_p \boldsymbol{u} T + \boldsymbol{F}^D + \frac{1}{2} \varrho \boldsymbol{u}^2 \boldsymbol{u} \right) = \varrho \boldsymbol{u} \cdot \boldsymbol{g}^{\text{eff}} . \tag{42}$$

Here we have excluded the frictional heating (41) as a source of convection. It has been suggested that this term influences the dynamics of the Earth's core. We shall also neglect radiative diffusion and mechanical energy flux, but will allow the flow to be turbulent, the immediate consequence being

$$\nabla \cdot (\bar{\varrho}\, \overline{\boldsymbol{u}' T'} + \overline{\varrho' T'} \bar{\boldsymbol{u}} + \overline{\varrho' T' \boldsymbol{u}'}) = (\bar{\varrho} \bar{\boldsymbol{u}} + \overline{\varrho' \boldsymbol{u}'}) \cdot (C_p^{-1} \boldsymbol{g}^{\text{eff}} - \nabla T) . \tag{43}$$

The final bracket represents the superadiabatic rate which in this book will always be denoted by

$$\boldsymbol{\beta} = C_p^{-1} \boldsymbol{g}^{\text{eff}} - \nabla T . \tag{44}$$

Several different notations for this gradient excess may be found in the literature:

$$\boldsymbol{\beta} = \nabla T^{\text{ad}} - \nabla T = \Delta \nabla T = H_p^{-1} T (\nabla - \nabla^{\text{ad}})$$
$$= (1 - \gamma^{-1})\, p^{-1} T \nabla p - \nabla T , \tag{45}$$

where H_p is the (necessarily positive) pressure scale height. The integral of β_r over r is the so-called "potential temperature".

Using the relationship

$$T\,\mathrm{d}S = C_p\,\mathrm{d}T - \mathrm{d}p/\varrho\,, \tag{46}$$

one can also express $\boldsymbol{\beta}$ in terms of the specific entropy, S:

$$\boldsymbol{\beta} = -C_p^{-1} T\,\nabla S\,, \tag{47}$$

with S, for an ideal gas, being given by

$$S = C_v \ln (p\varrho^{-\gamma})\,. \tag{48}$$

Since the entropy decreases as one moves outwards in a convectively unstable layer, the component of $\boldsymbol{\beta}$ vertically upwards is positive but, since clearly $\boldsymbol{\beta} = 0$ in an adiabatically stratified media (S = constant), this vertical component of $\boldsymbol{\beta}$, though positive, is rather small. It is

$$\beta_r = R^{-1}(-\mathrm{d}T/\mathrm{d}x - \Gamma) > 0\,, \tag{49}$$

where $\Gamma = Rg/C_p$ is the adiabatic temperature gradient. When evaluated, using the values appropriate to the Sun, namely

$$g = 2.7 \cdot 10^4\ \mathrm{cm/s^2}\,, \qquad C_p = 4 \cdot 10^8\ \mathrm{ergs/gK},$$

Γ proves to be of order 10^6 K, corresponding to an adiabatic temperature gradient of 10^{-4} K/cm. The magnitude of the temperature excess is much smaller, namely about $2 \cdot 10^{-10}$ K/cm. Noting that in unstable layers the vector $\boldsymbol{\beta}$ points outwards, we conclude that the actual radial temperature profile must be steeper than the adiabatic. That is also the essence of the Schwarzschild criterion.

A look at density fluctuations ϱ' also proves to be worthwhile. Most mixing-length theories are formulated in terms of rising and falling fluid elements with a length scale L and vertical velocity u'. The elements are assumed to travel at their terminal velocity, the buoyant driving forces being balanced by the drag $u'^2/\frac{1}{2}L$. Hence, for the r.m.s. fluctuations, we have

$$\varrho'/\varrho = 2u'^2/gL, \tag{50}$$

from which the vertical eddy mass flow $\overline{\varrho' u'}$ can be estimated, since

$$\overline{\varrho' u'}/\varrho = 2\overline{u'^3}/gL\,. \tag{51}$$

Table VIII.iv. Velocity and density fluctuations in the SCZ (Staude, 1977)

x	u'	L	ϱ'/ϱ	u_T
1	$6 \cdot 10^4$	$6 \cdot 10^8$	$4 \cdot 10^{-4}$	20
0.9	$1 \cdot 10^4$	$7 \cdot 10^9$	$1 \cdot 10^{-6}$	10^{-2}
0.75	$0.8 \cdot 10^4$	$2 \cdot 10^{10}$	$2 \cdot 10^{-7}$	$2 \cdot 10^{-3}$

This quantity is an "eddy velocity" u_T, but it proves to be very small. Table VIII.iv. contains the values (in c.g.s. units) for three well-separated layers in the SCZ model of Staude (1977). Even a very slow mean circulation exceeds the calculated eddy flow, u_T, so that terms such as $\overline{\varrho' \boldsymbol{u}'}$ or $\overline{\varrho' \boldsymbol{u}' T'}$ may be neglected. The consequence is

$$\nabla \cdot (\bar{\varrho} \overline{\boldsymbol{u}' T'}) = \bar{\varrho} \bar{\boldsymbol{u}} \cdot [\boldsymbol{\beta} - \nabla(\overline{\varrho' T'}/\bar{\varrho})]. \tag{52}$$

The final term in this relation can be omitted from mixing-length arguments, too. Since $\varrho'/\varrho \approx -T'/T$, we can approximate the gradient term by

$$\nabla(\overline{\varrho' T'}/\bar{\varrho}) \simeq \overline{(\varrho'/\varrho)^2}\, \nabla \bar{T}, \tag{53}$$

a quantity which makes almost no contribution in (52). Our final expression describing the heat flux in a convective domain is therefore

$$\nabla \cdot (\boldsymbol{F}^{\text{conv}} + \boldsymbol{F}^D) = \varrho C_p \bar{\boldsymbol{u}} \boldsymbol{\beta}, \tag{54}$$

where

$$\boldsymbol{F}^{\text{conv}} = \varrho C_p \overline{\boldsymbol{u}' T'}. \tag{55}$$

Thus, the second-order correlation between velocity and temperature fluctuations is decisive in determining turbulent heat transport. Obviously hot rising, as well as cool sinking, eddies carry thermal energy through a convective region. In contrast to, for example, the TMT this is the transport of merely a scalar quantity. The desired correlation is, therefore, a vector quantity rather than a tensor.

If, as in a non-rotating star, gravity is the only preferred direction, the eddy heat transport will be radial. Historically the first ansatz for this situation was a diffusive law, namely

$$\overline{\boldsymbol{u}' T'} = -\chi_T\, \nabla \bar{T}. \tag{56}$$

Yet, due to the heat of compression, correlations between temperature and velocity will occur even in a (say) isothermal layer unless it is unstratified. The simple diffusive ansatz (56) can therefore never be correct. In reality, the eddy heat flux must vanish in an adiabatically stratified media, and so

$$\overline{\boldsymbol{u}' T'} = -\chi_T (\nabla T - C_p^{-1} \boldsymbol{g}) = -\chi_T C_p^{-1} T\, \nabla S = \chi_T \boldsymbol{\beta}. \tag{57}$$

The eddy diffusivity χ_T possesses the dimension cm^2 s^{-1} and is most likely to be of order Lu', where L is the mixing length and u' the r.m.s. turbulent velocity.

Wasiutyński (1946) introduced a formulation which, though still local, was based on a tensorial conductivity,

$$\begin{aligned} F_i^{\text{conv}} &= \varrho C_p \overline{u_i' T'} = -\varrho C_p \chi_{ij} (T_{,j} - C_p^{-1} g_j) = \varrho C_p \chi_{ij} \beta_j \\ &= \varrho C_p (\chi \cdot \boldsymbol{\beta})_i. \end{aligned} \tag{58a}$$

Written in component form, we have

$$F_r^{\text{conv}} = \varrho C_p(\chi_{rr}\beta_r + \chi_{r\theta}\beta_\theta), \quad F_\theta^{\text{conv}} = \varrho C_p(\chi_{\theta r}\beta_r + \chi_{\theta\theta}\beta_\theta). \tag{58b}$$

If rotation is involved, the eddy conductivity tensor may contain the following elementary tensors ($\mathring{\boldsymbol{g}}$ being the unit radial vector)

$$\begin{aligned} &\delta_{ij}, \quad \mathring{g}_i\mathring{g}_j, \quad (\varepsilon_{ipl}\mathring{g}_j + \varepsilon_{jpl}\mathring{g}_i)\,\mathring{g}_p\Omega_l, \\ &(\mathring{\boldsymbol{g}}\cdot\boldsymbol{\Omega})^2\,\delta_{ij}, \quad (\mathring{\boldsymbol{g}}\cdot\boldsymbol{\Omega})^2\,\mathring{g}_i\mathring{g}_j, \quad (\mathring{\boldsymbol{g}}\cdot\boldsymbol{\Omega})\,(\mathring{g}_i\Omega_j + \mathring{g}_j\Omega_i), \quad \Omega_i\Omega_j, \end{aligned} \tag{59}$$

all of which are polar combinations, symmetric in i and j. The tensor χ_{ij} is, however, not necessarily symmetric so that the terms

$$\begin{aligned} &\varepsilon_{ijk}\Omega_k, \quad (\mathring{\boldsymbol{g}}\cdot\boldsymbol{\Omega})\,\varepsilon_{ijk}\mathring{g}_k, \quad (\varepsilon_{ipl}\mathring{g}_j - \varepsilon_{jpl}\mathring{g}_i)\,\mathring{g}_p\Omega_l, \\ &(\mathring{\boldsymbol{g}}\cdot\boldsymbol{\Omega})\,(\mathring{g}_i\Omega_j - \mathring{g}_j\Omega_i) \end{aligned} \tag{60}$$

may also appear. In contrast to the terms in (59), some of those in (60) are linear in the angular velocity Ω. Therefore, like the α-effect and the $V^{(0)}$-effect, they are first order effects. Only the last term in (60) is inconsistent with the Onsager symmetry conditions of thermodynamics which, contrary to what is often believed, do not require a strict symmetry of the conductivity tensor. When an axial vector (e.g. $\boldsymbol{\Omega}$) is present the correct formulation of Onsager's principle is

$$\chi_{ij}(\boldsymbol{\Omega}) = \chi_{ji}(-\boldsymbol{\Omega}) \tag{61}$$

(cf. de Groot, 1952) and this is clearly satisfied by the first term of (60). This, in turn, leads to the eddy heat flux

$$F_i^{\text{conv}} = -\varepsilon_{ijk}(\overline{T}_{,j} - C_p^{-1}g_j)\,\Omega_k, \tag{62}$$

which is essentially longitudinal. This actually forges a link to a calculation by Gough (1978) in which he wrote down the form

$$\boldsymbol{F}^{\text{conv}} = F^{(0)}\cdot(1, 0, \varepsilon) \quad \text{with} \quad \varepsilon \sim \Omega\sin\theta. \tag{63}$$

Yet the actual effect of the non-vertical part, ε, seems to be relatively small since the divergence of (62) is

$$-\varepsilon_{jki}\overline{T}_{,j}\Omega_{k,i}, \tag{64}$$

which vanishes if the gradients are only radial.

Some recent papers have studied the astrophysical implications of a scalar eddy conductivity that is latitude-dependent, an assumed consequence of the influence of rotation upon turbulence. This effect can arise only from those tensors in (59) that involve the Kronecker delta, i.e.

$$\chi_{ij} = \cdots\,\delta_{ij} + \cdots\,(\mathring{\boldsymbol{g}}\cdot\boldsymbol{\Omega})^2\,\delta_{ij}. \tag{65}$$

Although this ansatz is quadratic in Ω, it should nevertheless be looked at in greater detail to determine whether it is consistent with a more thorough analysis of rotating turbulence. To this end we will, in Chapter 17, investigate the eddy heat flux in the framework of the second-order correlation approximation. One of our results will be that, in some cases, the eddy conductivity tensor may, to a sufficiently good approximation, be replaced by the one-point velocity correlation (i.e. Reynolds stress) tensor. In such circumstances, the ansatz (65) is tantamount to supposing that all cross-correlations of the turbulence field are zero. Such a turbulence could never generate differential rotation via the Reynolds stres, or via the Λ-effect, which amounts to the same thing.

The conductivity tensor may be represented by a sum of elementary tensors

$$\begin{aligned}\chi_{ij} = {} & \chi_{00}\delta_{ij} + \chi_{01}\mathring{g}_i\mathring{g}_j + \chi_{10}\varepsilon_{ijk}\Omega_k + \chi_{11}(\mathring{\boldsymbol{g}}\boldsymbol{\Omega})\,\varepsilon_{ijk}\mathring{g}_k + \chi_{12}(\varepsilon_{ipl}\mathring{g}_j \\ & - \varepsilon_{jpl}\mathring{g}_i)\,\mathring{g}_p\Omega_l + \chi_{20}\Omega^2\delta_{ij} + \chi_{21}\Omega^2\mathring{g}_i\mathring{g}_j + \chi_{22}(\mathring{\boldsymbol{g}}\boldsymbol{\Omega})^2\,\delta_{ij} \\ & + \chi_{23}(\mathring{\boldsymbol{g}}\boldsymbol{\Omega})^2\,\mathring{g}_i\mathring{g}_j + \chi_{24}\Omega_i\Omega_j + \chi_{25}(\mathring{\boldsymbol{g}}\boldsymbol{\Omega})\,(\mathring{g}_i\Omega_j + \mathring{g}_j\Omega_i),\end{aligned} \tag{66}$$

where all the terms of (59) which fulfill the Onsager relations have been included. If $\boldsymbol{\Omega}$ is expressed in spherical coordinates [see $(A4)_1$], we obtain the components

$$\begin{aligned}\chi_{rr} &= \chi_{00} + \chi_{01} + (\chi_{20} + \chi_{21})\,\Omega^2 + (\chi_{22} + \chi_{23} + \chi_{24} + 2\chi_{25})\,\Omega^2\cos^2\theta, \\ \chi_{\theta\theta} &= \chi_{00} + \chi_{20}\Omega^2 + (\chi_{22}\cos^2\theta + \chi_{24}\sin^2\theta)\,\Omega^2, \\ \chi_{\phi\phi} &= \chi_{00} + \chi_{20}\Omega^2 + \chi_{22}\Omega^2\cos^2\theta, \\ \chi_{r\theta} &= \chi_{\theta r} = -(\chi_{24} + \chi_{25})\,\Omega^2\sin\theta\cos\theta, \\ \chi_{r\phi} &= -\chi_{\phi r} = (\chi_{10} + \chi_{12})\,\Omega\sin\theta, \\ \chi_{\theta\phi} &= -\chi_{\phi\theta} = (\chi_{10} + \chi_{11})\,\Omega\cos\theta.\end{aligned} \tag{67}$$

Apart from the asymmetry, the last two expressions remind us of the structure of the Λ-effect. They can, however, be ignored for all axisymmetric temperature problems for which β_ϕ vanishes.

In order to produce expressions consistent with the concept of a "latitude-dependent conductivity", one must, without any compelling reason, set

$$\chi_{23} = 0, \qquad \chi_{24} = -\chi_{25}. \tag{68}$$

For this special choice of parameters the relevant expressions are

$$\begin{aligned}\chi_{rr} &= \cdots + (\chi_{22} + \chi_{25})\,\Omega^2\cos^2\theta, \\ \chi_{\theta\theta} &= \cdots + (\chi_{22} + \chi_{25})\,\Omega^2\cos^2\theta, \\ \chi_{r\theta} &= \chi_{\theta r} = 0.\end{aligned} \tag{69}$$

Needless to say, there is no clear argument in favor of omitting the off-diagonal element, $\chi_{r\theta}$, which, in conjunction with the vertical temperature gradient, *will* produce a latitudinally varying heat flux. Since this effect should be strong, the associated temperature field should also be latitude-dependent. In view of the observed uniformity of temperature (and its vertical gradient, the flux) over the solar surface, this paradox constitutes the "solar heat flux problem".

A few words should be added concerning the $\chi_{r\phi}$ term or, equivalently, (62). Its existence requires that there should be correlations between the longitudinal velocity and the temperature fluctuations:

$$\overline{u_\phi' T'} = \cdots \Omega \beta_r \sin\theta . \tag{70}$$

The effect is linear in Ω, as is also the Λ-effect [i.e. its fundamental mode, $V^{(0)}$] and the α-effect of MHD.

For future reference a representation of the conductivity tensor will now be given. We shall choose a notation that is somewhat more general than that used in the expression derived earlier. We shall also restrict ourselves to the meridional components of the tensor:

$$\frac{\chi_{ij}}{\chi_T} = \begin{pmatrix} 1 & 0 \\ 0 & s \end{pmatrix} - \sum_l \begin{pmatrix} VV^{(l)} \sin^{2l}\theta & VH^{(l)} \sin^{2l-1}\theta\cos\theta \\ HV^{(l)} \sin^{2l-1}\theta\cos\theta & HH^{(l)} \sin^{2l}\theta \end{pmatrix}. \tag{71}$$

Comparing this with (67), we see that

$$\begin{aligned} \chi_T &= \chi_{00} + \chi_{01}, \qquad s = \chi_{00}/(\chi_{00} + \chi_{01}), \\ VV^{(0)} &= -\chi_T^{-1}\Omega^2(\chi_{20} + \chi_{21} + \chi_{22} + \chi_{23} + \chi_{24} + 2\chi_{25}), \\ VV^{(1)} &= \chi_T^{-1}\Omega^2(\chi_{22} + \chi_{23} + \chi_{24} + 2\chi_{25}), \\ HV^{(1)} &= VH^{(1)} = \chi_T^{-1}\Omega^2(\chi_{24} + \chi_{25}), \\ HH^{(1)} &= \chi_T^{-1}\Omega^2(\chi_{22} - \chi_{24}). \end{aligned} \tag{72}$$

Generally we consider that rotation is responsible for the last part of (71), which is therefore of higher order than the leading "standard" part:

$$\chi_{ij}^{(0)} = \chi_T \begin{pmatrix} 1 & 0 \\ 0 & s \end{pmatrix}. \tag{73}$$

As usual we denote the rotation terms of the tensor by a superposed tilde, so that

$$\tilde{\chi}_{ij} = -\chi_T \begin{pmatrix} \sum\limits_{l=0} VV^{(l)} \sin^{2l}\theta & \sum\limits_{l=1} VH^{(l)} \sin^{2l-1}\theta\cos\theta \\ \sum\limits_{l=1} HV^{(l)} \sin^{2l-1}\theta\cos\theta & \sum\limits_{l=0} HH^{(l)} \sin^{2l}\theta \end{pmatrix}. \tag{74}$$

In addition to our understanding that this tensor is of a higher order than the undisturbed tensor (73), there are also the numerically important constraints:

$$|\boldsymbol{V}\boldsymbol{V}^{(1)}|, \ldots |\boldsymbol{H}\boldsymbol{H}^{(1)}| \ll 1. \tag{75}$$

8.3. The Temperature Equation

Let us now turn to the basic equation (54) which includes rotation. All the quantities, apart from C_p, may be divided into two parts,

$$\varrho = \varrho^{(0)} + \tilde{\varrho}, \qquad T = T^{(0)} + \tilde{T}, \qquad \text{etc.}$$

where the influence of rotation is confined to the terms having tildes. As is evident from the underlying physics, the circulation is rotationally created, and $\tilde{\boldsymbol{F}}^D$ can be set to zero. The zeroth-order and first-order equations are

$$\nabla \cdot (\boldsymbol{F}^{\text{conv},0} + \boldsymbol{F}^{D,0}) = 0, \qquad \nabla \cdot \tilde{\boldsymbol{F}}^{\text{conv}} = C_p \varrho^{(0)} \tilde{\boldsymbol{u}} \cdot \boldsymbol{\beta}^{(0)}. \tag{76}$$

The zeroth-order part of β has only a radial component so that $(76)_2$ simplifies to

$$\nabla \cdot [\tilde{\varrho}\beta\mathring{\boldsymbol{g}} + \varrho^{(0)}(\chi_T{}^{-1}\tilde{\chi} \cdot \mathring{\boldsymbol{g}})\,\beta + T_\odot{}^{-1}R\varrho^{(0)}\tilde{\boldsymbol{\beta}}] = n(n+1)\,r^{-2}\chi_T{}^{-1}A_n\beta P_n, \tag{77}$$

where we have introduced

$$\boldsymbol{\beta}^{(0)} = (T_\odot\beta/R)\,\mathring{\boldsymbol{g}} \quad \text{with} \quad \beta = -T_\odot{}^{-1}(\mathrm{d}T^{(0)}/\mathrm{d}x + \Gamma). \tag{78}$$

The zeroth-order equation can be integrated directly, to give

$$\varrho^{(0)}C_p\chi_T R^{-1}T_\odot\beta + F_r{}^D = \text{constant}/r^2, \tag{79}$$

where the constant of integration is determined by the boundary condition. If this condition requires that all energy leaves the solar surface as electromagnetic radiation, then the Stefan-Boltzmann law should apply and we have, at $r = R$,

$$F_r{}^{(0)} = \sigma T_\odot{}^4, \qquad \tilde{F}_r = 4\sigma T_\odot{}^3\tilde{T}, \tag{80}$$

to zeroth and first order respectively. Thus at the surface we have

$$\varrho_\odot C_p\chi_T R^{-1}T_\odot\beta + F_r{}^D = \sigma T_\odot{}^4, \tag{81}$$

so that the complete solution of $(76)_1$ reads

$$\beta = (\varrho_\odot f^*/x^2\varrho^{(0)}) - (F_r{}^D R/\varrho^{(0)}C_p\chi_T T_\odot), \tag{82}$$

or, equivalently,

$$\mathrm{d}T^{(0)}/\mathrm{d}x = (F_r{}^D R/\varrho^{(0)}C_p\chi_T) - (\varrho_\odot T_\odot f^*/x^2\varrho^{(0)}) - \Gamma. \tag{83}$$

Clearly this relation connects the temperature and density (both undisturbed), and a second relation between these quantities follows from hydrostatic equilibrium.

Substituting (33) in (83), we at once obtain

$$\mathrm{d}T^{(0)}/\mathrm{d}x = -\chi_T(\chi_T + \chi_D)^{-1}\,(\Gamma + T_{\odot}f^*\varrho_{\odot}/x^2\varrho^{(0)}) \tag{84}$$

and then, using (78), we find

$$\beta = [\chi_T f^*(x^2\varrho^{(0)}/\varrho_{\odot})^{-1} - \chi_D\Gamma T_{\odot}^{-1}]/(\chi_T + \chi_D)\,. \tag{85}$$

As expected, the inclusion of radiative energy transport reduces the (positive) superadiabatic gradient β. The extent of this reduction depends on the ratio of the two χ's. For the radiative conductivity we employ the usual expression for optically thick media (diffusion approximation),

$$\chi_D = 16\sigma T^3/3\varkappa\varrho^2 C_p\,, \tag{86}$$

and then the χ ratio, becomes

$$f = \chi_D/\chi_T = 16\sigma T^3/3\varkappa\varrho^2 C_p\chi_T\,, \tag{87}$$

whose value at the surface is

$$f_{\odot} = 16f^*/3\varkappa\varrho_{\odot}R\,, \tag{88}$$

which can easily be estimated. Assuming Kramers' law for the opacity, and using c.g.s. units, we have

$$\varkappa = 0.34\,(1 + 6\cdot 10^{24}\varrho T^{-3.5})\,, \tag{89}$$

and a surface density of 10^{-6} g/cm^3 leads to an optical depth $\varkappa\varrho R$ of about 10^5, so that $f_{\odot}$ is considerably smaller than f^*. The likelihood that radiation plays a prominent role in the rotation problem is slight. The region in which radiation dominates is extremely thin. At the level where the convective velocities are maximal (50—100 km below the photosphere), the ratio f is already less than 0.1 and convection transports 75% of the energy. These numbers have been taken from the tables of Baker and Temesvary, as well as those of Staude. Staude's model gives $f^* \approx 800$ at the level where convection sets in. In our approach it has always been assumed in making estimates that $\nu_T \approx u'L$, and the results do not then support the idea that radiation above the top of the SCZ is fundamental to the rotation problem.

There may exist another important application of these formulations. Imagine a non-turbulent thin layer (of width $\approx$ 100 km) above the SCZ, in which conduction is mainly through radiation. The layer is strongly stratified and has a large f^*. To estimate f^* we write

$$f^* = \sigma T_{\odot}^{\,3}R/\varrho_{\odot}\chi_D C_p = 3\varkappa\varrho_{\odot}R/16 \simeq (2-10)\cdot 10^4\,, \tag{90}$$

and evidently $f^* \gg 1$. The outermost layer of the solar model of Baker and Temesvary has a very strong stratification G, namely

$$G = -\mathrm{d} \ln \varrho / \mathrm{d} \ln x \simeq 3000. \tag{91}$$

Since this is smaller than f^*, the response function (18) can be used to describe the screening of any temperature disturbances at the bottom of a layer from the top of the layer. For a thin layer, this screening is effective if $f^*\mathrm{d}$ greatly exceeds unity. For a layer of (say) 70 km depth, we have $\mathrm{d} = 10^{-4}$ so that $f^*\mathrm{d} \approx 2 - 10$. Thus we expect a remarkably large screening considering how very thin the radiating layer is. Table VIII.v. shows the response (17) for $n = 2$ and $x_i = 0.9999$. It is confirmed that the parameters used above should lead to a surprisingly strong screening.

Table VIII.v. Temperature response for a very thin layer ($x_i = 0.9999$, $n = 2$).

f^* ↓ $G \rightarrow$	10^3	10^4	10^5
10^4	0.21	0.28	0.71
10^5	0.03	0.04	0.20

Let us return to the temperature equation. For the case in which β is given, the first-order equation (77) can be studied in greater detail. Using the harmonic expansion (3), we find from (77) that

$$\begin{aligned}\frac{R}{T_\odot} \nabla \cdot (\varrho^{(0)} \tilde{\beta}) = & \left[\frac{n(n+1)\, \beta A_n}{\chi_T r^2} - \frac{1}{r^2} \frac{\partial}{\partial r} (r^2 \tilde{\varrho}_n \beta)\right] P_n \\ & + \sum_l \left[\frac{1}{r^2} \frac{\partial}{\partial r} (r^2 \beta \varrho^{(0)} V V^{(l)}) \sin^{2l} \theta \right. \\ & \left. + \frac{H V^{(l)} \beta \varrho^{(0)}}{r \sin \theta} \frac{\partial}{\partial \theta} (\cos \theta \cdot \sin^{2l} \theta)\right].\end{aligned} \tag{92}$$

The entire right-hand side of (92) can be written in Legendre polynomials by using

$$\sin^{2l} \theta = Q_n^{(l)} P_n, \tag{93}$$

Table VIII.vi. Numerical values for $Q_n^{(l)}$ as defined in Equation (93)

l ↓ $n \rightarrow$	0	2	4
0	1	0	0
1	2/3	−2/3	0
2	8/15	−16/21	8/35

where the first few coefficients Q are shown in Table VIII.vi. They also occur in the relation

$$(\sin\theta)^{-1}\,\mathrm{d}(\cos\theta\,\sin^{2l}\theta)/\mathrm{d}\theta = -[n(n+1)/2l]\,Q_n^{(l)}P_n, \tag{94}$$

which we shall need in the derivations below. From (92) we obtain

$$\frac{R}{T_\odot}\nabla\cdot(\varrho^{(0)}\tilde{\boldsymbol{\beta}}) = \left[\frac{n(n+1)\,\beta A_n}{\chi_T r^2} - \frac{1}{r^2}\frac{\mathrm{d}}{\mathrm{d}r}(r^2\beta\tilde{\varrho}_n)\right.$$
$$\left.+\frac{1}{r^2}\frac{\mathrm{d}}{\mathrm{d}r}(r^2\varrho^{(0)}\beta VV^{(l)})\,Q_n^{(l)} - \frac{n(n+1)}{2lr}HV^{(l)}\beta\varrho^{(0)}Q_n^{(l)}\right]P_n, \tag{95}$$

so that now we have only to specify the rotationally created superadiabatic excess $\tilde{\boldsymbol{\beta}}$, which is

$$\tilde{\boldsymbol{\beta}} = (\boldsymbol{g}^{\mathrm{eff}} - \boldsymbol{g})\,C_p^{-1} - \nabla\tilde{T}, \tag{96}$$

where the difference $(\boldsymbol{g}^{\mathrm{eff}} - \boldsymbol{g})$ is due to the centrifugal force,

$$\boldsymbol{g}^{\mathrm{eff}} - \boldsymbol{g} = \Omega^2\mathbf{s}. \tag{97}$$

Hence we have

$$\tilde{\boldsymbol{\beta}} = \begin{pmatrix} r\Omega^2\sin^2\theta/C_p - \partial\tilde{T}/\partial r \\ r\Omega^2\sin\theta\cos\theta/C_p - r^{-1}\,\partial\tilde{T}/\partial\theta \\ 0 \end{pmatrix}, \tag{98}$$

so that for a rigid rotation, (95) finally becomes

$$\frac{1}{\varrho^{(0)}}\frac{\mathrm{d}}{\mathrm{d}x}\left(x^2\varrho^{(0)}\frac{\mathrm{d}\tilde{T}_n}{\mathrm{d}x}\right) - n(n+1)\,\tilde{T}_n + \frac{n(n+1)\,A_n\beta T_\odot}{\chi_T\varrho^{(0)}R} - \frac{T_\odot}{\varrho^{(0)}}\frac{\mathrm{d}}{\mathrm{d}x}(x^2\beta\tilde{\varrho}_n)$$
$$= \frac{2}{3}\frac{\Omega_\odot{}^2R^2x^2}{C_p}\left[\left(3 + \frac{\mathrm{d}\ln\varrho^{(0)}}{\mathrm{d}\ln x}\right)\delta_{n0} - \frac{\mathrm{d}\ln\varrho^{(0)}}{\mathrm{d}\ln x}\delta_{n2}\right]$$
$$+\left[-\frac{1}{\varrho^{(0)}}\frac{\mathrm{d}}{\mathrm{d}x}(x^2\varrho^{(0)}\beta VV^{(l)}) + \frac{n(n+1)}{2l}HV^{(l)}\beta x\right]Q_n^{(l)}T_\odot. \tag{99}$$

This equation describes the influence of rotation on the temperature distribution. The coefficients $VV^{(l)}$ and $HV^{(l)}$ appear to play quite similar roles — at least for small l.

The effect of the centrifugal force, which we included for completeness, will certainly be of minor interest.

We shall conclude the present Chapter with a brief discussion of what happens to the latitudinal temperature profile when we add a prescribed meridional flow, and also a density variation $\tilde{\varrho}$. For simplicity, the mean density will be assumed uniform: $\varrho^{(0)} \equiv \bar{\varrho}$. Hence, the governing equation is

$$x^2t_n'' + 2xt_n' - n(n+1)\,t_n = F_n, \tag{100}$$

where

$$F_n = -[n(n+1)\,A_n\beta/R\chi_T\bar{\varrho}] + \mathrm{d}(x^2 d_n\beta)/\mathrm{d}x, \tag{101}$$

and the normalized functions are given by $t_n = \tilde{T}_n/T_\odot$ and $d_n = \tilde{\varrho}_n/\bar{\varrho}$. The superadiabatic excess β, given by (82), is positive and may be approximated by

$$\beta = f^*/x^2. \tag{102}$$

The solution of (100) is simply

$$t_n = C_1 x^n + C_2 x^{-n-1} + (2n+1)^{-1} \int_x^1 F(\xi)\,(\xi^n x^{-n-1} - x^n \xi^{-n-1})\,\mathrm{d}\xi, \tag{103}$$

where the arbitrary coefficients have to be determined from two boundary conditions. In this analysis we choose constraints on the radial component of the heat flux,

$$F_r \sim -\partial t/\partial x + \beta d, \tag{104}$$

which must be zero at both the top and the base of the zone. Of course, the condition at $x = 1$ is not identical to the Stefan-Boltzmann law, which we usually favor, but the resulting differences are too small to affect the validity of our general conclusions.

From (103) we obtain

$$\frac{\mathrm{d}t_n}{\mathrm{d}x} = nx^{n-1}\,C_1 - (n+1)\,x^{-n-2}\,C_2 - (2n+1)^{-1} \int_x^1 F(\xi) \times [(n+1)\,x^{-n-2}\xi^n + nx^{n-1}\xi^{-n-1}]\,\mathrm{d}\xi, \tag{105}$$

so that, applying (104), we find that C_1 and C_2 must satisfy the conditions

$$nC_1 - (n+1)\,C_2 = f^* d_n(1),$$

$$nx_i^{n+1}C_1 - (n+1)\,x_i^{-n}C_2 \tag{106}$$

$$= f^* d_n(x_i) + (2n+1)^{-1} \int_{x_i}^1 F(\xi)\,[(n+1)\,x_i^{-n}\xi^n + nx_i^{n+1}\xi^{-n-1}]\,\mathrm{d}\xi.$$

As we are mainly interested in the surface profile of the temperature, i.e. $t_n(1) = C_1 + C_2$, we need to know only this sum of the constants. A short calculation yields

$$t_n(1) = f^* \frac{(2n+1)\,d_n(x_i) - nx_i^{n+1}d_n(1) - (n+1)\,x_i^{-n}d_n(1)}{n(n+1)\,(x_i^{n+1} - x_i^{-n})}$$

$$+ \frac{\int_{x_i}^1 F(\xi)\,[(n+1)\,\xi^n x_i^{-n} + n\xi^{-n-1}x_i^{n+1}]\,\mathrm{d}\xi}{n(n+1)\,(x_i^{n+1} - x_i^{-n})}, \tag{107}$$

in which we can now substitute the function (101). There are two different contributions to (107), viz: meridional flow and density distortion. The contribution from the former is very simple:

$$t_n(1) = \frac{f^*}{R_{\odot}\chi_T\bar{\varrho}} \frac{\int\limits_{x_i}^{1} [(n+1)\,\xi^{n-2}x_i^{-n} + nx_i^{n+1}\xi^{-n-3}]\,A_n\,d\xi}{x_i^{-n} - x^{n+1}} + \cdots. \qquad (108)$$

Since the denominator is positive-definite we see that $t_n(1)$ is positive for positive A. The case $n = 2$ is particularly impressive. For this, $\delta t \equiv t_{\text{pole}} - t_{\text{eq}} \approx 1.5\,t_2$, so that, for positive t_2, the poles are warmer than the equator. According to (108), this occurs when A is positive, corresponding to a circulation that is clockwise in the northern hemisphere. This result is not surprising because a flow that is upward at the poles transports more heat to the surface there.

Now consider the part of (101) created by the density deviation. A partial integration procedure gives a more useful result:

$$t_n(1) = \cdots + f^* \frac{\int\limits_{x_i}^{1} d_n(\xi)\,(\xi^{n-1}x_i^{-n} - \xi^{-n-2}x_i^{n+1})\,d\xi}{x_i^{-n} - x_i^{n+1}}. \qquad (109)$$

Because the kernel of the integral contains a minus sign, no definitive statement can be made about the sign of the effect. We can conclude only that its magnitude is small in comparison with the effect of the circulation. For constant d_n we immediately find from (107) that

$$t_n(1) = \frac{f^*d_n}{n(n+1)} \frac{n + 1 + nx_i^{2n+1} - (2n+1)\,x_i^n}{1 - x_i^{2n+1}} \simeq \frac{f^*d_n\,\mathrm{d}}{2}, \qquad (110)$$

where the final approximation holds for shallow zones. In this case the signs of d_n and t_n appear to coincide, and the denser the poles are the warmer they become. In an ideal gas, of course, the polar pressure must also be higher. It is hard, however, to estimate the order of magnitude of this effect without knowing what the real density stratification is.

Chapter 9. The Pole-Equator Temperature Difference

9.1. The Simplest Model

It is now possible to study the solar heat flux problem using the results of the previous Chapter. The arbitrarily postulated disturbances studied there can be set aside in favor of a more general approach. In addition to a possible meridional flow, the θ dependencies of the generalized heat fluxes,

$$\begin{aligned} F_r &= \varrho C_p \chi_T \left\{\beta_r - \sum_l (VV^{(l)} \sin\theta\beta_r + VH^{(l)} \cos\theta\beta_\theta) \sin^{2l-1}\theta\right\}, \\ F_\theta &= \varrho C_p \chi_T \left\{\beta_\theta - \sum_l (HV^{(l)} \cos\theta\beta_r + HH^{(l)} \sin\theta\beta_\theta) \sin^{2l-1}\theta\right\}, \end{aligned} \tag{1}$$

are also sources of non-uniform surface temperatures. The situation now becomes clearer: the rotationally induced modes produce pole-equator differences in all cases. The only question is whether these differences are observable, i. e. whether they exceed 5K. By studying suitable models we hope to clarify the physical picture. Even the most primitive model, described in this section, must take into account the sizeable heat flux that passes through the top and bottom of the convection zone. At the bottom, we assume the heat flux to be completely uniform over the spherical surface:

$$F_r(x_i) = j = \text{constant}. \tag{2}$$

In contrast, the same energy leaves the solar surface in the form of electromagnetic radiation, so that the radial heat flux is related to the temperature by Stefan's law $F_r(1) = \sigma T^4(1)$. Since we are looking for relatively small latitudinal temperature differences, we may use the ansatz $T = T^{(0)} + \tilde{T}$, and consider $\tilde{T}$ to be so small that $\tilde{F}_r = 4\sigma T^{(0)3}\tilde{T}$ on the surface. We have then

$$\beta_r^{(0)} = T_\odot f^*/Rx^2d, \qquad \beta_\theta^{(0)} = 0, \tag{3}$$

hence

$$\tilde{F}_r = -\chi_T \varrho C_p R^{-1}(\partial\tilde{T}_n/\partial x + VV^{(l)} f^* Q_n^{(l)} T_\odot x^{-2} d^{-1})\, P_n. \tag{4}$$

Initially, we shall assume that the mean density is uniform ($d = 1$). The equation governing the temperature (8.99) then becomes

$$x^2\tilde{T}_n'' + 2x\tilde{T}_n' - n(n+1)\,\tilde{T}_n = [n(n+1)/2lx]\,HV^{(l)}f^*Q_n^{(l)}T_\odot\,, \tag{5}$$

whose solution simply is

$$\tilde{T}_n = C_n^1x^n + C_n^2x^{-n-1} - (1/2lx)\,HV^{(l)}f^*Q_n^{(l)}T_\odot\,. \tag{6}$$

The arbitrary coefficients C_n^1 and C_n^2 can be determined from the boundary conditions. The bottom constraint (2), i.e. $\tilde{F}_r = 0$, leads to

$$nx_i^{n+1}C_n^1 - (n+1)\,x_i^{-n}C_n^2 = -\left(\frac{1}{2l}\,HV^{(l)} + VV^{(l)}\right)f^*Q_n^{(l)}T_\odot\,, \tag{7}$$

and condition (8.80) at the top yields

$$\begin{aligned}&(4f^* + n)\,C_n^1 + (4f^* - n - 1)\,C_n^2\\ &= \left[\frac{1}{2l}\,(4f^* - 1)\,HV^{(l)} - VV^{(l)}\right]f^*Q_n^{(l)}T_\odot\,.\end{aligned} \tag{8}$$

Using these relations the coefficients C_n^1 and C_n^2 can be determined. Solution (6) is then known. We are interested only in the profile of the *surface* temperature

$$\tilde{T}_n(1) = C_n^1 + C_n^2 - HV^{(l)}f^*Q_n^{(l)}T_\odot/2l\,, \tag{9}$$

for which we obtain

$$\tilde{T}_n(1) = [\phi(1)\,VV^{(l)} + \Psi(1)\,HV^{(l)}]Q_n^{(l)}T_\odot\,, \tag{10}$$

where we have written

$$\begin{aligned}\phi(1) &= -\frac{2n + 1 - nx_i^{n+1} - (n+1)\,x_i^{-n}}{n(n + 1 - 4f^*)\,x_i^{n+1} - (n+1)\,(n + 4f^*)\,x_i^{-n}}\,f^*,\\ \Psi(1) &= -\frac{2n + 1 + n^2x_i^{n+1} - (n+1)^2\,x_i^{-n}}{n(n + 1 - 4f^*)\,x_i^{n+1} - (n+1)\,(n + 4f^*)\,x_i^{-n}}\,\frac{f^*}{2l}\,.\end{aligned} \tag{11}$$

If we adopt the thin-shell approximation, $\mathsf{d} = 1 - x_i \ll 1$, we may reduce (11) to

$$\begin{aligned}\phi(1) &\simeq -n(n+1)\,\mathsf{d}^2f^*/2[n(n+1)\,\mathsf{d} + 4f^*],\\ \Psi(1) &\simeq -n(n+1)\,\mathsf{d}f^*/2l[n(n+1)\,\mathsf{d} + 4f^*].\end{aligned} \tag{12}$$

The limiting forms of these results are, for small f^* ($f^* \ll \mathsf{d}$),

$$\phi(1) \simeq -\mathsf{d}f^*/2, \qquad \Psi(1) \simeq -f^*/2l\,, \tag{13}$$

and, for large f^* ($f^* \gg \mathsf{d}$),

$$\phi(1) \simeq -n(n+1)\,\mathsf{d}^2/8, \qquad \Psi(1) \simeq -\,n\,(n+1)\,\mathsf{d}/8l\,. \tag{14}$$

In these limits the temperature profiles are described by

$$\tilde{T}_n(1)/T_\odot \simeq \begin{cases} -\dfrac{1}{2} f^*(\mathsf{d}VV^{(l)} + l^{-1}HV^{(l)})\, Q_n^{(l)}, & f^* \ll \mathsf{d}, \\ -\,\mathsf{d}(\mathsf{d}VV^{(l)} + l^{-1}HV^{(l)})\,[n(n+1)/8]\, Q_n^{(l)}, & f^* \gg \mathsf{d}. \end{cases} \tag{15}$$

We see that, in the first approximation for very small f^*, the temperature modes due to VV and HV are strongly reduced. In § 8.1 however, we had good reason to favor $f^* = 100$, this large value being due to the small surface density of about 10^{-6} g/cm³. Had we used instead the density at the bottom of the layer (about 10^{-1} g/cm³), f^* would have been reduced to 10^{-3}, and it might then have been considered to satisfy the condition, $f^* \ll \mathsf{d}$, of smallness. Then the values $(15)_1$ of the temperature ratio would have been the relevant ones, and they are small. Consequently, if it is really permissible to adopt the value of f^* that corresponds to the base of the layer, the temperature profile should be relatively flat.

Because of the fundamental relation

$$\delta F_r/F_r \simeq 4\delta T/T \qquad \text{at} \qquad x = 1\,, \tag{16}$$

the result $(15)_1$ also provides the profile of the emergent heat flux. The difference between the two profiles consists only of the factor 4. This remark becomes apposite if we now apply a boundary condition other than the Stefan-Boltzmann law. Let us assume that the surface temperature is uniform but that the surface heat flux is not (Gilman, 1978). In this case, (8) must be replaced by

$$C_n^1 + C_n^2 = HV^{(l)} f^* Q_n^{(l)} T_\odot/2l\,. \tag{17}$$

Then

$$\begin{Bmatrix} C_n^1 \\ C_n^2 \end{Bmatrix} = \begin{Bmatrix} 1 - x_i^{-n} \\ x_i^{n+1} - 1 \end{Bmatrix} \frac{HV^{(l)} f^* Q_n^{(l)} T_\odot/2l}{x_i^{n+1} - x_i^{-n}}\,, \tag{18}$$

so that, in accordance with (4), a rotationally induced heat flux results:

$$\frac{\tilde{F}_r(1)}{\sigma T_\odot^4} = -\left[\frac{2n+1-(n+1)\,x_i^{-n} - n x_i^{n+1}}{x_i^{n+1} - x_i^{-n}} \frac{HV^{(l)}}{2l} + VV^{(l)}\right] Q_n^{(l)} P_n\,. \tag{19}$$

The thin-shell approximation yields

$$\tilde{F}_r(1)/\sigma T_\odot^4 = -\left[VV^{(l)} + \frac{1}{4l}\, n(n+1)\, \mathsf{d} HV^{(l)}\right] Q_n^{(l)} P_n\,. \tag{20}$$

By means of (15) and (20), the efficacy of modes VV and HV in generating heat flux surface profiles can be compared for various boundary conditions.

The following coefficients in the associated series expansions occur:

mode	Boltzmann		non-Boltzmann
	$f^* \ll 1$	$f^* \gg 1$	
$VV^{(1)}$	$2f^*\mathrm{d}$	$n(n+1)\,\mathrm{d}^2/2$	1
$HV^{(1)}$	$2f^*$	$n(n+1)\,\mathrm{d}/2$	$n(n+1)\,\mathrm{d}/4$

It can easily be seen that the smallest pole-equator differences in the heat flux occur when the σT^4-law is applied *and* when

$$f^* \equiv \sigma R T_\odot^3/\varrho_\odot C_p \chi_T \ll \mathrm{d}. \tag{21}$$

It is, then, the cooperation of two effects that prevents a considerable differential heat flux or temperature from occurring: solar energy must be radiated via electromagnetic waves *and* the eddy diffusivity must be sufficiently large. The physical explanation is the same as for the screening process considered in § 8.1: heat a small area of a large but thin plate from below and measure the temperature distribution on its surface. This will be almost uniform only if the plate is made from a material that possesses a high conductivity, which itself depends on the intensity of the non-uniform heat source, so that a condition of the form (21) is fulfilled. The same principle also applies to the frying-pan; otherwise one of the steaks could easily burn while its neighbor remained uncooked!

We shall now say a few words about a very special problem: $\tilde{F}_r = 0$ cannot be used as the bottom boundary condition for the latitude-independent part of $\tilde{F}_r$. This follows from the energy equation, in conjunction with the Stefan-Boltzmann condition at the surface. If no θ-dependence exists, we have $\tilde{F}_\theta = 0$ and the energy equation is reduced to $\mathrm{d}(r^2\tilde{F}_r)/\mathrm{d}r = 0$, so that $\tilde{F}_r \sim r^{-2}$. If $\tilde{F}_r$ vanishes at the bottom, then it must vanish everywhere, i.e.

$$\tilde{F}_r \equiv -\chi_T \varrho C_p R^{-1}(\partial \tilde{T}_0/\partial x + VV^{(l)} f^* Q_0^{(l)} T_\odot/x^2\, d) = 0. \tag{22}$$

This first order differential equation can be solved only if we have one further boundary condition, and this condition must come from the Stefan-Boltzmann law. Since $\tilde{F}_r$ vanishes everywhere and $\tilde{F}_r = 4\sigma T_\odot^3 \tilde{T}$ at the surface, we have to ensure that $\tilde{T}_0$ vanishes there. Hence, the two conditions which determine the temperature $\tilde{T}_0$ are $\tilde{F}_r(x_i) = \tilde{T}_0(1) = 0$. All the parameters have been determined; no free choice remains. The temperature is completely decided by the physics of the problem. Consequently the functions $\phi(1)$ and $\Psi(1)$ in (10) both vanish for $n = 0$ so that $\tilde{T}_0$ does indeed vanish at the surface.

Another enlightening proof of the same point is based on the energy integral over the convection zone, i.e.

$$x_i^2 j = \frac{1}{2}\int_0^\pi F_r(1)\sin\theta\,d\theta = \frac{1}{2}\sigma\int_0^\pi (T^{(0)} + \tilde{T})^4 \sin\theta\,d\theta. \quad (23)$$

The kernel of the integral can be expanded as

$$x_i^2 j = \sigma T_\odot^4 + 2\sigma T_\odot^3 \int_0^\pi \tilde{T}_n P_n \sin\theta\,d\theta, \quad (24)$$

hence

$$x_i^2 j = \sigma T_\odot^4 + 4\sigma T_\odot^3 \tilde{T}_0(1). \quad (25)$$

Thus, if we postulate $x_i^2 j = \sigma T^4{}_\odot$, the vanishing of $\tilde{T}_0(1)$ follows directly Note the appropriateness of the Legendre polynomial series (8.3) in carrying out the integration (24).

9.2. Two-Shell Model with Constant but Different Densities

Since it fails to allow for the rapid outward decrease of the density, the model just discussed may be too simple. We recall that a "flux disturbance" which is initiated at too shallow a depth below the surface is only slightly reduced by the small conductivity of the intervening material (see Table VIII.i.). This "surface difficulty" is not, however, relevant, to the mechanism we consider here, for which the outer layers of the convection zone are not influential. The rotationally created functions, VV and HV, vary in proportion to $(2\Omega\tau_c)^2$, which itself becomes very small near the surface since the correlation (or life-) time τ_c decreases strongly in the outward direction. Thus, the part of the conductivity tensor which is due to rotation decreases more rapidly than does the small-scale eddy conductivity. The heat flux in this relatively thin layer is simply

$$F_r = -\varrho_\odot C_p \chi_T R^{-1}(\partial T/\partial x + \Gamma), \quad F_\theta = -\varrho_\odot C_p \chi_T R^{-1}\, x^{-1}\, \partial T/\partial\theta, \quad (26)$$

in contrast to the more complex expressions (1) which are appropriate for the bulk of the SCZ where actually $f^* \Rightarrow f_i \ll 1$. One is, therefore, tempted to construct a model of two shells each with constant but different density.

Having provided the motivation, we may immediately formulate solutions for the temperature in both shells:

$$\tilde{T}_n^s = C_n^1 x^n + C_n^2 x^{-n-1}, \quad (27)$$

$$\tilde{T}_n^i = C_n^3 x^n + C_n^4 x^{-n-1} - HV^{(l)} f_i Q_n^{(l)} T_\odot / 2\,lx, \quad (28)$$

where

$$f_i = j x_i^2 R / \varrho_i \chi_T C_p T_\odot, \quad (29)$$

j being the heat flux entering the convection zone from below. To a good approximation, we have $j = \sigma x_i^{-2} T_\odot^4$ with $T_\odot \equiv T^{(0)}(1)$, which for the Sun yields $j = 9.73 \cdot 10^{10}$ c.g.s., corresponding to a surface temperature of 5385 K. Adopting these values, we find that f_i then has the (small) value of $2 \cdot 10^{-3}$.

In order to formulate the boundary conditions we must know the heat fluxes at top and bottom of the convection zone. The first expression has to be set equal to $4\sigma T_{\odot}^3 \tilde{T}_n(1)$, whereas the second one vanishes. The two free coefficients C_n^1 and C_n^3 in (27) and (28) can thus be determined, and the temperatures become

$$\tilde{T}_n^s = \{x^{-n-1} + [(n+1-4f^*)/(n+4f^*)]\, x^n\}\, C_n^2, \tag{30}$$

and

$$\tilde{T}_n^i = \{x^{-n-1} + [(n+1)/n]\, x_i^{-2n-1} x^n\}\, C_n^4 - \left[\frac{1}{2}(lx)^{-1} H V^{(l)} + \left(\frac{1}{2} l^{-1} H V^{(l)} + V V^{(l)}\right) n^{-1} x_i^{-n-1} x^n\right] f_i Q_n^{(l)} T_{\odot}, \tag{31}$$

where f^* (= 100, as usual) represents the surface value of the parameter f. Both the remaining coefficients can be determined by the conditions, still to be formulated, at the interface between the two shells.

Let the outer shell be $1 \geqq x \geqq x_h \equiv 1 - h$, where $h \ll d$. At the interface, both the temperature and the radial heat flux must be continuous, i.e.

$$\tilde{T}^i(x_h) = \tilde{T}^s(x_h), \tag{32}$$

and

$$\tilde{F}_r^i(x_h) = \tilde{F}_r^s(x_h): \tag{33}$$

The lateral component of the heat flux has a discontinuity since $\nabla \cdot \tilde{F} = 0$ only requires that the radial component of $\tilde{F}$ be continuous. Due to conditions (32) and (33), we have two equations for the two coefficients C_n^2 and C_n^4. As we are only interested in the surface profile of the temperature,

$$\tilde{T}_n^s(1) = [(2n+1)/(n+4f^*)]\, C_n^2, \tag{34}$$

we, fortunately, only need find the coefficient C_n^2. This coefficient may be deduced, and the expression for the surface temperature then becomes

$$\tilde{T}_n(1) = \frac{(2n+1)\, f_i f^* Q_n^{(l)} T_{\odot}}{N} \left\{\left[2n + 1 - n\left(\frac{x_i}{x_h}\right)^{n+1} - (n+1)\left(\frac{x_i}{x_h}\right)^{-n}\right] \times V V^{(l)} + \left[2n + 1 + n^2\left(\frac{x_i}{x_h}\right)^{n+1} - (n+1)^2\left(\frac{x_i}{x_h}\right)^{-n}\right] \frac{H V^{(l)}}{2l}\right\}, \tag{35}$$

where the denominator, N, is

$$N = \Big\{(n+1)\, x_i^{-n}\big[n(n+1-4f^*)(f^* - f_i)\, x_h^{2n+1} + (n+4f^*) \times \big(nf^* + (n+1) f_i\big)\big] - n x_i^{n+1}\big[(n+1)(n+4f^*)(f^* - f_i)\, x_h^{-2n-1} + (n+1-4f^*)\big(nf_i + (n+1) f^*\big)\big]\Big\}. \tag{36}$$

As is to be expected, these expressions reduce to results (10) and (11) for the one-shell model if $x_h \to 1$ and $f_i = f^*$.

Here we are particularly interested in the case $\varrho_i \gg \varrho_\odot$, i.e. $f_i \ll f^*$. The inner density, ϱ_i, exceeds the surface value, $\varrho_\odot$, by more than 5 orders of magnitude. By this approximation, the denominator N may be simplified to

$$N = n(n+1) f^* \{[(n+1-4f^*) x_h^{2n+1} + n + 4f^*] x_i^{-n} - [(n+4f^*) x_h^{-2n-1} + n + 1 - 4f^*] x_i^{n+1}\}, \tag{37}$$

and, if in addition we assume $h \ll 1$, then

$$N = (2n+1) n(n+1) f^* \{[1 - (n+1-4f^*) h] x_i^{-n} - [1 + (n+4f^*) h] x_i^{n+1}\}. \tag{38}$$

This expression can be reduced still further if d is also taken to be small:

$$N \simeq (2n+1)^2 n(n+1) f^* (\mathsf{d} - h + 4f^* \mathsf{d} h). \tag{39}$$

By adopting a similar approach for the numerator in (35), we are finally led to the results

$$\phi(1) = -\frac{\mathsf{d}(\mathsf{d} - 2h) f_i}{2(\mathsf{d} - h + 4f^* \mathsf{d}h)}, \quad \Psi(1) = -\frac{(\mathsf{d} - h) f_i}{2l(\mathsf{d} - h + 4f^* \mathsf{d}h)}, \tag{40}$$

where ϕ and Ψ have the same meaning as in (10).

For $h \to 0$ the expressions are, of course, identical with the result $(15)_1$ of the one-shell model. The "passive" role of the outer layer is clear from (40). The thickness of the remaining "active" layer is $\mathsf{d} - h$, an almost trivial result. If, however, f^* is large, if indeed it exceeds $1/h$, then the resulting expressions,

$$\phi(1) = -f_i \mathsf{d}/8f^*h, \quad \Psi(1) = -f_i/8lf^*h, \tag{41}$$

are significantly smaller than the corresponding values for the one-shell model. The magnitude of this reduction is of the order of f^*h. This finding agrees precisely with the arguments that led to (8.18). As for the outer layer of the SCZ, for which the estimates $h = 0.1$, $f^* = 100$ are not too unreasonable, we can state that there is an additional smoothing of the temperature difference by one order of magnitude. Finally, if we compare the variation in the surface heat flux, due to (say) VV alone, as predicted by our model,

$$\tilde{F}_r/F_r^{(0)} \simeq -(f_i \mathsf{d}/2f^*h)\, VV^{(l)} Q_n^{(l)} P_n \approx -(10^{-3} - 10^{-4})\, VV^{(l)} Q_n^{(l)} P_n, \tag{42}$$

with that predicted by the "wrong" model having uniform surface temperatures,

$$\tilde{F}_r/F_r^{(0)} \simeq -VV^{(l)} Q_n^{(l)} P_n, \tag{43}$$

we obtain an idea of the effectiveness of the SCZ to create *and* to screen pole-equator differences.

After this exposition of the basic physics and its corroboration through the results of simple models, it might be desirable to confirm our findings with a final, more sophisticated, model in which the density stratification is continuously varying. Such models would enable us to make better comparisons with computer results than is possible with models whose structures vary discontinuously. We have designed such a model in an original paper (Rüdiger, 1982) and there also the numerical results have been discussed. They are in accordance with those of the model with two shells. In summary we may note the following conclusions:

(i) Positive $VV^{(1)}$ and/or $HV^{(1)}$ produce poles warmer than the equator (positive pole-equator difference);
(ii) In unstratified zones the HV-effect is stronger than the VV-effect;
(iii) The opposite is true in strongly stratified media;
(iv) If the outward decrease in correlation time is included, the pole-equator temperature differences become nearly equal, and rather small. Even if the turbulence coefficients are of order unity, the pole-equator differences are reduced to less than 1% and are unobservable;
(v) For the realistic case of large f^* the outward decrease in correlation time *must* be included in order to obtain small pole-equator temperature differences.

Chapter 10. A Complete Model

10.1. The Zero-Order Equations

After laying the groundwork in previous chapters, we are now in a position to establish a complete solar model that includes the full, mutual interaction of mean-field hydro- and thermo-dynamics. Most of the governing equations which we shall need have already been derived, including also the zeroth-order equations, which determine the structure of the convection zone for zero rotation. When considered on the large scale, these properly belong to hydrostatics rather than hydrodynamics. Ignoring rotation, we may write the Reynolds equation (5.41) as

$$\nabla p = \varrho \boldsymbol{g} - \nabla \cdot (\varrho Q), \tag{1}$$

and the energy equation (8.54) as

$$\nabla \cdot (\boldsymbol{F}^{\text{conv}} + \boldsymbol{F}^{D}) = 0, \tag{2}$$

where

$$\boldsymbol{F}^{\text{conv}} = \varrho C_p \chi_T \boldsymbol{\beta} \tag{3}$$

($\boldsymbol{F}^{D} = 0$, if radiation is still neglected, see § 8.3). Converting these expressions to spherical coordinates, we obtain the system

$$\begin{aligned} \mathrm{d}p/\mathrm{d}r &= -\varrho g - \mathrm{d}(\varrho \overline{u_r^{(0)2}})/\mathrm{d}r + r^{-1}\varrho(\overline{u_\theta^{(0)2}} + \overline{u_\phi^{(0)2}} - 2\overline{u_r^{(0)2}}), \\ (\mathrm{d}/\mathrm{d}r)\,&[r^2\varrho(\mathrm{d}T/\mathrm{d}r + C_p^{-1}g)] = 0, \end{aligned} \tag{4}$$

provided the product $C_p\chi_T$ is taken to be independent of the depth. As usual, we normalize the equation by writing

$$r = Rx, \quad \varrho = \varrho_\odot d, \quad T = T_\odot t. \tag{5}$$

We suppose the medium to be an ideal gas: $p = (\mathfrak{R}/\mu)\,\varrho T$. As a further simplification we also neglect the turbulent pressure and find

$$(dt)' = -\gamma\, dg^*, \tag{6}$$

where in (6) we have introduced the normalized gravity

$$g^* = \mu g \cdot R/\gamma \mathcal{R} T_\odot = (\gamma - 1)^{-1} \Gamma / T_\odot . \tag{7}$$

In a similar way, we find from the energy equation $(4)_2$ that

$$t' + (\gamma - 1)\, g^* = \text{constant}/x^2\, d , \tag{8}$$

the constant being determined by using the Stefan-Boltzmann law as an outer boundary condition. This requires

$$(C_p^{-1} g + \mathrm{d}T/\mathrm{d}r)|_1 = -\sigma T_\odot^{\,4} / \varrho_\odot C_p \chi_T , \tag{9}$$

or, if we normalize further,

$$t'(1) + (\gamma - 1)\, g^* = -f^* , \tag{10}$$

where the normalized energy flux, or adiabaticity excess, f^* is given by (8.10).

The temperature decreases outwards more rapidly in convection zones than in adiabatic regions. Hence, the second equation for the SCZ proves to be

$$dt' = -x^{-2} f^* + (1 - \gamma)\, g^*\, d . \tag{11}$$

To this relation must be added the hydrostatic equation (6), in the form

$$t\, d' = x^{-2} f^* - g^*\, d , \tag{12}$$

as well as the boundary conditions

$$d(1) = t(1) = 1 . \tag{13}$$

The system (11)—(13) governs the zeroth-order problem, the defining parameters being g^* and f^*. The boundary conditions (13) follow simply, of course, from the definitions (5).

To assess the orders of magnitude of g^* and f^*, we shall analyze two different models of the SCZ. The first one, developed by Belvedere and Paternò (1977), gives the surface values

$$T_\odot = 6200\,\mathrm{K}, \quad \varrho_\odot = 2 \cdot 10^{-7}\,\mathrm{g \cdot cm^{-3}}, \quad \chi_T = 5 \cdot 10^{13}\,\mathrm{cm^2\, s^{-1}} , \tag{14}$$

the atomic constants being $\gamma = 5/3$ and $\mu = 0.6$. We at once obtain $g^* = 1500$ and $f^* = 200$ so that, for the values of density and temperature at the inner surface, we have $d_i = 2600$ and $t_i = 220$.

In order to provide a general overall view of the models that can be described by (11)—(13), we present in Table X.i the values d_i and t_i at the base of the SCZ for various values of g^* and f^*. Since t_i is practically inde-

Table X.i. The values of d_i (and t_i) at the base of the SCZ for various values of f^* and $g^*(x_i = 0.7)$.

$g^* \rightarrow$ / $f^* \downarrow$	1000	2000	4000
100	2,200	7,800	28,000
200	1,600	6,600	26,000
400	700	4,500	21,000
t_i	200	400	800

pendent of f^*, only its variation with g^* is shown. It is clear that $t_i \approx g^*/5$, and we note that values derived by Belvedere and Paternò can be reproduced with ease.

The situation is different for the more detailed model of the SCZ derived by Staude (1977). This gives surface values of

$$T_\odot = 6200\,\mathrm{K}, \quad \varrho_\odot = 3 \cdot 10^{-7}\,\mathrm{g \cdot cm^{-3}}, \quad \chi_T = 3 \cdot 10^{13}\,\mathrm{cm^2\,s^{-1}}. \tag{15}$$

If we adopt a mean specific heat of $5 \cdot 10^{-8}$ c.g.s. for this model, we obtain

$$g^* = 1400, \quad f^* = 200, \tag{16}$$

values which are very close to those of Belvedere and Paternò. It is, however, not possible to reproduce the values $d_i \approx 3 \cdot 10^5$, $t_i \approx 8500$ for the base of the SCZ because of the more complicated physics which is built into the Staude model. The same comments are pertinent also for the SCZ model of Baker and Temesváry (1966).

The particular model with which we shall work is defined by

$$g^* = 1500, \quad f^* = 200, \tag{17}$$

Table X.ii. The zeroth-order model: values of the density, temperature, density gradient and normalized mixing-length in the SCZ at given depths.

x	d	t	G	$-dx/d(\ln p)$
1	1	1	1300	$4.00 \cdot 10^{-4}$
0.95	258	51	27.7	$2.04 \cdot 10^{-2}$
0.90	714	101	13.3	$4.04 \cdot 10^{-2}$
0.85	1302	152	8.42	$6.08 \cdot 10^{-2}$
0.80	1997	202	5.96	$8.08 \cdot 10^{-2}$
0.75	2789	252	4.47	$1.01 \cdot 10^{-1}$
0.70	3655	301	3.48	$1.21 \cdot 10^{-1}$

which, with $\gamma = 5/3$, leads to the radial profiles shown in Table X.ii. The final column of Table X.ii. gives the pressure scale height, a variable which is usually identified with the mixing length. It will play a role later when we have to parameterize the rotationally created eddy coefficients. Our model covers a density range of 3—4 orders of magnitude and a somewhat smaller temperature interval. Because of limitations of computer storage, we had to be content with this. Fig. 10.1 compares the density profile with that of an adiabat.

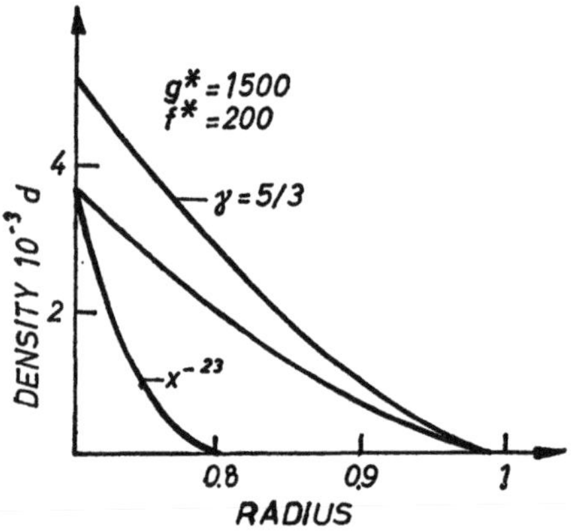

Fig. 10.1. The normalized density profile of the zeroth-order model compared with the adiabatic stratification, $d = [1 + (\gamma - 1)\, g^*(x - 1)]^{[1/\gamma - 1]}$. The potential temperature, i.e. the difference $T - T^{ad}$, is even smaller. A profile x^{-23} is also shown.

It is a simple matter to estimate the parameters linked with rotation. First, there is the Taylor number

$$Ta = \Omega_\odot^2 R^4 / \nu_T^2 , \tag{18}$$

which measures the solar rotation rate, $\Omega_\odot = 2.6 \cdot 10^{-6}\,\mathrm{s}^{-1}$. To evaluate (18), we need also the eddy viscosity, or equivalently the turbulent Prandtl number. Belvedere and Paternó postulated $\nu_T = 2 \cdot 10^{12}\,\mathrm{cm}^2/\mathrm{s}$ so that their Taylor number is approximately $Ta = 4 \cdot 10^7$. Staude's model yields an eddy viscosity only via the Prandtl number, which we usually take to be (Eschrich and Rüdiger, 1983).

$$Pr = 0.33 : \tag{19}$$

The corresponding Taylor number is $Ta = 10^6$.

As is clear from its definition, the Taylor number links the Coriolis force to the Reynolds equation, whereas the pressure is coupled to the Reynolds equation through another quantity which essentially defines a normalized sound velocity. When defined by

$$T^* = (\gamma \mathcal{R} R^2/\mu \nu_T^2)\, T_\odot, \tag{20}$$

this quantity can be interpreted as the square of a Reynolds number based on the speed of sound or, more simply, as a normalized surface temperature. Substituting the numerical values given by Belvedere and Paternó, we obtain the relatively large value of $T^* = 2.5 \cdot 10^9$, while the Staude model, implies only $T^* = 7 \cdot 10^7$ Both these models of the SCZ give the same value, 0.015, for the (square of the) Mach number of the rotation:

$$M = Ta/T^*. \tag{21}$$

10.2. Rotation Equations

Let us proceed to those parts of the equations that owe their existence to rotation. These may be obtained by subtracting the undisturbed state studied above from the complete system. As usual all physical quantities are split into two parts, only the ones with tildes being associated with rotation:

$$\bar{\varrho} = \varrho^{(0)} + \tilde{\varrho}, \quad \bar{T} = T^{(0)} + \tilde{T}, \text{ etc.} \tag{22}$$

The quantities superscripted $^{(0)}$ correspond to those discussed in § 10.1. Although the procedure just outlined can be applied more generally, we shall use it as a perturbation method by supposing that quantities beneath tildes are so small that their squares and products are negligible. This leads to the equations

$$\begin{aligned} \nabla \tilde{p} &= \tilde{\varrho} \boldsymbol{g}^{\mathrm{eff},(0)} + \varrho^{(0)} \tilde{\boldsymbol{g}}^{\mathrm{eff}} - \nabla \cdot (\tilde{\varrho} Q^{(0)}) - \nabla \cdot (\varrho^{(0)} \tilde{Q}) + 2\varrho^{(0)} \bar{\boldsymbol{u}} \times \boldsymbol{\Omega}, \\ \nabla \cdot [\tilde{\varrho}(\chi^{(0)} \cdot \boldsymbol{\beta}^{(0)}) &+ \varrho^{(0)}(\tilde{\chi} \cdot \boldsymbol{\beta}^{(0)}) + \varrho^{(0)}(\chi^{(0)} \cdot \tilde{\boldsymbol{\beta}})] = \varrho^{(0)} \bar{\boldsymbol{u}} \cdot \boldsymbol{\beta}^{(0)}, \end{aligned} \tag{23}$$

and

$$\nabla \cdot (\varrho^{(0)} \bar{\boldsymbol{u}}) = 0, \tag{24}$$

if the flow is steady and the medium is inviscid. Again $\boldsymbol{\beta}$ is the super-adiabatic gradient, i.e. $\boldsymbol{\beta} = \boldsymbol{g}^{\mathrm{eff}} C_p^{-1} - \nabla \bar{T}$, with $\boldsymbol{g}^{\mathrm{eff}} = \boldsymbol{g} + \Omega^2 \boldsymbol{s}$. The vector $\boldsymbol{s}$ is perpendicular to the rotation axis; (A4) gives its components in spherical coordinates.

In the meridional components of the Reynolds equation $(23)_1$ we shall: (i) neglect the term $\varrho Q_{ij}^{(0)}$ and (ii) simplify the correlation tensor Q_{ij} by setting

$$Q_{ij} \to Q_{ij}^d = -\nu_T(\bar{u}_{i,j} + \bar{u}_{j,i}). \tag{25}$$

The first point again concerns the turbulent pressure which we ignored in the zeroth-order equations; the second point was dealt with in Chapter 5.

On adopting these two assumptions, we can derive two equations, similar to (5.43) taken in conjunction with (5.55):

$$\begin{aligned} \partial\tilde{p}/\partial r &= -\tilde{\varrho}g + \varrho^{(0)}r \sin^2\theta\, \Omega^2 + \cdots, \\ \partial\tilde{p}/\partial\theta &= \varrho^{(0)}r^2 \sin\theta \cos\theta\, \Omega^2 + \cdots; \end{aligned} \tag{26}$$

The associated ϕ-component is, of course, identical in content with

$$\nabla \cdot (\varrho^{(0)}r^2 \sin^2\theta\, \Omega\bar{\boldsymbol{u}} + \varrho^{(0)}r \sin\theta\, \overline{u_\phi'\boldsymbol{u}'}) = 0, \tag{27}$$

where the tilde over $\overline{u_\phi'\boldsymbol{u}'}$ has been omitted since it has no zeroth-order part [cf. (6.6)]. For our model we shall again use the representations (6.24) and (6.32).

The heat equation $(23)_2$ can be written down immediately by simply inserting the definition

$$\boldsymbol{\beta}^{(0)} = \begin{pmatrix} -C_p^{-1}g - \mathrm{d}T^{(0)}/\mathrm{d}r \\ 0 \\ 0 \end{pmatrix} = \frac{T_\odot}{R} \begin{pmatrix} \beta \\ 0 \\ 0 \end{pmatrix}, \tag{28}$$

and using (8.98). Expressions for the χ-tensor are provided by (8.73) and (8.74).

As is our wont we shall, as often as permissible, treat the basic rotation rate, Ω, as a constant,

$$\Omega = \Omega_\odot + \tilde{\Omega}. \tag{29}$$

Here $\Omega_\odot$ is the prescribed constant value, and $\tilde{\Omega}$ is the non-uniform part representing the differential rotation. This arises only in the eddy viscosity term in the rotation equation (27) and in the centrifugal force in the Reynolds equation, where we again set

$$\Omega^2 = \Omega_\odot(\Omega_\odot + 2\tilde{\Omega}). \tag{30}$$

The leading order part of (30) sets up an inhomogeneity in the system of equations, as do the turbulence effects $\boldsymbol{V}$, $\boldsymbol{H}$, $\boldsymbol{VV}$ and $\boldsymbol{HV}$. There are therefore five different "sources" of differential rotation to consider, each of which is in principle able to act on its own.

Our approach is, of course, only an approximate one, but conceptually it is ideal since it separates effects that would otherwise be coupled. In an accurate model, the real inhomogeneities would occur, not in the equations, but in the boundary conditions: we might, for instance, define $\Omega_\odot$ to be Ω (1,90°). The mathematical intricacies of such a model are, however, too demanding for us at the moment.

Let us now adopt our usual expansions in latitude:

$$\{\tilde{T}, \tilde{\varrho}, \tilde{p}\} = \{\tilde{T}_n, \tilde{\varrho}_n, \tilde{p}_n\} P_n. \tag{31}$$

If we again apply the ideal gas approximation

$$\tilde{p} = \mu^{-1}\mathcal{R}(\varrho^{(0)}\tilde{T} + \tilde{\varrho}T^{(0)}), \tag{32}$$

we finally arrive at

$$\begin{aligned}\frac{\mathcal{R}}{\mu}(\varrho^{(0)}\tilde{T}_n + \tilde{\varrho}_n T^{(0)})' &= -\tilde{\varrho}_n g + \frac{2}{3}\varrho^{(0)} r \Omega_\odot^2 (\delta_{n0} - \delta_{n2}) \\ &\quad + 2\Omega_\odot \varrho^{(0)} r \left[\frac{(n+1)(n+2)}{2n+3}\Omega_n - \frac{n(n-1)}{2n-1}\Omega_{n-2}\right] \\ &\quad + \nu_T r^{-2} n(n+1)\{A_n'' - (\ln \varrho^{(0)})' A_n' \\ &\quad - [r^{-2}n(n+1) + 2(\ln \varrho^{(0)})''] A_n\},\end{aligned} \tag{33}$$

and

$$\begin{aligned}&-\frac{\mathcal{R}}{\mu}(\varrho^{(0)}\tilde{T}_n + \tilde{\varrho}_n T^{(0)}) \\ &= \frac{1}{3}\varrho^{(0)} r^2 \delta_{n2}\Omega_\odot^2 + 2\Omega_\odot r^2 \varrho^{(0)}\left(\frac{n+2}{2n+3}\Omega_n + \frac{n-1}{2n-1}\Omega_{n-2}\right) \\ &\quad + \nu_T\left\{-A_n''' + (\ln \varrho^{(0)})' A_n'' + \left[\frac{n(n+1)}{r^2} + 2\frac{(\ln \varrho^{(0)})'}{r} + (\ln \varrho^{(0)})''\right] A_n' \right. \\ &\quad \left. - \frac{2n(n+1)}{r^3} A_n\right\}\end{aligned} \tag{34}$$

($n = 2, 4, \ldots$) for the meridional equations. We have already met these equations in Chapter 5, where we also discussed the equation that can be derived by eliminating the pressure p_n between (33) and (34).

When $n = 0$, (33) requires particularly careful handling since it incorporates an exceptional relation for which we have no boundary condition. There is a fundamental arbitrariness in all $n = 0$ equations associated with the temperature, $\tilde{T}_0$, and the rotation, $\tilde{\Omega}_0$. This arbitrariness arises from the definition of surface values. It can only be removed by considering the entire Sun, and not merely its surface layers as is done here. For simplicity, we have set $\tilde{\varrho}_0 = \tilde{T}_0 = \tilde{\Omega}_0 = 0$ at $x = 1$ in our calculations. These are precisely the quantities for which surface values ($\varrho_\odot, T_\odot, \Omega_\odot$) are to be prescribed.

Our rotation equation is obtained by combining (5.6) and (6.30) to yield, as can be easily verified,

$$r^2\tilde{\Omega}_n'' + [4 + (\ln \varrho^{(0)})'r]\, r\tilde{\Omega}_n' - n(n+3)\, \tilde{\Omega}_n$$
$$+ \frac{2\Omega_\odot}{\nu_T \varrho^{(0)}} \left[\frac{(n+2)\,(n+3)}{2n+5} \frac{A_{n+2}}{r} - \frac{n(n+1)}{2n+1} \frac{A_n}{r} \right.$$
$$\left. + \frac{n+3}{2n+5} A'_{n+2} + \frac{n}{2n+1} A_n' \right]$$
$$= \Omega_\odot \left\{ [3 + r(\ln \varrho^{(0)})']\, V^{(l)} + rV^{(l)\prime} - \frac{n(n+3)}{2l} H^{(l)} \right\} \frac{q_{n+1}^{(l)}}{(n+1)\,(n+2)}. \tag{35}$$

The behavior of this equation for either meridional flow or the Λ-effect has been discussed in Chapters 5 and 6. We shall now study the interactions between the diffusion, the mean flow and Λ. In order to separate the different contributions, the computer program had to be constructed in a way that allowed the meridional circulation to be removed.

The energy equation has already been presented in previous Chapters. Here we shall consider, in addition to the density variations, the effects of circulation and centrifugal force. It then becomes

$$r^2\tilde{T}_n'' + [2 + (\ln \varrho^{(0)})'r]\, r\tilde{T}_n' - n(n+1)\, \tilde{T}_n + n(n+1)\, (\beta_r^{(0)}/\varrho^{(0)}\chi_T)\, A_n$$
$$+ (\beta_r^{(0)}/\varrho^{(0)})\, r^2[-\tilde{\varrho}_n' + (\ln \varrho^{(0)})'\tilde{\varrho}_n] = \frac{2}{3} C_p^{-1} r^2 \Omega_\odot^2 \{[3 + r(\ln \varrho^{(0)})']\, \delta_{n0}$$
$$- r(\ln \varrho^{(0)})'\delta_{n2}\} + [-rVV^{(l)\prime} + \frac{1}{2} l^{-1} n(n+1)\, HV^{(l)}]\, Q_n^{(l)} r\beta_r^{(0)}. \tag{36}$$

It is convenient to use normalized relations in the numerical computations. To this end we introduce, in addition to (5), the definitions

$$\tilde{\Omega} = \Omega_\odot \omega, \quad A = \varrho_\odot R \nu_T a. \tag{37}$$

The density stratification may again be described by

$$G = -\mathrm{d}(\ln \varrho^{(0)})/\mathrm{d}(\ln r). \tag{38}$$

Normalization of the circulation yields

$$\bar{u}_\theta|_1 \equiv -R^{-1}\nu_T a_n'(1)\, P_n^1, \tag{39}$$

which (for $n = 2$) gives, in mid-latitudes,

$$\bar{u}_\theta \simeq -a_2'(1) \;\; [\mathrm{m/s}]. \tag{40}$$

Summing up, our "complete model" is defined by the system of equations

$$T^*(d^{(0)}\tilde{t}_n + t^{(0)}\tilde{d}_n)'/\gamma = -g^*T^*\tilde{d}_n + \frac{2}{3}\,d^{(0)}xTa\ (\delta_{n0} - \delta_{n2})$$

$$+ 2d^{(0)}x\left[\frac{(n+1)(n+2)}{2n+3}\,\omega_n - \frac{n(n-1)}{2n-1}\,\omega_{n-2}\right]Ta$$

$$+ \frac{n(n+1)}{x^2}\left\{a_n'' + \frac{Ga_n'}{x} - [n(n+1) + 2G - 2G'x]\,\frac{a_n}{x^2}\right\}, \tag{41}$$

$$T^*(d^{(0)}\tilde{t}_n + t^{(0)}\,\tilde{d}_n)/\gamma = -\frac{1}{3}\,x^2d^{(0)}\,\delta_{2n}Ta - 2x^2d^{(0)}\,\{(n+2)\,\omega_n/(2n+3)$$

$$+ (n-1)\,\omega_{n-2}/(2n-1)\}\,Ta + a_n''' + Ga_n''/x$$

$$+ \big(G + xG' - n(n+1)\big)\,a_n'/x^2 + 2n\,(n+1)\,a_n/x^3, \tag{42}$$

$$x^2\omega_n'' + (4-G)\,x\omega_n' - n(n+3)\,\omega_n$$

$$+ \frac{2}{d^{(0)}}\left(\frac{(n+2)(n+3)}{2n+5}\,\frac{a_{n+2}}{x} - \frac{n(n+1)}{2n+1}\,\frac{a_n}{x} + \frac{n+3}{2n+5}\,a_{n+2}' + \frac{n}{2n+1}\,a_n'\right)$$

$$= \big((3-G)\,V^{(l)} + xV^{(l)\prime} - n(n+3)\,H^{(l)}/2l\big)\,q_{n+1}^{(l)}/(n+1)\,(n+2), \tag{43}$$

$$x^2\tilde{t}_n'' + (2-G)\,x\tilde{t}_n' - n(n+1)\,\tilde{t}_n + n(n+1)\,Pr\,\frac{\beta a_n}{d^{(0)}} - \frac{\beta x}{d^{(0)}}\,(x\tilde{d}_n' + G\tilde{d}_n)$$

$$= \frac{2}{3}\,\big((3-G)\,\delta_{n0} + G\delta_{n2}\big)\,x^2(\gamma - 1)\,M$$

$$+ x\,\big(-xVV^{(l)\prime} + n(n+1)\,HV^{(l)}/2l\big)\,Q_n^{(l)}\beta. \tag{44}$$

As has already been mentioned, the quantity (21), i.e. $M = \mu\Omega_\odot{}^2R^2/(\gamma\mathcal{R}T_\odot)$ is the Mach number of the rotation, i.e. the ratio between the linear rotation velocity and the speed of sound. It is clearly smaller than unity. A normalized superadiabatic rate β has also been introduced in (44). The solution of the zeroth-order equation $(4)_2$ yields

$$\beta = f^*/x^2\,d^{(0)}. \tag{45}$$

Most of the boundary conditions for the above system can be formulated quite easily. The only complication concerns the perturbed heat flux, which we assume vanishes over the base of the SCZ while being consistent with the electromagnetic radiation law at the top, i.e.

$$\tilde{F}_r(x_i) = 0, \quad \tilde{F}_r(1) = 4\sigma T_\odot{}^3\tilde{T}(1). \tag{46}$$

From (8.58) we immediately obtain

$$\tilde{F}_r = \varrho_\odot C_p\left(\chi^{(0)} \cdot \beta^{(0)}\tilde{d} + \tilde{\chi} \cdot \beta^{(0)}d^{(0)} + \chi^{(0)} \cdot \tilde{\beta}\, d^{(0)}\right) \mathring{g}\,, \tag{47}$$

or, with (28) and (45),

$$\tilde{F}_r = \frac{\varrho_\odot C_p \chi_T T_\odot}{R}\left(\frac{\tilde{d}f^*}{d^{(0)}x^2} + \frac{\tilde{\chi}_{rr}f^*}{\chi_T x^2} + \frac{Rd^{(0)}}{T_\odot}\tilde{\beta}_r\right). \tag{48}$$

If one substitutes $\chi_{rr} = -\chi_T V V^{(l)} Q_n^{(l)} P_n$ from (8.74) and uses (8.98), one finds

$$\tilde{F}_r = \frac{\varrho_\odot C_p \chi_T T_\odot}{R} \times\left\{\frac{f^*\tilde{d}_n}{d^{(0)}x^2} - \frac{f^* V V^{(l)} Q_n^{(l)}}{x^2} + d^{(0)}\left[\frac{2}{3}\,x(\gamma - 1)\,M(\delta_{n0} - \delta_{n2}) - \tilde{t}_n'\right]\right\} P_n\,, \tag{49}$$

an expression which, once again, includes the Mach number M. From (49) we obtain the two conditions: at $x = 1$

$$f^*\tilde{d}_n - 4f^*\tilde{t}_n - \tilde{t}_n' = f^* V V^{(l)} Q_n^{(l)} + \frac{2}{3}(\gamma - 1)\,M\delta_{n2}\,, \quad n > 0; \tag{50}$$

and for $x = x_i$

$$\frac{f^*\tilde{d}_n}{d_i x_i^2} - d_i\tilde{t}_n' = \frac{f^*}{x_i^2} V V^{(l)} Q_n^{(l)} - \frac{2}{3}\,x_i(\gamma - 1)\,M(\delta_{n0} - \delta_{n2})\,d_i\,, n \geqq 0. \tag{51}$$

The missing condition for $x = 1$ and $n = 0$ is simply $\tilde{t}_0(1) = 0$ if $f^* \neq 0$.

The remaining boundary conditions are quite standard, being

$$\bar{u}_r = Q_{r\theta} = Q_{r\phi} = 0 \quad \text{at} \quad x = x_i, 1\,, \tag{52}$$

and they are chosen to ensure that the surfaces are stress-free. Consequently, we have at $x = x_i$ and $x = 1$,

$$a_n = x^2 a_n'' + (G - 2)\,x a_n' = x\omega_n' - V^{(l)} q_{n+1}^{(l)}/(n + 1)\,(n + 2) = 0. \tag{53}$$

Using these conditions, we can for $n = 0$ immediately carry out the first of the two integrations necessary to determine each of ω_0 and $\tilde{t}_0$.

These relations, together with the boundary conditions $\omega_0(1) = \tilde{t}_0(1) = 0$, will be solved numerically in the next Chapter.

Chapter 11. Differential Rotation, Meridional Circulation and Pole-equator Temperature Difference: Results.

Mathematically speaking, the over-all rotation problem which we have formulated may be thought of as the search for a solution to an inhomogeneous algebraic system of equations. Its scope is wider than that of merely finding the rotation profile alone; rather, it is the study of the rotationally created latitude-dependence of an entire set of physical quantities such as angular velocity, meridional flow, temperature and density. By its inhomogeneity, the system ensures that a nontrivial solution exists having, in particular, a latitudinally varying rotation profile. It cannot, of course, predict the input parameters of the theory, such as the basic angular velocity; this must be prescribed, e.g. by choosing Ω to be $\Omega(x_0)$ at the latitude of some particular point, x_0 on the solar surface. Another quantity that has to be supplied is the heat flux which enters the convection zone from below.

In our formulation, we have assumed the rotation to be nearly rigid, so that the basic rotation is represented by the constant leading term $\Omega_{\odot}$, of a series and not by the angular velocity at one particular point. Five quantities, $V^{(0)}$, $V^{(1)}$, $H^{(1)}$, $VV^{(1)}$, and $HV^{(1)}$, have been introduced to represent the most essential properties of the turbulence, and these are basic to the latitudinal profiles we wish to explain. They all owe their existence to the effects of the basic rotation on an anisotropic (but spherically symmetric) turbulence, and they vanish as $\Omega \to 0$. As a first approximation we are justified when computing the Λ-effect in taking the angular velocity to be constant (and equal to $\Omega_{\odot}$). As a result, the corresponding terms that represent these turbulent effects enter the basic equations as inhomogeneities on which the results depend linearly.

By "results" we refer to the $n = 2$ components of the surface values of the angular velocity, temperature, density and other variables. The coefficients in the next order equation ($n = 4$) will often be ignored as we can only apply the coefficients for $l = 1$, i.e. $V^{(1)}$, $H^{(1)}$, $VV^{(1)}$ and $HV^{(1)}$. This is simply a matter of consistency.

The results may be compressed into an output "vector",

$$\mathcal{O} = (\omega'_{\mathrm{eq}}, \omega_2, a_2', t_2, d_2), \tag{1}$$

evaluated on the surface. There, because of the boundary conditions, we have

$$\omega'_{eq} \equiv \omega_0' - 1.5\,\omega_2'|_1 = V^{(0)}(1) + V^{(1)}(1). \tag{2}$$

All the quantities in (1) are observable in principle, although measurements of the "differential" density are still lacking. The meridional flow at the surface is represented by a_2' as in (10.40). Negative $\omega_2(1)$ corresponds to equatorial acceleration.

The input parameters of the turbulence that generates the output values, $\mathcal{O}$, will be representated by another "vector",

$$\mathcal{T} = \left(V^{(0)}(x_i),\, V^{(1)}(x_i),\, H^{(1)}(x_i),\, VV^{(1)}(x_i),\, HV^{(1)}(x_i)\right), \tag{3}$$

which is related to $\mathcal{O}$ through a matrix $\mathcal{A}$ by

$$\mathcal{O} = \mathcal{M} + \mathcal{A}\mathcal{T}, \tag{4}$$

here

$$\mathcal{A} = \begin{pmatrix} \partial\omega'_{eq}(1)/\partial V^{(0)}(x_i) & \dots & \partial\omega'_{eq}(1)/\partial HV^{(1)}(x_i) \\ \vdots & & \vdots \\ \partial d_2(1)/\partial V^{(0)}(x_i) & \dots & \partial d_2(1)/\partial HV^{(1)}(x_i) \end{pmatrix}.$$

In (4), we have allowed for the fact that a nontrivial output arises even when $\mathcal{T} = 0$. The "vector" $\mathcal{M}$ recognizes the effects of the centrifugal force of the basic rotation, assumed rigid. Its main ingredient is a meridional flow, a part of the baroclinic flow, which excites a differential rotation, a temperature variation, and so forth. All aspects of this non-turbulent effect vary with the Mach number and vanish as $M \to 0$. It is for this reason that we denote the effect by $\mathcal{M}$ in (4).

$\mathcal{A}$ and $\mathcal{M}$ are totally numerical products of a computer program. Provided the matrix $\mathcal{A}$ is of rank 5, it can be inverted to determine the turbulence parameters that will produce the observed $\mathcal{O}$. One can use the $\mathcal{T}$ finally obtained to calculate some sixth quantity, the value of which can then be considered to be a prediction of the theory. We shall present here results deriv ed from evaluations of $\mathcal{A}$ for several different models, in which the turbulence parameters comprising $\mathcal{T}$ are either independent of depth or, for the higher order terms, show an outward decrease proportional to $x^{-\lambda}$:

$$\begin{Bmatrix} V^{(1)} \\ H^{(1)} \\ VV^{(1)} \\ HV^{(1)} \end{Bmatrix} = \begin{Bmatrix} V_i^{(1)} \\ H_i^{(1)} \\ VV_i^{(1)} \\ HV_i^{(1)} \end{Bmatrix} (x_i/x)^{\lambda}. \tag{5}$$

The closing section of this chapter deals with SCZ models whose outermost layers are independent of $V^{(1)}, \dots, HV^{(1)}$. While $V^{(0)}$ is independent of the depth, the higher order terms vanish in the interval $x_j \leqq x \leqq 1$, where $x_j > x_i$. Section 11.1 deals only with turbulence effects that are depth-independent, i.e. $\lambda = 0$.

11.1. A Simple Model

We shall define our 'standard model' by the set of thermodynamic quantities

$$f^* = 200, \quad g^* = 1500, \quad Pr = 0.33, \quad T^* = 5 \cdot 10^7. \tag{6}$$

The associated rotation rate may be represented by the Taylor number

$$Ta = 5 \cdot 10^5, \tag{7}$$

which is neither very large nor very small. While a more slowly rotating model will also be considered later, more rapidly rotating models failed because of a "fast-rotation-instability" to which the program was prone (cf. § 11.4). The corresponding Mach number of the basic rotation is 0.01.

Initially the program computed the non-turbulent effects, i.e. $\mathcal{M}$:

$$\mathcal{M} = (0, 1.06 \cdot 10^{-3}\,(\mathrm{II}), 1.48 \cdot 10^{1}\,(\mathrm{II}), -1.29 \cdot 10^{-3}, -2.80 \cdot 10^{-3}). \tag{8}$$

The roman numericals here indicate the number of nodes the functions possesses within the region $(x_i, 1)$, e.g. in the case of a', the number of cells of the meridional flow on any radius. We ascertained that, in this case of no turbulence, all components of $\mathcal{O}$ are rather small. In particular the rotation profile generated by the nonturbulent part of baroclinic flow is very flat, although surface flow velocities attain values of more than 10 m/s. The qualitative picture, as exhibited by the computer, is:

(i) a very slight super-rotation;
(ii) an equatorial deceleration at the surface, and an acceleration in the bulk of the SCZ;
(iii) a poleward flow faster than 10 m/s at both the top and the bottom of the SCZ;
(iv) in general, an equatorial plane that is warmer and denser than average throughout the entire SCZ, both effects amounting to less than 2%.

Turning now to the effects of turbulence, we first recall the result that, if the meridional flow is ignored, then $V^{(1)}$ and/or $H^{(1)}$ must maintain the rotation profile unaided, i.e.

$$\omega_2(1) = -(2.9\,V^{(1)} + 6.7 H^{(1)}) \cdot 10^{-2}. \tag{9}$$

This is the result of a computer calculation undertaken for comparison purposes, but it is also consistent with the picture developed in Chapter 6. Positive $V^{(1)}$ and $H^{(1)}$ generate equatorial acceleration.

The results of our complete model may be written as a matrix:

$$\mathcal{A} = \begin{pmatrix} 1 & 1 & 0 & 0 & 0 \\ 3.13 \cdot 10^{-3} & \boxed{-2.54 \cdot 10^{-2}\,(\mathrm{I}) \quad -8.26 \cdot 10^{-2}} & & -7.82 \cdot 10^{-1} & 6.79 \cdot 10^{-1} \\ -4.28\,(\mathrm{II}) & 5.35\,(\mathrm{IV}) & 2.43\,(\mathrm{II}) & 1.38 \cdot 10^{3}\,(\mathrm{II}) & 2.09 \cdot 10^{2}\,(\mathrm{III}) \\ 6.60 \cdot 10^{-5}\,(\mathrm{II}) & -3.25 \cdot 10^{-5}\,(\mathrm{I}) & 2.76 \cdot 10^{-4} & \boxed{8.03 \cdot 10^{-2}\,(\mathrm{I})} & -4.29 \cdot 10^{-3} \\ 1.43 \cdot 10^{-4} & 5.59 \cdot 10^{-5} & 5.93 \cdot 10^{-4}\,(\mathrm{I}) & -5.21 \cdot 10^{-3}\,(\mathrm{II}) & -9.65 \cdot 10^{-3}\,(\mathrm{II}) \end{pmatrix}. \qquad (10)$$

Each column represents the response of one of the turbulence parameters $V^{(0)}, \ldots, HV^{(1)}$. Note that the values of $\omega_2(1)$ resulting from $V^{(1)}$ and $H^{(1)}$ [which are surrounded in the second line of (10) by a rectangular box] are close to the values (9) for a pure Reynolds stress. We find the meridional flow plays only a minor role here, and it is always poleward for the higher-order modes $V^{(1)}, \ldots, HV^{(1)}$. Strikingly enough, the number of cells in the meridional flow varies from II to IV. Positive $V^{(1)}$, $H^{(1)}$ and $VV^{(1)}$, as well as negative $V^{(0)}$ and $HV^{(1)}$, generate equatorial acceleration. As the first column is of particular interest since it is related to the Biermann-Kippenhahn approach. Positive values of $V^{(0)}$ are needed in order to describe Kippenhahn's model of anisotropic viscosity and dominating horizontal "Austausch". For positive $V^{(0)}$ we also find super-rotation, equatorward surface flow and, surprisingly, equatorial deceleration. Again, this can be attributed to the presence of two cells. The lower cell streams equatorward at the bottom of the convection zone and, if this motion is strong enough, there results, quite understandably, an equatorial deceleration.

The latitudinal profiles of temperature and density remain flat in this model. The expected combination of equatorward surface flow [$a_2'(1) < 0$] and equatorial acceleration [$\omega_2(1) < 0$] has been obtained only from the calculations for (negative) $HV^{(1)}$, but this was a case where III cells arose. For the other cases, with an even number of cells (II or IV), we found the opposite combination: poleward flow and equatorial acceleration.

A noticeable pole-equator difference in the temperature and/or density is only produced by the $VV^{(1)}$-effect [see the entry boxed on the third row of

(10)]. This agrees with the result of a numerical study by Arlt and Rüdiger (1984) who demonstrated that a (small) "differential" temperature could in principle be produced by $VV^{(1)}$ in a strongly stratified media. Only the outward decrease in $VV^{(1)}$, resulting from the outward shortening of the lifetimes of the convective elements, reduced the pole-equator difference to unobservably small values.

To allow a more detailed discussion of the results (8) and (10), we will consider them in parallel with the observations, though we cannot do this for the "differential density", which has so far not been measured. We do, however, know one fact about the horizontal Reynolds stress $Q_{\theta\phi}$, namely that the coefficient $w^{(1)}$ is of order unity. In accordance with (2.20), (6.23) and with (3.17), we have, in general,

$$w^{(1)}(1) = 15\,\omega_2(1) + H_i^{(1)}x_i^{\lambda}. \tag{11}$$

Clearly, this result for $w^{(1)}$ plays an independent role only when $H^{(1)}$ does not vanish at the surface. Otherwise $w^{(1)}$ would always duplicate the rotation law, $\omega_2(1)$.

Let the new "observable vector" $\mathcal{O}'$ be

$$\mathcal{O}' = \begin{pmatrix} \omega'_{\rm eq} \\ \omega_2 \\ a_2' \\ t_2 \\ w^{(1)} \end{pmatrix} \doteq \begin{pmatrix} -1 \\ -0.04 \\ 0 \\ 0 \\ 1 \end{pmatrix}, \tag{12}$$

where the postulate $\bar{u}_\theta(1) = 0$ has been made because of the uncertainty in the observational results. Flows, $\bar{u}_\theta(1)$, of ± 10 m/s hardly change the results, as can be easily ascertained from (8) and (10). The matrix $\mathcal{A}'$ corresponding to $\mathcal{O}'$ is the same as $\mathcal{A}$ apart from the fifth line which is replaced by the $w^{(1)}$ defined in (11), i.e.

$$(-4.87 \cdot 10^{-2},\ -0.38,\ -0.24,\ -1.17 \cdot 10^{1},\ 1.02 \cdot 10^{1}). \tag{13}$$

Correspondingly, the fifth component of $\mathcal{M}'$ is $1.56 \cdot 10^{-2}$, and its other components are as in (8). The predominantly negative signs of these new results demonstrate that it may be no easy task to produce the positive correlation sought.

The turbulence parameters $\mathcal{T}'$ that reproduce the observations is the solution of the relation $\mathcal{O}' = \mathcal{M} + \mathcal{A}'\mathcal{T}'$, and this turns out to be

$$\mathcal{T}' = (2.38,\ -3.38,\ 1.6,\ 7.07 \cdot 10^{-3},\ 4.92 \cdot 10^{-3}). \tag{14}$$

As in Chapter 6, $H^{(1)}$ is positive and $V^{(1)}$ negative. Also $V^{(0)}$ is positive. The values of the V's imply the desired sub-rotation of the surface layers, their order of magnitude being 1. The values of the eddy-heat flux quantities are, however, striking. Their smallness reflects the fact that they need only be small to evoke a relatively large response, as can be seen in the fourth and fifth columns of $\mathcal{A}$.

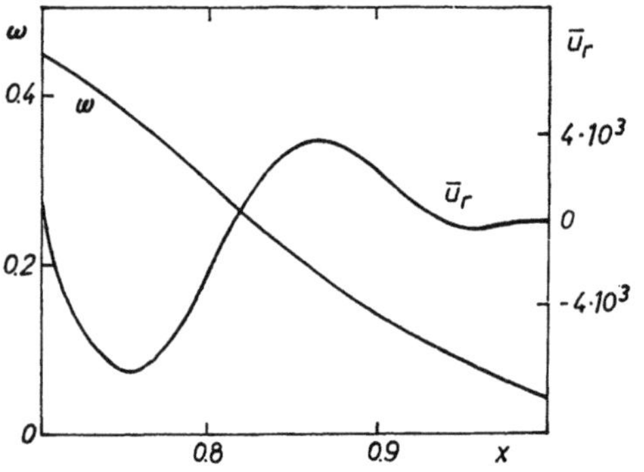

Fig. 11.1. The depth-dependence of the normalized angular velocity and the radial drift beneath the equator for uniform Λ, i.e. $\lambda = 0$. The SCZ is subrotating and has three cells of meridional flow. Turbulence coefficients are as specified in (14).

We also re-computed the model with the turbulence specified by (14). Of course, the required observation vector emerged. In Fig. 11.1 we give, in addition, the radial profiles of the angular velocity and radial drift, $\bar{u}_r$ at the equator

$$\omega = \omega_0 - 1.5\,\omega_2 + 1.875\,\omega_4, \qquad x^2 d^{(0)}\bar{u}_r \simeq -3\,a_2 + 7.5\,a_4. \tag{15}$$

For the pole-equator density difference, produced by the turbulence (14), we found that $d_2 = -1.4 \cdot 10^{-3}$. Hence, the equator should be denser than the poles but only by about 0.2%. This prediction follows from our simple standard model having uniform turbulence quantities $V^{(0)}, \ldots, HV^{(1)}$.

We have also considered a more slowly rotating convection zone. In order to isolate the Taylor number dependence, Ta was reduced by a factor 10: $Ta = 5 \cdot 10^4$. The resulting $\mathcal{M}$ is

$$\mathcal{M} = (0,\ 2.76 \cdot 10^{-4},\ -2.60 \cdot 10^{-1},\ -9.32 \cdot 10^{-5},\ -2.09 \cdot 10^{-4}). \tag{16}$$

Comparing this with (8) we see that the values (16) are smaller but that not all are smaller by a factor of 10. The surface flow even changes its direction. Obviously, the system is more complex than we thought. This is confirmed by the solutions produced by the individual $V^{(0)}, V^{(1)}, \ldots, HV^{(1)}$ and which together yield the matrix

$$\mathcal{A} = \begin{pmatrix} 1 & 1 & 0 & 0 & 0 \\ -2.91\cdot10^{-4} & \boxed{-2.91\cdot10^{-2} \quad -6.79\cdot10^{-2}} & & -4.84\cdot10^{-1} & 6.12\cdot10^{-1} \\ 2.09\cdot10^{-1} & 2.19\cdot10^{-1} & 4.38\cdot10^{-1} & 6.61\cdot10^{+1} & -2.59\cdot10^{+1} \\ 7.13\cdot10^{-5} & 7.46\cdot10^{-5} & 1.90\cdot10^{-5} & \boxed{1.04\cdot10^{-1}} & 3.01\cdot10^{-3} \\ -2.51\cdot10^{-4} & -1.62\cdot10^{-4} & 3.28\cdot10^{-5} & 4.30\cdot10^{-2} & 5.15\cdot10^{-3} \end{pmatrix}. \tag{17}$$

Here, and in all the other such matrices, we have marked important components, such as $\mathcal{A}_{44} = \partial t_2(1)/\partial[VV_i(1)]$ in the present case, by enclosing them in boxes. Note that some of the entries are even larger than in (10). The interaction between rotation and circulation is rather complicated. In particular, the differential rotation at the surface, $\omega_2(1)$, is scarcely changed by reducing the angular velocity. The turbulence "vector", $\mathcal{T}'$, which generates the observations $\mathcal{O}'$, is very similar to (14):

$$\mathcal{T}' = (0.6,\ -1.6,\ 1.6,\ 2.6\cdot10^{-4},\ 3.0\cdot10^{-2}). \tag{18}$$

11.2. Moderately Inhomogeneous Models

In the following work we set $\lambda = 4$, 8 and 15, so allowing the higher-order turbulence quantities to decrease moderately in the outward direction. In particular, we consider different radial distributions for the quantities $V^{(1)}, \ldots, HV^{(1)}$. Table XI.i. summarizes the models numerically. Surface values decrease by a factor of about 5 as one moves from one value of λ to the next. The final column is proportional to the fourth power of the mixing length (see § 11.3). Profiles with large λ tend to drive the effects of turbulence down to the base of the SCZ. For comparison purposes, we again provide rotation laws for the case when the meridional flow is artificially ignored:

Table XI.i. The depth-dependence $x^{-\lambda}$ for various values of λ

$\lambda \rightarrow$ / $x \downarrow$	0	4	8	15	"L^4"
0.7	1	1	1	1	1
0.8	1	0.59	0.34	0.13	0.203
0.9	1	0.37	0.13	0.023	0.013
1	1	0.24	0.058	0.0047	$1.2\cdot10^{-10}$

$$\omega_2(1) = -\begin{Bmatrix}1.2\\0.59\\0.22\end{Bmatrix} 10^{-2} V_i^{(1)} - \begin{Bmatrix}4.0\\2.7\\1.6\end{Bmatrix} 10^{-2} H_i^{(1)} \quad \text{for} \quad \lambda = \begin{Bmatrix}4\\8\\15\end{Bmatrix}. \tag{19}$$

Clearly the efficiency of the Λ-effect is reduced, for increasing λ though not by very much. This means that the differential rotation, which is generated in the volume of the convection zone, is not strongly screened by the outer, more passive layers.

The matrices for two of the values of λ are:

for $\lambda = 4$

$$\mathcal{A} = \begin{pmatrix} 1 & 2.40 \cdot 10^{-1} & 0 & 0 & 0 \\ 3.13 \cdot 10^{-3} & \boxed{-5.96 \cdot 10^{-3}} & \boxed{-5.06 \cdot 10^{-2}} & -2.87 \cdot 10^{-1} & 4.09 \cdot 10^{-1} \\ -4.28 & 9.63 & 9.83 \cdot 10^{-1} & 6.00 \cdot 10^{2} & -5.32 \cdot 10^{1} \\ 6.60 \cdot 10^{-5} & -1.90 \cdot 10^{-4} & 1.80 \cdot 10^{-4} & \boxed{1.53 \cdot 10^{-2}} & 1.03 \cdot 10^{-4} \\ 1.43 \cdot 10^{-4} & -3.90 \cdot 10^{-4} & 3.85 \cdot 10^{-4} & -9.72 \cdot 10^{-3} & -6.14 \cdot 10^{-5} \end{pmatrix}, \tag{20}$$

or $\lambda = 8$

$$\mathcal{A} = \begin{pmatrix} 1 & 5.76 \cdot 10^{-2} & 0 & 0 & 0 \\ 3.13 \cdot 10^{-3} & \boxed{9.36 \cdot 10^{-5}} & \boxed{-3.43 \cdot 10^{-2}} & -1.14 \cdot 10^{-1} & 2.74 \cdot 10^{-1} \\ -4.28 & 1.19 \cdot 10^{1} & -2.87 \cdot 10^{-1} & 2.38 \cdot 10^{2} & -9.88 \cdot 10^{1} \\ 6.60 \cdot 10^{-5} & -2.56 \cdot 10^{-4} & 1.41 \cdot 10^{-4} & \boxed{2.27 \cdot 10^{-3}} & 1.05 \cdot 10^{-3} \\ 1.43 \cdot 10^{-4} & -5.27 \cdot 10^{-4} & 3.01 \cdot 10^{-4} & -5.22 \cdot 10^{-3} & 2.04 \cdot 10^{-3} \end{pmatrix}. \tag{21}$$

Salient differences between (20) and (21) and the matrix (10) should be noted:

(i) as λ increases, the term $\mathcal{A}_{22}$ becomes positive but its value remains small;
(ii) $\mathcal{A}_{34}$ decreases with increasing λ. Obviously the reduced $VV^{(1)}$ generates only a slow surface flow;
(iii) $\mathcal{A}_{35}$ changes sign between $\lambda = 0$ and $\lambda = 4$. The flow for constant $HV^{(1)}$ is directed towards the pole, in contrast to that for non-zero λ;
(iv) increasing λ reduces the value of $\mathcal{A}_{44}$. The profile of the surface temperature becomes increasingly flat [see also Arlt and Rüdiger, 1984].

Note also the smallness of $\mathcal{A}_{12}$ in (20) and (21) compared with its value in the matrix (10), demonstrating that the radial gradient of Ω is now mainly due to $V^{(0)}$.

In interpretating these results we follow the above scenario. As usual, we normalize with $w^{(1)}(1) = 1$, and we then obtain the values for the various turbulence parameters shown in Table XI.ii. This table exhibits four main features of particular interest:

(i) except for $V_i^{(0)}$, all entries increase with increasing λ;
(ii) $H_i^{(1)}$, $VV_i^{(1)}$ and $HV_i^{(1)}$ are always positive, the product $V_i^{(0)}V_i^{(1)}$ always negative;
(iii) between $\lambda = 0$ and $\lambda = 4$ (or, more precisely, for some λ in $1 \leqq \lambda \leqq 2$), $V_i^{(0)}$ and $V_i^{(1)}$ simultaneously change sign; these quantities do not preserve their signs individually, even though their product does so;
(iv) the predicted values of the excess, $d_2(1)$, in surface density remain very small for all λ.

Table XI.ii. The turbulence $\mathcal{T}$ for three cases of outwardly decreasing correlation time

λ	0	4	8
$V_i^{(0)}$	2.4	-1.20	-2.41
$V_i^{(1)}$	-3.4	0.832	24.5
$H_i^{(1)}$	1.6	6.66	27.8
$VV_i^{(1)}$	0.007	0.0126	0.107
$HV_i^{(1)}$	0.005	0.755	3.39
$d_2(1)$	$-1.1 \cdot 10^{-3}$	$-9.2 \cdot 10^{-4}$	$-1.3 \cdot 10^{-3}$

The principal cause of (i) is (11). Were the outward decrease great, the surface value of $H_i^{(1)}$ would be too small to produce the required $w^{(1)}$, unless $H_i^{(1)}$ were rather large. Inspecting (11) we see that

$$H_i^{(1)} = 1.6\, x_i^{-\lambda}, \tag{22}$$

so that $H_i^{(1)}$ increases continuously with increasing λ. The other components of $\mathcal{T}'$ also follow this trend. As the turbulence coefficients $V^{(0)}, \ldots, HV^{(1)}$ are, of course, expected to be of order unity, even at the base of the convection zone, λ cannot be very great. Consequently the correlation time cannot dramatically decrease outward, and the observation that w is non-negative reveals a basic property of the sub-photospheric turbulence. Once again, the inclusion of the $Q_{\theta\phi}$ observations appears to be a demanding test for any theory to meet. If instead, disregarding $Q_{\theta\phi}$, we tentatively satisfy the condition $d_2(1) = 0$, the resulting turbulence $\mathcal{T}$, e.g. for $\lambda = 8$, is

$$\mathcal{T} = (-1.53,\ 9.26,\ 15.63,\ 0.155,\ 1.92), \tag{23}$$

which is "smaller" than the $\mathcal{T}'$ given above for the same value of λ. That there is no effective screening of the rotation profile has already been appreciated in the discussion of (19).

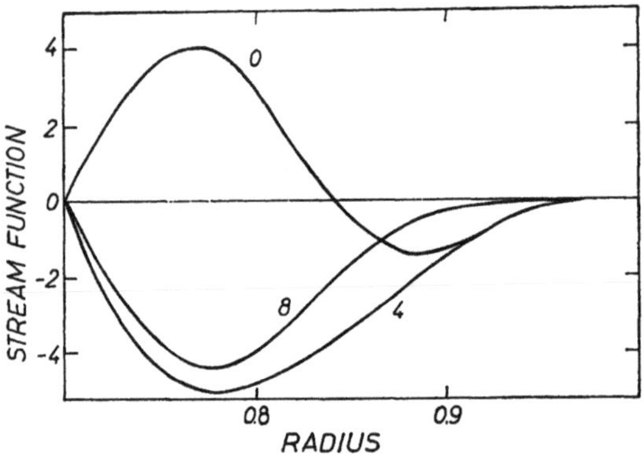

Fig. 11.2. The stream functions a_2 for $\lambda = 0$, $\lambda = 4$ and $\lambda = 8$. The curves for $\lambda \neq 0$ do not possess genuine scales, and no general solutions exists.

The most exciting feature of Table XI.ii is the different signs required for the given V-coefficients for small λ and for large λ. Obviously, we have two kinds of solutions: small-λ type and large-λ type. Their main difference is illustrated in Fig. 11.2, where the stream function, a_2, is given for various values of λ. It appears that along any radius the solution for small λ has two cells, whereas the solutions for large λ have only one. Fig. 11.3, however, tells a different story. It displays the quantity $x^2\, d\bar{u}_r$ on the equatorial plane, as given by $(15)_2$, and we see that, in general, there are three cells. For small and large λ the flow is in opposite directions (cf. Fig. 11.4). Another property concerns the radial rotation law in the outer regions. We have, by definition, sub-rotation in the upper SCZ, but a super-rotating regime develops in the lower SCZ as λ is increased (cf. Fig. 11.5). In particular, the radial rotation law for $\lambda = 8$ has a profile resembling the one reported recently by Duvall *et al.* (1984).

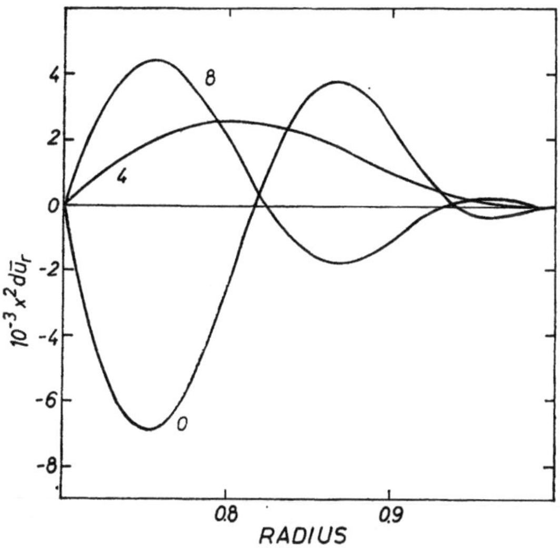

Fig. 11.3. The depth-dependence of $x^2\, d\bar{u}_r$ in the equatorial region for various values of λ. The three cases behave completely differently. For moderate λ there is only one cell, otherwise three cells occur.

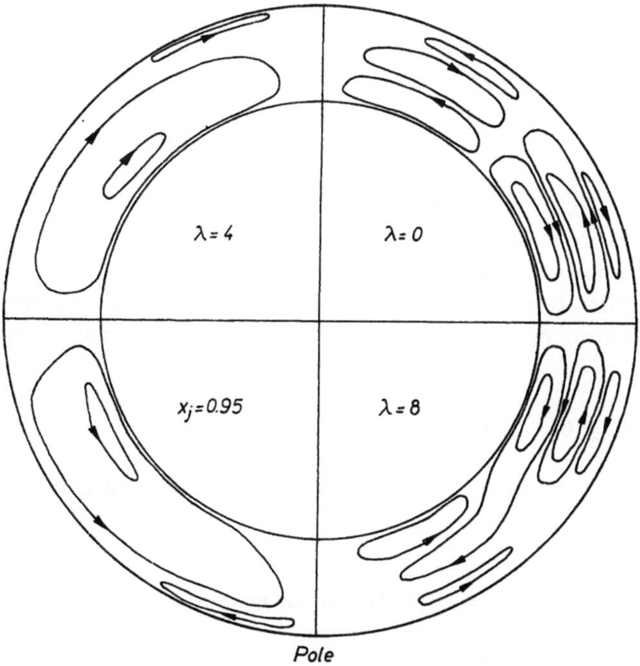

Fig. 11.4. The circulation system in the SCZ for models with small, moderate and large values of λ. The model "$x_j = 0.95$" is dealt with in § 11.3.

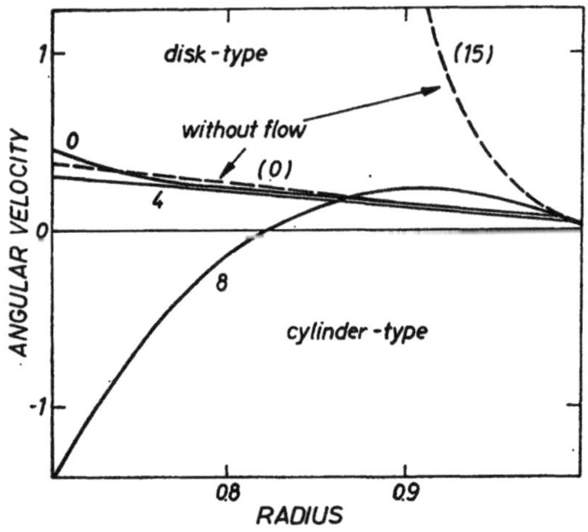

Fig. 11.5. The rotation laws beneath the equator for $\lambda = 0$, $\lambda = 4$ and $\lambda = 8$ (the labels on the curves). The broken lines give the rotation laws for $\lambda = 0$ and $\lambda = 15$ when the meridional flow is artificially neglected. Note in particular that, for $\lambda = 8$, only the outermost layer is sub-rotating. The bulk of the SCZ has a slower rotation rate.

It is no easy task to decide which type of solution is the right one. As long as the radial rotation law has yet to be finally determined, as long as the outward decrease in the higher modes of Λ and χ, i.e. the correlation time, is unknown, the mean-field concept provides no indisputable evidence in favor of any one particular solution. Another observational constraint concerns the depth-dependence of the differential rotation. If we accept the idea that the sunspots are anchored in subsurface layers, then their rotation should reflect the rotation of those layers. As young spots show an equatorial acceleration very similar to that of the photosphere, though with a somewhat stronger maximum, we may tentatively apply the additional observational condition

$$(\partial \omega_2/\partial x)|_1 \gtrsim 0 \tag{24}$$

to the solutions. The corresponding values for our models are given in Table XI.iii.

Table XI.iii. The depth-dependence of $\omega_2'(1)$ for various models

λ	0	1	4	8
$\omega_2'(1)$	0.45	0.36	−0.026	−0.14

Comparison with (24) clearly favors models of modest λ.

We now find ourselves in the following situation:

(i) models in which the turbulent coefficients decrease markedly with radius make predictions that agree with the current "measurements" of the depth-dependence of Ω,

(ii) coefficients that decrease only weakly yield agreement with the better established measurements of the depth-dependence of the equatorial acceleration.

The basic uncertainty concerns the depth-dependence of the correlation times. The arguments of mixing length theory favor a strong outward decrease. The $Q_{\theta\phi}$-measurements, however, tell another story and favor nearly uniform $H^{(1)}$, i.e. nearly constant correlation times. The measured $Q_{\theta\phi}$-profile can only arise if $H^{(1)}$ is quite large. If sunspots really are deep phenomena, the measured $Q_{\theta\phi}$-correlations must belong to those deeper layers, for example to $x_w = 0.95$. We therefore suggest replacing condition (11) by

$$1 = 15\,\omega_2(x_w) + H_i^{(1)} x_i^{\lambda} x_w^{-\lambda}, \tag{25}$$

which, however, has no great effect because $\omega_2(x_w) \approx \omega_2(1)$ unless λ is very large. Since both these seem to be true, the effect of modification (25) on the resulting $H_i^{(1)}$ remains small.

A direct comparison of the numerical results for models with and without meridional circulation is worthwhile. The main question to answer concerns the effect of a meridional flow on the equatorial acceleration generated at the surface. If this flow is ignored, an analysis of (10.43) yields the results

$$\omega_2(1) = -\begin{Bmatrix} 2.9 \\ 0.59 \\ 0.22 \end{Bmatrix} 10^{-2} V_i^{(1)} - \begin{Bmatrix} 6.7 \\ 2.7 \\ 1.6 \end{Bmatrix} 10^{-2} H_i^{(1)} \quad \text{for} \quad \lambda = \begin{Bmatrix} 0 \\ 8, \\ 15 \end{Bmatrix} \tag{26}$$

which have been already given, in part, in (19). The complete system of equations (i.e. with flow) yields

$$\omega_2(1) = -\begin{Bmatrix} 2.53 \cdot 10^{-2} \\ -3.56 \cdot 10^{-4} \\ -2.34 \cdot 10^{-3} \end{Bmatrix} V_i^{(1)} - \begin{Bmatrix} 8.26 \\ 3.43 \\ 2.08 \end{Bmatrix} 10^{-2} H_i^{(1)} \quad \text{for} \quad \lambda = \begin{Bmatrix} 0 \\ 8. \\ 15 \end{Bmatrix} \tag{27}$$

Clearly the coefficients of $H_i^{(1)}$ are hardly influenced at all by the omission of meridional flow. The $H^{(1)}$-effect appears to be able to generate the rotation law in a quite direct manner, rather than via meridional flow, and the value of λ does not play an essential role. Yet this is not true for the vertical TMT: For small λ the differences between the two approaches are still small but not, however, for larger λ. With increasing λ the flux mode $V^{(1)}$ loses its power and the meridional flow creates the differential rotation. Thus, the equatorial acceleration can even develop into a deceleration.

11.3. Models in which $V^{(1)}\ldots HV^{(1)}$ Decrease Strongly

The simplest theory makes use of only the fundamental mode $V^{(0)}$ of the Λ-effect (the "Biermann-Kippenhahn theory"). It ignores the higher modes both in the TMT and in the eddy heat flux. For the standard model one finds the general relations

$$\omega'_{\text{eq}} = V^{(0)}, \quad \omega_2 = (1.1 + 3.1\, V^{(0)})\, 10^{-3}, \qquad a_2' = 15 - 4.3\, V^{(0)}, \tag{28}$$

and, almost trivially, $w^{(1)} = 0.017 + 0.047\, V^{(0)}$. If we use the second relation to determine the value of $V^{(0)}$ corresponding to the observed $\omega_2 = -0.04$, we obtain the negative (!) value

$$V^{(0)} = -13. \tag{29}$$

This result is the opposite of Kippenhahn's finding. Negative $V^{(0)}$ arises from the domination of the vertical "Austausch" over the horizontal "Austausch". The first and the third relations in (28) yield $\omega'_{\text{eq}} = -13$ and $a_2' = 64$. Note that here sub-rotation and poleward flow prevail, rather than super-rotation and equatorward flow that Kippenhahn found in his original paper. Fig. 11.6 gives the flow pattern of the resulting meridional circulation, and it is this pattern which creates the desired equatorial acceleration. However, a surface flow of about 50 m/s, as well as a negative Ward coefficient, $w^{(1)} \approx -0.6$, clearly contradict the observations. We also worked with non-uniform $V^{(0)}$ coefficients, the gradient being taken to be constant. As a first example we set $V_i^{(0)} = 4\, V^{(0)}(1)$ and in a second example the assumption was $V_i^{(0)} = -2 V^{(0)}(1)$. The amplitude of $V^{(0)}$ was chosen so

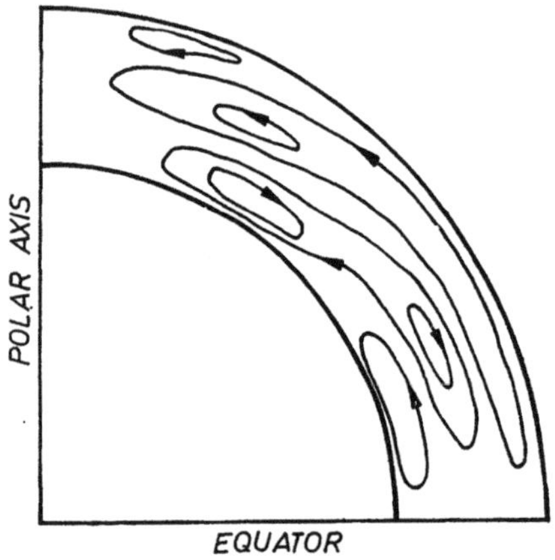

Fig. 11.6. The flow pattern for the Biermann-Kippenhahn theory in which, by a suitable specification of the free parameter $V^{(0)}$, the observed equatorial acceleration is generated by the meridional circulation.

that the hoped-for equatorial acceleration occurred. Then the values of $V_i^{(0)}$ at the base of the SCZ were always approximately -10 and the surface drift $a_2'(1)$ was about 80. Clearly, the latter is also too large to be consistent with the observations.

In short, the one-parameter theory does not work, and other turbulence quantities must be included. On the other hand, we believe the higher modes $V^{(1)}, \ldots, HV^{(1)}$ vanish in the surface layers of the SCZ. They should certainly exhibit a strong outward decrease. The true profiles, however, are unknown. In the last section we used normalized profiles, such as $x^{-\lambda}$, but other choices are also possible. In particular, the modes may, to all intents and purposes, vanish in the surface layers but remain finite at the base of the SCZ. Two different profiles have been used in the following investigation.

The concept behind the first model is similar to that of mixing length theory. The idea is that $V^{(1)}, \ldots, HV^{(1)}$ vary as $(\tau_c \Omega)^2$, where the turnover time is taken to be the correlation time: $\tau_c \approx L/u'$. So we derive

$$(\tau_c \Omega_\odot)^2 \approx L^4 Ta/R^4, \tag{30}$$

and we choose for L the pressure scale height:

$$L/R = -\mathrm{d}x/\mathrm{d}(\ln p) = t/\gamma g^* . \tag{31}$$

The last column of Table X.ii. displays the corresponding values for the zeroth-order SCZ model. With suitable normalization we may write (see Table XI.i)

$$\left(V^{(1)}(x), \ldots, HV^{(1)}(x)\right) = (V_i^{(1)}, \ldots, HV_i^{(1)})\,(t/t_i)^4 . \tag{32}$$

We suggest that the turbulence coefficients decrease outwards extremely strongly (Fig. 11.7). In practical terms, they are pushed, in their entirety, to the base of the SCZ, and they may therefore be expected to influence the

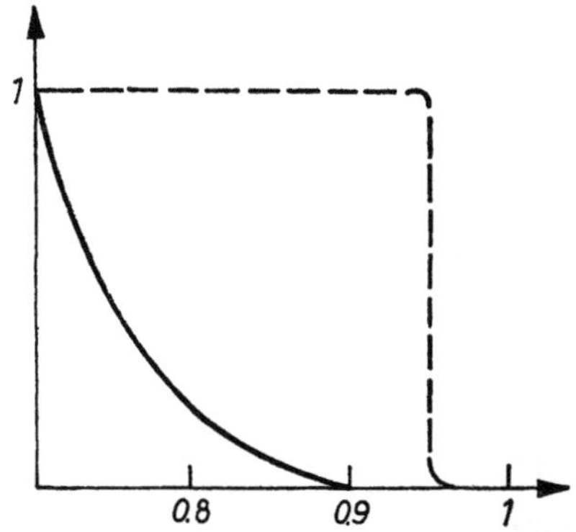

Fig. 11.7. A representation of our two models that include a strong outward decrease of the functions $V^{(1)}, \ldots, HV^{(1)}$. The solid line is for the "mixing-length model", with a mixing length that varies as x^{-15}; the broken line is for the "two-shell model".

results very little. The matrix $\mathcal{A}$ exhibits the surface values:

$$\mathcal{A} = \begin{pmatrix} 1 & 0 & 0 & 0 & 0 \\ 3.13 \cdot 10^{-3} & \boxed{3.35 \cdot 10^{-3}} & \boxed{-2.55 \cdot 10^{-2}} & -2.84 \cdot 10^{-2} & 2.01 \cdot 10^{-1} \\ -4.28 & 1.88 \cdot 10^{1} & -1.30 & 2.06 \cdot 10^{1} & -1.12 \cdot 10^{2} \\ 6.60 \cdot 10^{-5} & -4.43 \cdot 10^{-4} & 1.27 \cdot 10^{-4} & \boxed{-2.12 \cdot 10^{-4}} & 1.42 \cdot 10^{-3} \\ 1.43 \cdot 10^{-4} & -6.80 \cdot 10^{-4} & 2.69 \cdot 10^{-4} & -3.58 \cdot 10^{-4} & 2.91 \cdot 10^{-3} \end{pmatrix}. \tag{33}$$

A comparison with (10) reveals that the values are reduced, though not very greatly. The main result of the calculations is that the decrease in the differential rotation and the surface flow is merely of one order of magnitude. As we know, the coefficient $\mathcal{A}_{22}$ plays an exception role, since its value appears to change from negative to positive values if $V^{(1)}$ decreases markedly towards the surface. One result must again be emphasized: it is not possible in the mixing length concept to explain Ward's observation, $w^{(1)}(1) \approx 1$. This is because we have no $H^{(1)}$ in the surface layer, and the relation (11) therefore degenerates to $w^{(1)}(1) = 15\,\omega_2(1) \approx -0.6$. The observation of non-negative $w^{(1)}$ coefficients seems to exclude mixing length concepts from the Λ-effect!

Our second example is a model having a two-shell structure (see again Fig. 11.7). At $x_j = 0.95$ (say), the $V^{(1)}, \ldots, HV^{(1)}$ decrease sharply to very small values, so that in the outermost layer ($x_j \leqq x \leqq 1$) only the fundamental mode $V^{(0)}$ exists. Again we have an outer passive layer which overlies a shell with "giant granulation". The resulting matrix is

$$\mathcal{A} = \begin{pmatrix} 1 & 0 & 0 & 0 & 0 \\ . & \boxed{-1.74 \cdot 10^{-2}\,(\mathrm{I})} & \boxed{-8.01 \cdot 10^{-2}} & -5.54 \cdot 10^{-1} & 6.47 \cdot 10^{-1} \\ . & -2.78\,(\mathrm{II}) & 3.11\,(\mathrm{II}) & 2.22 \cdot 10^{3}\,(\mathrm{II}) & -1.09 \cdot 10^{2}\,(\mathrm{II}) \\ . & 8.74 \cdot 10^{-5}\,(\mathrm{I}) & 2.54 \cdot 10^{-4} & \boxed{-3.94 \cdot 10^{-2}\,(\mathrm{I})} & -1.32 \cdot 10^{-4} \\ . & 1.84 \cdot 10^{-4} & 5.46 \cdot 10^{-4}\,(\mathrm{I}) & -8.09 \cdot 10^{-2}\,(\mathrm{II}) & -4.41 \cdot 10^{-4}\,(\mathrm{I}) \end{pmatrix}. \tag{34}$$

Note the relatively large value of the $\mathcal{A}_{44}$ component. Of course, the Ward condition $w^{(1)} \approx 1$ cannot be satisfied directly at the surface since $H^{(1)}$ vanishes there. This value can be attained only below x_j, e.g. at $x_w = 0.94$. There we can easily fulfill the condition:

$$H^{(1)}(x_w) + 15\omega_2(x_w) = 1. \tag{35}$$

In other words, we assume that the measured $Q_{\theta\phi}$-correlations are essentially characteristic of the deeper layers associated with sunspots. As a consequence of this idea, we obtain the turbulence "vector"

$$\mathcal{T}' = (-0.980, -32.0, 3.63, -0.0803, -0.537), \tag{36}$$

whose components differ strongly from those obtained previously for the $x^{-\lambda}$-models. In particular, negative signs now dominate. Comparing the values with our earlier results, as shown in Table XI.ii, this feature appears surprising. Only the signs of $V^{(0)}$ and $H^{(1)}$ remain unchanged.

One must at once wonder about the cause of this unexpected phenomenon. Is it a consequence of the remarkable value for $\mathcal{A}_{44}$? The answer is, "Yes". As a test, we artificially changed the value of this coefficient, first to zero and then to the positive value 0.04. All the higher turbulence coefficients $V^{(1)}, \ldots, HV^{(1)}$ then became positive. Obviously, the temperature profile generated by $VV^{(1)}$ plays a sensitive role in our system. If $\mathcal{A}_{44}$ is reduced by any physical mechanism, for example through radiation, then the shell model given here will also work with positive turbulence coefficients $V^{(1)}, \ldots, HV^{(1)}$.

11.4. A New Instability for Fast Rotation?

Surprisingly enough, in certain parameters ranges, the homogeneous part of our system of equations does have a solution. For various combinations of T^* and Ta, the computer indicated that the determinant of the homogeneous system is zero. In other words, an eigensolution exists for which a differential rotation, a meridional circulation, and a "differential" density and temperature, are all generated even without external sources. This is precisely the main property of an instability: the linearized equations show bifurcation, i.e. they possess a solution, other than the trivial one, if a certain parameter exceeds a critical value. It may be worth setting down here, for separate discussion, the equations of the homogeneous problem. All high-order turbulence effects are ignored. The equations for the flow, the density and the temperature are then

$$\begin{aligned}
&\partial \varrho u_i/\partial t + \mu^{-1}\mathcal{R}\, \partial(\varrho T)/\partial x_i = \varrho g_i^{\text{eff}} + \nu_T\, \partial[\varrho(\bar{u}_{i,j} + \bar{u}_{j,i})]/\partial x_j, \\
&\frac{\partial}{\partial t}(\varrho s^2\Omega) + \nabla \cdot [\varrho s^2(\Omega\bar{u} - \nu_T\nabla\Omega)] = 0,
\end{aligned} \tag{37}$$

$$\frac{\partial \varrho}{\partial t} + \nabla \cdot (\varrho \bar{\boldsymbol{u}}) = 0,$$
$$\frac{1}{\gamma} \frac{\partial \varrho T}{\partial t} + \chi_T \nabla \cdot (\varrho \boldsymbol{\beta}) = \varrho \bar{\boldsymbol{u}} \boldsymbol{\beta}. \tag{37}$$

If we divide the equations in the usual way, we obtain an undisturbed part [cf. (10.11) and (10.12)] and a rotational part. The latter may be expanded in a series of Legendre polynomials. For the nonlinear expressions we need two types of Gaunt integrals [cf. Rüdiger *et al.* (1986) for details]:

$$P_n P_m^1 = \sum a_{nmk} P_k^1, \quad \cot \theta P_n^1 P_m^1 = \sum c_{nmk} P_k^1. \tag{38}$$

The numerical determination of a_{nmk} and c_{nmk} is an almost trivial task.

For reasons of completeness we give the entire time-dependent system, the temporal variable being $\tau = t\nu_T/R^2$:

$$\begin{aligned}
n(n+1)\, x^{-2} \dot{a}_n = & -T^* \gamma^{-1} (d^{(0)} \bar{t}_n + t^{(0)} \bar{d}_n)' - T^* g^* \bar{d}_n + n(n+1)\, x^{-2} \\
& \times \{a_n'' + G x^{-1} a_n' - [n(n+1) + 2G - 2xG']\, x^{-2} a_n\} \\
& + x d^{(0)} (2n+1)\, Ta \sum [m(m+1)/(2m+1)] \\
& \times a_{nkm} \omega_{k-1} \omega_{m-1},
\end{aligned}$$

$$\begin{aligned}
\dot{a}_n' = & -T^* \gamma^{-1} (d^{(0)} \bar{t}_n + t^{(0)} \bar{d}_n) + a_n''' + G x^{-1} a_n'' \\
& + [G + xG' - n(n+1)]\, x^{-2} a_n' \\
& + 2n(n+1)\, x^{-3} a_n - Ta\, x^2 d^{(0} \sum c_{kmn} \omega_{k-1} \omega_{m-1},
\end{aligned} \tag{39}$$

$$\begin{aligned}
x^2 \dot{\omega}_n = & \, x^2 \omega_n'' + (4 - G)\, x \omega_n' - n(n+3)\, \omega_n \\
& - (1/d^{(0)}) \sum [(m+1)(m+2)\, a_{(n+1)k(n+1)} a_{m+1} (2x^{-1} \omega_{k-1} + \omega_{k-1}') \\
& - k(k+1)\, a_{k(m+1)(n+1)} a'_{m+1} \omega_{k-1}],
\end{aligned}$$

$$\begin{aligned}
\gamma^{-1} Pr x^2 \dot{\bar{t}}_n = & \, x^2 \bar{t}_n'' + (2 - G)\, x \bar{t}_n' - n(n+1)\, \bar{t}_n \\
& + [n(n+1)\, Pr\, \beta a_n - \beta x (x \bar{d}_n' + G \bar{d}_n)]/d^{(0)} \\
& - (\gamma - 1)\, M x^2 \sum \left[\frac{(2n+1)\, m(m+1)}{2m+1} a_{nkm} \frac{(x^3 d^{(0)} \omega_{k-1} \omega_{m-1})'}{x^2 d^{(0)}} \right. \\
& \left. + n(n+1)\, c_{kmn} \omega_{k-1} \omega_{m-1} \right].
\end{aligned}$$

The system is a time-dependent, nonlinear description of the rotationally generated baroclinic flow, with the turbulence effects $V, \ldots, VH$ not included. Naturally, it also contains the steady linearized version (10.41)—(10.44) considered earlier. A special choice of parameters must be made, which recognizes our linearization of the Coriolis and centrifugal forces. In order to reproduce the linear system we must set $\omega_k = (2 - \delta_{m1})\,\delta_{k0}$ in $(39)_1$ and $(39)_2$, and write $\omega_k = \delta_{k0}$, $\omega_m = \delta_{m0}$ in $(39)_3$ and $(39)_4$. We know that this linearized system [taken in conjunction with the related boundary conditions (10.50)—(10.53)] possesses eigensolutions. Also a depth-dependence of gravity of the form

$$g = g_\odot / x^2 \tag{40}$$

can be included. The essential parameters of the homogeneous system are T^* and Ta. For each Taylor number, a normalized surface temperature T^* exists above which the system becomes unstable. A similar phenomenon related to the effect of changing the Prandtl number has already been described by Schmidt (1982) and Chan *et al.* (1987).

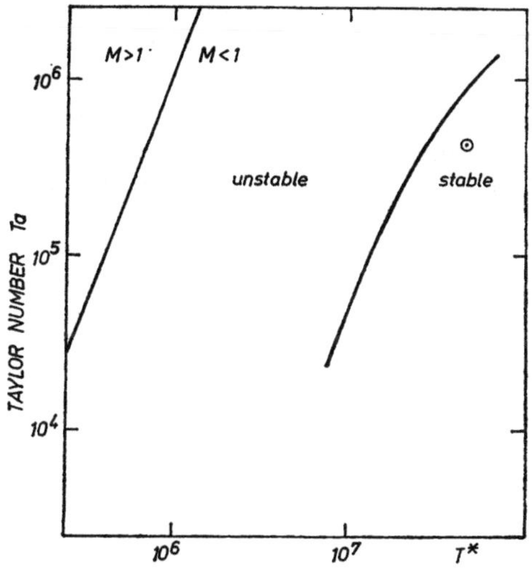

Fig. 11.8. The homogeneous part of the full equation possesses a solution for the case shown here. Thus for a given surface temperature T^* an eigenvalue for the Taylor number exists. The stable region is on the slow rotation side. The Sun (⊙) is probably slow enough to be in the subcritical area. M is the Mach number of the rotation.

We found that our standard model, defined by (6) and (7), lies in a stable region of parameter space (cf. Fig. 11.8). If it had not done so, the solutions of the inhomogeneous system which we presented above would not have made sense. Nevertheless, it seems that eigensolutions can exist even for modest values of the Taylor numbers that decrease if the assumed T^* is reduced. The corresponding Mach numbers thus become smaller and smaller. We are, therefore, led to the conclusion that, for a given rotation rate, cool stars are unstable and hot stars are stable. In other words, for a given surface temperature the slow rotators are stable and the fast rotators are unstable. This is strongly reminiscent of the Vaughan-Preston gap for solar-type stars which was discovered in 1980. Yet we must be careful as our basic approximation can strictly only be applied to slow rotators.

To discover whether this fast-rotation instability also exists for large Ω is a matter for future research to settle. Also the time derivative ought to be restored to the theory to allow for possible oscillations. In addition, we need a detailed analytical investigation of a simplified system in order to clarify the basic physics. A note of caution, however, should be sounded about the dangers of over-emphasizing the importance of the phenomenon. Since, strictly speaking, the disturbance $\tilde{\Omega}$ is no longer small compared with Ω near the onset of the "instability", the linearized equations are no longer pertinent. The apparent self-excitation occurs for a model which cannot be properly justified. A similar situation was discussed in Chapter 7 in connection with "negative eddy viscosity". We expect, of course, physical quantities to be finite everywhere when the non-linearity is fully included. At our present state of knowledge, we fear the instability is a mere artifact, that would disappear were more realistic models considered. Whether the solutions have any physical peculiarity along the eigenvalue branch of Fig. 11.8 is a matter that can only be decided by studying the non-linear equations.

Chapter 12. Lorentz Force and Differential Rotation

12.1. Introduction

In a paper published in 1976, Stix (1976b) discussed the role played by differential rotation in the solar dynamo. He considered $\alpha\Omega$-dynamos with surfaces of isorotation that were cylindrical. Such models must meet two conditions, one for each of the two fundamental observational facts that have to be explained:

(i) $\partial\Omega/\partial r > 0$ (super-rotation), in order to ensure equatorial acceleration;
(ii) $\alpha\ \partial\Omega/\partial r < 0$ in the northern hemisphere, in order to produce the correct butterfly diagram.

These demand that α be negative in the northern hemisphere, even though its positivity was established convincingly in an early paper by Krause (1967). Accepting both conditions (i) and (ii), Stix found that the magnetic field components $\overline{B}_r$ and $\overline{B}_\phi$ evolved practically in phase, in contradiction with the observations. He presented a butterfly diagram with contour lines of radial field, derived from magnetograms measured at the Mt. Wilson Observatory and dating back to 1959 (see Figure 12.1). In reality, $\overline{B}_r$ and $\overline{B}_\phi$ are completely out of phase, i.e. one has a phase lag of about π with respect to the other. The product $\overline{B}_r\overline{B}_\phi$ is mostly negative. This additional observational constraint is easily satisfied by models with sub-rotation ($\partial\Omega/\partial r < 0$) and positive α (in the northern hemisphere). This has been established by the integration of such models by Steenbeck and Krause (1969), Deinzer and Stix (1971) and Roberts (1972). Stix was compelled to reject his own models with cylindrical surfaces of isorotation because of their unrealistic predictions.

Models with sub-rotation have negative $\overline{B}_r\overline{B}_\phi$ near the active region i.e. where the toroidal field is a maximum. This is illustrated, for example, on page 230 of the book by Krause and Rädler (1980). Model 1 of Yoshimura (1981), that is the one with the "open" boundary conditions, also gives negative $\overline{B}_r\overline{B}_\phi$ near the active regions. In contrast Schüssler (1981) proposed a flux tube model having positive $\overline{B}_r\overline{B}_\phi$ both in, and south of, the

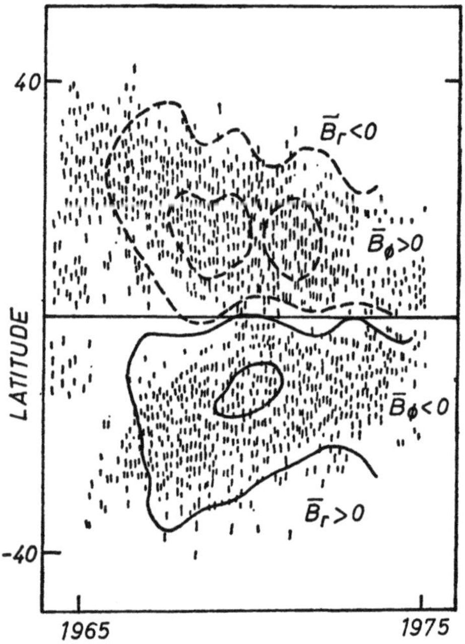

Fig. 12.1. Butterfly diagram and isolines of the radial field (solid lines show positive field; dashed lines, negative). The toroidal field is positive in the northern hemisphere and negative in the southern hemisphere. $\bar{B}_r\bar{B}_\phi$ proves to be negative (after Stix, 1976b).

active region in the northern hemisphere. As we know this also conflicts with direct observation of the solar surface. Yet Schüssler was aiming in quite another direction. He proposed that the product $\bar{B}_r\bar{B}_\phi$ should be regarded as the toroidal component of the Lorentz force, in order to explain the solar torsional oscillation discovered by Howard and La Bonte (1980). They had found, after subtracting the mean differential rotation from the observed angular velocity, a systematic flow structure, Ω_{cyc}, remained:

$$\bar{\Omega} = \tilde{\Omega} + \Omega_{\text{cyc}} = A + B\cos^2\theta + C\cos^4\theta + \Omega_{\text{cyc}}. \qquad (1)$$

The flow pattern $\Omega_{\text{cyc}}(\theta)$ consisted of alternating bands of fast and slow rotation that drifted towards the equator; the cycle was repeated approximately every 22 years. The center of magnetic activity was generally located between a pair of bands for which the more rapidly moving band lies on the equatorward side of the center. In other words, close to an activity belt $\partial\Omega_{\text{cyc}}/\partial\theta > 0$. The linear velocity characteristically had a value of only 5 m/s (see Chapter 3). There is also an important phase relation: the more rapidly rotating zone was born in a polar region during a minimum and arrived at the equator at the next minimum but one.

Contrary to the views of Howard and La Bonte, i.e. the discoverers of this phenomenon, Schüssler (1981), Yoshimura (1981) and the present author all believe that we see in a torsional oscillation the feedback of the cyclic magnetic field on the mean flow. We should bear in mind that electromagnetic induction by differential rotation is the most effective mechanism for generating toroidal fields in the solar dynamo. And we should pay heed to Lenz's rule: an effect tries to weaken its cause.

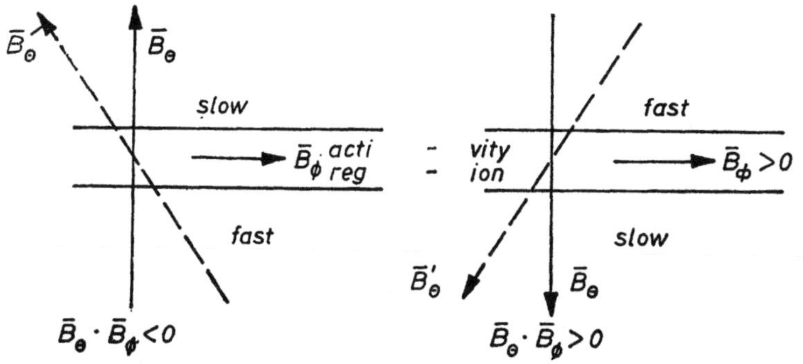

Fig. 12.2. Two configurations that fulfil Lenz's rule. The applied toroidal field is positive. A zonal flow shears the poloidal fields to produce toroidal fields that oppose the original toroidal field. If the *fast* region lies below the activity zone the product $\bar{B}_\theta \bar{B}_\phi$ is negative.

It may be worth examining Fig. 12.2 to elucidate this point further. Suppose that in an active region with $\bar{B}_\phi > 0$, a polewardly directed latitudinal field, $\bar{B}_\theta < 0$, exists. When the Lorentz force generates a rotation with $\partial \Omega_{cyc}/\partial \theta > 0$ (i.e. with a fast zone between the equator and the activity region), the direction in which B_θ is sheared zonally is such as to weaken the original toroidal field. Since $\bar{B}_\theta \bar{B}_\phi$ is indeed negative in the dynamo model of Steenbeck and Krause mentioned earlier, the occurrence of fast zones between activity regions seems to have been explained.

The activity zones are, of course, critical for the dynamo mechanism, since it is within these that the radial differential rotation (for this type of dynamo) produces its toroidal field from the radial component of $\bar{B}$ ($\bar{B}_r \bar{B}_\phi < 0$ for sub-rotation). If, at the same time, a latitudinal differential rotation is present, having an equatorial acceleration resembling the one observed, dynamo excitation is obstructed (cf. Table 10 of Steenbeck and Krause, 1969). Sub-rotation and equatorial acceleration do not act in the same way for, according to Lenz's rule, dynamo regeneration is impeded if the magnetic feedback on the flow increases the latitudinal gradient of angular velocity in the active region. Hence, in the northern hemisphere we expect a slow zone north of the active latitude and a fast zone south of it. But it is precisely in this case that the gradient $\partial \Omega/\partial \theta$ is increased there, and the dynamo is seriously obstructed. It may be seen that the basic property of torsional oscillations is a direct consequence of applying Lenz's rule to simple dynamo models.

The observed torsional oscillations also show that the dynamo principally functions through the depth-dependence, and not the latitude-dependence, of the differential rotation. Were latitude dependence responsible, the torsional oscillations would tend to smooth out the latitudinal rotation profile (near the active region) in obedience to Lenz's rule, i.e. to obstruct the dynamo. In his paper, Schüssler (1981) presented calculations for the azimuthal component of the Lorentz force and its relation to the maximum toroidal field. In this way he obtained the isolines of the Lorentz force, and these indicated that in the northern hemisphere the zonal acceleration dominates in latitudes south of the centre of activity. "Since the flow reacts quasi-simultaneously to changes in the Lorentz force, the isolines can also be taken as velocity contours."

The phase relation between force and flow depends on the ratio between inertial and viscous forces. Inertia favors a phase lag of $\pi/2$ and viscosity invites synchronism. It is the ratio $\sigma^* = \sigma/\nu K^2$ (where σ is the frequency of the driving force) that determines the phase. If σ^* is large, the phase lag is large. What is the characteristic scale, $2\pi/K$, for the problem: is it the radial scale or the horizontal? The latter greatly exceeds the characteristic depth; a fast belt has a linear dimension of about 25° or $3 \cdot 10^{10}$ cm. Hence, with $\nu \approx 5 \cdot 10^{12}$ cm²/s and $\sigma = 8.7 \cdot 10^{-9}$ s^{-1}, we find

$$\sigma^* \approx 1,$$

corresponding to a phase lag of (say) 45°. If instead, following Schüssler, we adopt a radial scale of about 10,000 km, the ratio proves to be very small: $\sigma^* \approx 10^{-3}$.

In order to illustrate the physical configuration, we may construct a simple model. Let us assume that an external force drives a two-dimensional flow, $u(y, z)$, in a channel, $0 \leqq z \leqq Z$. The force is in the x-direction and is independent of x and z (see Fig. 12.3). The equation of motion is

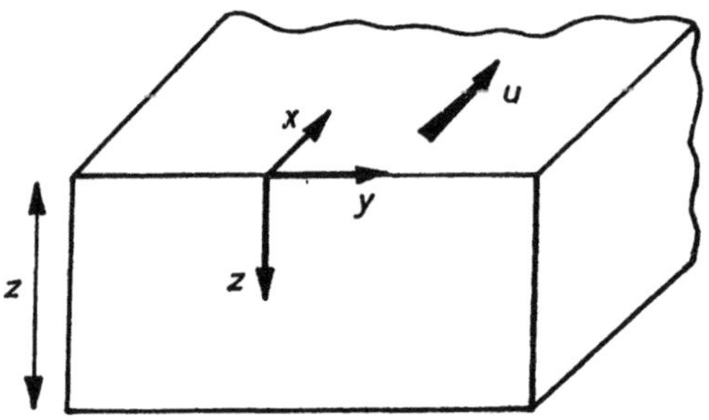

Fig. 12.3. The geometry of the simple channel flow discussed in the text.

simply

$$\frac{\partial u}{\partial t} - \nu \left(\frac{\partial^2 u}{\partial y^2} + \frac{\partial^2 u}{\partial z^2} + \frac{\partial (\ln \varrho)}{\partial z} \frac{\partial u}{\partial z} \right) = \frac{L}{\varrho}, \tag{2}$$

in which we shall set $L = \hat{L} \sin(\sigma t) \cos(k_y y)$, and $\varrho = \hat{\varrho} \exp(k_z z)$. Using the ansatz $u = [V(z) \sin \sigma t + U(z) \cos \sigma t] \cos(k_y y)$, we obtain

$$\begin{aligned} \nu k_z \, \mathrm{d}U/\mathrm{d}z + \nu \, \mathrm{d}^2U/\mathrm{d}z^2 - \nu k_y{}^2 U - \sigma V &= 0, \\ \nu k_z \, \mathrm{d}V/\mathrm{d}z + \nu \, \mathrm{d}^2V/\mathrm{d}z^2 - \nu k_y{}^2 V + \sigma U &= -\hat{\varrho}^{-1} \hat{L} \, \mathrm{e}^{-k_z z}. \end{aligned} \tag{3}$$

$U(0)$ and $V(0)$ together define a phase angle ϕ between the forces and the flow, namely $\phi = \arctan[U(0)/V(0)]$. With the normalization $z = \zeta Z$ and the abbreviations

$$\alpha = k_y Z, \quad \gamma = k_z Z, \quad \sigma^* = \sigma Z^2/\nu, \tag{4}$$

we find

$$\begin{aligned} &U^{IV} + 2\gamma U''' + (\gamma^2 - 2\alpha^2) U'' - 2\alpha^2 \gamma U' + (\alpha^4 + \sigma^{*2}) U = -\mathrm{e}^{-\gamma\zeta}, \\ &V = (U'' - \alpha^2 U + \gamma U')/\sigma^*, \end{aligned} \tag{5}$$

where the (unimportant) amplitude of the force has been set to unity. At one boundary we adopt the no-slip condition; the other boundary is stress-free:

$$U = V = 0 \quad \text{at} \quad \zeta = 1, \quad U' = V' = 0 \quad \text{at} \quad \zeta = 0. \tag{6a}$$

Thus, using (5) to eliminate V, we arrive at the simple boundary conditions

$$U = U'' + \gamma U' = 0 \quad \text{at} \quad \zeta = 1, \qquad U' = U''' + \gamma U'' \quad \text{at} \quad \zeta = 0. \tag{6b}$$

We are, of course, only interested in values at the free surface ($\zeta = 0$), and these are displayed for four models in Table XII.i. Note that the phase lag is in all cases small, showing that the magnetic force and the related flow are, indeed, quasi-simultaneous. Deep channels of stratified fluid exhibit minimal phase differences; shallow water shows the greatest phase differences. In the analytical part of his paper, Yoshimura (1981) employed similar concepts. If the poloidal field varies as $\sin \alpha$ and the toroidal field varies as $\sin(\alpha + \delta)$, then the Lorentz force varies as $L \sim \sin(2\alpha + \delta)$. The region of maximum activity corresponds to a node of the Lorentz force and the induced torsional oscillations are

$$\Omega \sim \sin(2\alpha + \delta - \delta') \quad \text{with} \quad \delta' = \arctan \sigma^*.$$

Since the observations reveal a node of Ω_{cyc} in the activity region, Yoshimura concluded that $\delta \approx 0$, so that σ^* must be small, as is true for the outer region of the SCZ.

Table XII.i. The phase lag ϕ for different channel models.

model	γ	a	σ^*	$\tan\phi = U(0)/V(0)$
non-stratified	0			
shallow		0.1	1	-0.4
			10^{-4}	$-4 \cdot 10^{-5}$
deep		10	1	-10^{-2}
			10^{-4}	-10^{-6}
stratified	5			
shallow		0.1	1	-0.07
			10^{-4}	$-7 \cdot 10^{-6}$
deep		10	1	$-8 \cdot 10^{-3}$
			10^{-4}	$-8 \cdot 10^{-3}$

So far it has all been clear and understandable. But the observations also show that the zone on the equatorward side of the activity region is *fast*. What is the real reason why the Lorentz force is positive there and not negative? Yoshimura's model demonstrates this, but does not explain it.

12.2. Toroidal Lorentz Force and Dynamo Models

Let us now discuss in more detail the toroidal component of the Lorentz force, i.e.

$$\mathcal{L}^T = \mathbf{e}.(\bar{\mathbf{j}} \times \overline{\mathbf{B}}) = \mathbf{e}.(\bar{\mathbf{j}}_P \times \overline{\mathbf{B}}_P) = \bar{j}_r \overline{B}_\theta - \bar{j}_\theta \overline{B}_r. \tag{7}$$

We shall particularly want to apply the results of the forthcoming analysis to different dynamo models in order to discover what the sign is of the force acting at low latitudes. If it is accelerating, we believe that the models will satisfy the observational constraint of the "torsional oscillations". More detailed statements can only be made by resorting to numerical computations, the preliminary results of which we shall present later on.

The toroidal Lorentz force (7) can also be written as

$$\begin{aligned}\mathcal{L}^T &= \frac{\overline{\mathbf{B}}}{\mu r \sin\theta} \cdot \nabla (r \sin\theta \overline{B}_\phi) \\ &= \frac{1}{\mu r \sin\theta} \left[\overline{B}_r \frac{\partial}{\partial r} (r \sin\theta \overline{B}_\phi) + \overline{B}_\theta \frac{\partial}{\partial\theta} (\sin\theta \overline{B}_\phi)\right].\end{aligned} \tag{8}$$

It is interesting to see the close relation, which is suggested by (8), between the current lines of $\bar{\mathbf{j}}_P$ and the isolines of the toroidal magnetic field B_ϕ. The current lines are defined by $\mathrm{d}r : r\,\mathrm{d}\theta = j_r : j_\theta$, so that with

$$\boldsymbol{j}_p = \mu^{-1}[(r \sin\theta)^{-1}\,\partial(\sin\theta \cdot B_\phi)/\partial\theta\,,\ -r^{-1}\,\partial(rB_\phi)/\partial r\,,\ 0]\,, \tag{9}$$

we immediately find $\mathrm{d}(r \sin\theta\, B_\phi) = 0$. It follows that current lines are the isolines of $r \sin\theta\, B_\phi$ which, at large distances from the rotation axis, is much the same as saying that they are isolines of the toroidal field B_ϕ. Since $\nabla \cdot \bar{\boldsymbol{j}} = 0$ everywhere, and $\bar{\boldsymbol{j}} = 0$ for $r > R$ the isolines of $\bar{\boldsymbol{j}}$ are closed, and lie entirely within the sphere $r = R$. Near maxima of $\bar{B}_\phi$, the current vanishes and near a node of $\bar{B}_\phi$ it is a local maximum, as (for example) on the equatorial axis, $\theta = \pi/2$. Clearly, there are three possible causes for a vanishing Lorentz force: $\bar{\boldsymbol{j}}$ vanishes, $\bar{\boldsymbol{B}}$ vanishes, or $\bar{\boldsymbol{j}}$ and $\bar{\boldsymbol{B}}$ are parallel. The first case is realized near an activity zone ($\bar{B}_\phi = \bar{B}_{\max}$), while the second occurs at a neutral point N of $\bar{\boldsymbol{B}}_P$. If, last but not least, the field lines of $\bar{\boldsymbol{B}}_P$ coincide with the isolines of $\bar{B}_\phi$, then $\bar{\boldsymbol{j}}_P$ and $\bar{\boldsymbol{B}}_P$ are nearly parallel, and the Lorentz force again vanishes.

Having dealt with the preliminaries, let us now turn to the special dynamo models of solar magnetism. In order to obtain an overview we ignore, for the moment, the sinusoidal time dependence of the Lorentz force and focus on one characteristic phase of the cycle, namely the moment when the dominant flux tube of the toroidal field has just displaced the remnant of its predecessor from the equatorial region. At about this time, a neutral point appears on the equatorial axis and begins to move towards the interior. This is also the moment when the θ-component of the magnetic field in the equatorial region changes sign.

In Fig. 12.4 we show the associated characteristic field configurations for various models. The first one was developed by Roberts (1972) and the other two are models 1 and 2 of Steenbeck and Krause (1969), the main

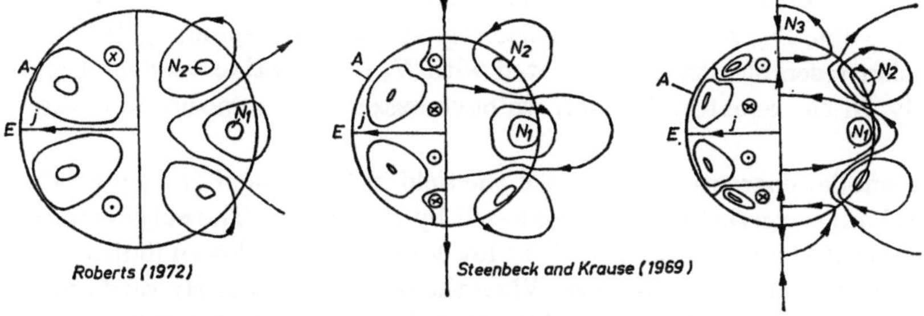

Fig. 12.4. Phase pictures for three solar dynamo models. The left-hand side of each diagram shows the isolines of the toroidal field, while the poloidal lines of force are on the right. A marks the active latitude, E the equator and N neutral points of the poloidal field. The poloidal currents ($\boldsymbol{j}$) approximately follow the toroidal field contours. In the northern hemisphere (low and middle latitudes) they flow clockwise.

differences between them being the thicknesses chosen for the SCZ. Roberts's model is the simplest one to study: α and $d\Omega/dr$ are completely depth-independent, i.e. $\mathsf{d} = 1$. Model 1 of Steenbeck and Krause applies to a rather deep turbulent region ($\mathsf{d} = 0.38$), unlike their second model for which $\mathsf{d} = 0.1$. None of these models has a latitudinal variation in rotation. It is the deep-zone models and not the shallow-zone model that agree with the observed phase relation between the global toroidal and poloidal fields. The sign of the general *polar* field of the Sun changed in 1957/58, in 1969/71 and in 1980, and the change always took place during maximal activity, or at a short time after sunspot maximum. The fields of Steenbeck and Krause's model 2 change sign during the minimum phase.

The surface region (E, A) between the equator (E) and an activity belt (A) is of the greatest interest. We suggest that at both boundaries of (E, A) the toroidal Lorentz force is very small, or even vanishes. The reason is to be found in the existence of the neutral point (N_1) in E and in the maximum of $\overline{B}$ in A. A Lorentz force, $\mathcal{L}^T$, that is positive in the region (E, A) would agree with the observations. In all three models considered, the toroidal fields are very similar between E and A. The corresponding poloidal current flows outwards in the equatorial plane. Near the surface it turns towards the poles so that in low northern latitudes we have, below the surface, a northward directed current. In the models having deep convection zones, this current interacts with poloidal fields that are nearly radial ($\overline{B}_r > 0$) in this region. An *accelerating* Lorentz force, $\mathcal{L}^T > 0$, results. The negative $\overline{B}_\phi$ is correlated with a positive $\overline{B}_r$, and also with a positive or vanishing $\overline{B}_\theta$. The products $\overline{B}_r\overline{B}_\phi$ and $\overline{B}_\phi\overline{B}_\theta$ are negative.

Quite a different situation arises for the third, shallow-zone model. Its poloidal field in low latitudes is in quite the opposite direction to those in the models just described. The Lorentz force is therefore decelerating and the products of the field components, which represent the angular momentum transport by the magnetic field, are positive. This great discrepancy arises because the fields in this model have smaller scales, and an additional neutral point, N_3, exists in the polar region. The corresponding field configuration resembles those associated with N_2 in the other models. Consequently, in the equatorial region we have exactly the opposite field structure.

The smaller characteristic scale of the field in this model is a by-product of the thinner convection zone possessed by this model. The surprising fact emerges that the Lorentz force, $\mathcal{L}^T$ at low latitudes provides an indication of how deep the convection zone is. When the observations of Howard and La Bonte and the meanfield theory of the turbulent dynamo are taken together, the results favor a relatively deep SCZ. Only such models can provide both the observed phase relation between $\overline{B}_r$ and $\overline{B}_\phi$ and the correct Lorentz force in low latitudes. Positive $\overline{B}_r\overline{B}_\phi$ indicates that the angular momentum flux is directed inwards; negative values that it is directed outwards.

Since the radial component of the electrical current vanishes near the surface, (7) can be approximated by

$$\mathcal{L}^T \simeq -\bar{j}_\theta \bar{B}_r \simeq (\mu r)^{-1} \bar{B}_r \, \partial(r\bar{B}_\phi)/\partial r \simeq -\bar{B}_{\phi,\max}\bar{B}_r/\mu \Delta r \,, \tag{10}$$

where we have set $\bar{B}_\phi \approx 0$ and $\partial\bar{B}_\phi/\partial r \approx -\bar{B}_{\phi,\max}/\Delta r$, i.e. the outward flux of angular momentum due to the magnetic field provides an accelerating force ($\bar{B}_r\bar{B}_\phi < 0$, as observed). It vanishes if $\bar{B}_r$ vanishes, this condition being fulfilled at the equator (N_1) and at N_2, a point which in the northern hemisphere is somewhat North of the activity belt (see Fig. 12.4 and 12.5). Approximation (10) also confirms the earlier result, that the Lorentz force for models with deep convection zones are accelerating. We stated that the product $\bar{B}_r\bar{B}_\phi$ is negative, and so (10) yields a positive L^T.

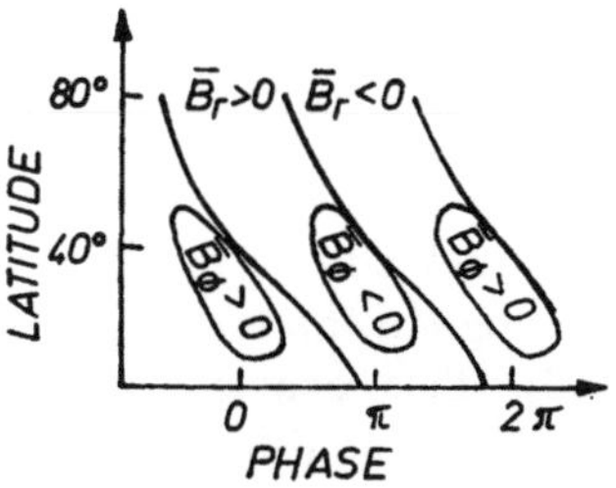

Fig. 12.5. Phase relationship between the r- and ϕ-components of the magnetic field at low latitudes for model 1 (deep convection zone) of Steenbeck and Krause (1969). The product of the two quantities is negative (outward transport of angular momentum). Compare the diagram with the observations illustrated in Fig. 12.1.

Another interesting point concerns the time taken for a single toroidal flux belt to move from pole to equator. Obviously, the duration of this process depends on the number of belts. If there were only one belt, it would take 11 years. A pattern with two belts requires 22 years, while for three belts the time would be 33 years, and so on. The observations prove that 22 years is correct, and that thus the toroidal field has two belts. The torsional oscillations therefore provide one further tool with which to probe the solar dynamo, a tool independent of sunspots statistics. It is the deep-zone models that generally have (at most) two belts at any one time. In contrast, the shallow-zone models (cf. Fig. 12.4c) possess three belts so that, were these relevant for the Sun, each would take 33 years to propagate from pole to equator. Again the evidence of the observations favor a relatively deep SCZ.

In their paper of 1982 La Bonte and Howard argued against the magnetic character of torsional oscillations. One of their observations is of exceptional weight, and deserves very careful consideration: the striking temporal constancy of the flow amplitude during an oscillation. This stands in sharp constrast to the obvious variations in the magnetic field strength. "The torsional velocities are constant to within 1 m/s throughout the 22 year transit from pole to equator." And this holds even though the magnetic tension "should produce large systematic variations in the torsional velocity amplitude as the sub-surface magnetic fields strengthen and decay during the magnetic cycle". Their opinion appears to be based on the implicit assumption that flow profiles with identical amplitudes are always necessarily due to forces having identical amplitudes. This, however, is untrue. This may be easily appreciated by considering the simplest of all possible arrangements: the balance between friction and external force. The latter is equal to the second spatial derivative, and hence does not depend on the amplitude of the flow alone; the "shape" of the flow determines the force amplitude required.

Let us consider a simple example. The force profile

$$L = \nu Y^{-2}(1 - 2y^2 Y^{-2}) \exp(-y^2 Y^{-2}), \tag{11}$$

reates for any Y the flow $u = \exp[(-y/Y)^2]$. Corresponding to any given Y, the external force creates a flow of unit maximum amplitude. Every change in the Y of the applied force results in a change in the width of the flow, never a change in its amplitude (Fig. 12.6). This example establishes a principle: it is a misunderstanding to believe that differing force amplitudes will create differing flow amplitudes.

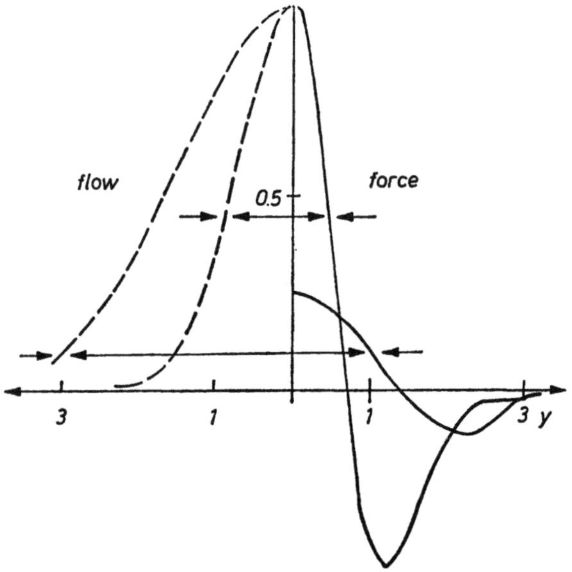

Fig. 12.6. Profiles of the external forces (on the right) and the flows they generate (on the left). Forces with strongly differing profiles generate flows of exactly the same amplitude.

In analogy with this example, although the Lorentz force undoubtedly varies during the cycle, and although the flow must necessarily react to this, the flow need *not* necessarily change its amplitude; the structure of the flow may alter instead. Indeed, "as the waves move towards the equator, the amount by which they extend across the latitudes becomes thinner, narrower". This might be their reaction to the temporal variations in the force. There *is* a variation in the flow, but in shape rather than in amplitude.

12.3. Hydrodynamics with Lorentz Force

There are alternating belts of Lorentz force at higher latitudes too. Yet in the neighborhood of N_2 it is not easy to predict the flow pattern correctly, and correspondingly the depth-dependence of the force cannot easily be inferred. Moreover, until now we have ignored the influence of the meridional circulation generated by the poloidal component of the Lorentz force. With these points in mind, it will perhaps be useful to set down in this section the formulae which govern the full problem of mean flow created by magnetic forces. In the final section of this Chapter we shall derive from the expressions below the surface flow patterns implied by certain dynamo models.

In rotating system of coordinates, the equation of (inviscid) flow under the influence of a magnetic field is

$$\varrho[\partial \boldsymbol{u}/\partial t + (\boldsymbol{u} \cdot \nabla)\, \boldsymbol{u}] = -\nabla p + \varrho \boldsymbol{g}^{\mathrm{eff}} + 2\varrho \boldsymbol{u} \times \boldsymbol{\Omega} + \mu^{-1}(\nabla \times \boldsymbol{B}) \times \boldsymbol{B}\,. \quad (12)$$

In turbulent media the mean flow, $\overline{\boldsymbol{u}}$, obeys the equation

$$\varrho[\partial \overline{\boldsymbol{u}}/\partial t + (\overline{\boldsymbol{u}} \cdot \nabla)\, \overline{\boldsymbol{u}}] = -\nabla \overline{p} + \overline{\varrho} \boldsymbol{g}^{\mathrm{eff}} + 2\varrho \overline{\boldsymbol{u}} \times \boldsymbol{\Omega} \\ - \nabla \cdot (\varrho Q) + \mu^{-1}(\nabla \times \overline{\boldsymbol{B}}) \times \overline{\boldsymbol{B}} + \mu^{-1}\nabla \cdot B\,, \quad (13)$$

with (8.97), where Q_{ij} is the correlation tensor and

$$B_{ij} = \overline{B_i'(\boldsymbol{x}, t)\, B_j'(\boldsymbol{x}, t)}\,. \quad (14)$$

Density fluctuations have been ignored, and the pressure contains a magnetic part. Here the old Boussinesq concept, i.e. (10.25), will be adopted and magnetic fluctuations will be ignored (see, however, Rüdiger *et al.*, 1986). The turbulence is then only parameterized by an eddy viscosity ν_T, which is of order $5 \cdot 10^{12}$ cm²/s. We shall also neglect the meridional components $\overline{u}_r$ and $\overline{u}_\theta$ when they appear at second or higher order. We then have

$$\frac{\partial}{\partial t}(\varrho \overline{u}_i) - 2\varrho \varepsilon_{ijk} \overline{u}_j \Omega_k + \frac{\partial}{\partial x_k}(\varrho Q_{ik}) \\ = \varrho g_i^{\mathrm{eff}} - \frac{\partial}{\partial x_i}\left(\overline{p} + \frac{\overline{B}^2}{2\mu}\right) + \frac{1}{\mu}\frac{\partial}{\partial x_k}(\overline{B}_i \overline{B}_k)\,, \quad (15)$$

where the magnetic terms will be regarded as the source of the motion.

The ϕ-component of (15) is identical with the conservation law for angular momentum and, like that law, it can be written in divergence form [see (4.15)]:

$$\partial(\varrho s^2 \Omega)/\partial t + \nabla \cdot \{\varrho s[\overline{u_\phi' \boldsymbol{u}'} + s\Omega \bar{\boldsymbol{u}} - (\mu\varrho)^{-1} \bar{B}_\phi \bar{\boldsymbol{B}}]\} = 0. \tag{16}$$

According to the Boussinesq ansatz (2.2) we obtain

$$\begin{aligned} &\partial(\varrho s^2 \Omega)/\partial t + \varrho \bar{\boldsymbol{u}} \cdot \nabla(s^2 \Omega) - \nu_T \nabla \cdot (\varrho s^2 \nabla \Omega) = L^T, \\ &L^T = \bar{\boldsymbol{B}} \cdot \nabla(s \bar{B}_\phi)/\mu. \end{aligned} \tag{17}$$

The solar magnetic cycle creates a slowly varying part in the source term, L^T. Thus we must again divide all quantities into two parts, e.g.

$$\Omega = \Omega^{(0)} + \Omega_{\text{cyc}}, \tag{18}$$

only one of which (here Ω_{cyc}) is time-dependent. In the case of the Sun, it is reasonable to assume that

$$u^{(0)} = L^{T,(0)} = \varrho_{\text{cyc}} = 0. \tag{19}$$

Magnetic buoyancy is still ignored, and therefore

$$r^2 \sin^2 \theta \varrho \dot{\Omega}_{\text{cyc}} + \varrho \boldsymbol{u}_{\text{cyc}} \cdot \nabla(r^2 \sin^2 \theta\, \Omega^{(0)}) - \nu_T \nabla \cdot (\varrho r^2 \sin^2 \theta\, \nabla \Omega_{\text{cyc}}) = L^T \tag{20}$$

For simplicity we shall use only the uniform, constant part of the angular velocity, i.e. we write $\Omega^{(0)} = \Omega_\odot$ so obtaining

$$\begin{aligned} r^2 \sin^2 \theta \varrho \dot{\Omega}_{\text{cyc}} + 2\Omega_\odot\, [r^{-1} \sin \theta\, \partial A_{\text{cyc}}/\partial\theta - \cos\theta\, \partial A_{\text{cyc}}/\partial r] \\ - \nu_T \nabla \cdot (\varrho r^2 \sin^2 \theta \nabla \Omega_{\text{cyc}}) = L^T_{\text{cyc}}, \end{aligned} \tag{21}$$

where the stream function A, defined as in (5.1), has been introduced.

If we now write

$$\begin{aligned} \Omega_{\text{cyc}} &= \sum_n \Omega_\odot [\omega^s_{n-1} \sin(\sigma t) + \omega^c_{n-1} \cos(\sigma t)]\, P_n^1/\sin\theta, \\ A_{\text{cyc}} &= \sum_n [A_n^s \sin(\sigma t) + A_n^c \cos(\sigma t)]\, P_n^1 \sin\theta, \\ L^T_{\text{cyc}} &= \Omega_\odot \varrho \nu_T \sum_n [L_n^{T,s} \sin(\sigma t) + L_n^{T,c} \cos(\sigma t)]\, P_n^1 \sin\theta, \end{aligned} \tag{22}$$

we finally obtain the two relations (cf. Rüdiger *et al.*, 1986)

$$\begin{aligned} &[+/-]\, \sigma^* x^2 \omega_n^{c/s} + \frac{2}{\nu_T \varrho R} \left[\frac{(n+2)(n+3)}{2n+5} \frac{A^{s/c}_{n+2}}{x} - \frac{n(n+1)}{2n+1} \frac{A_n^{s/c}}{x} \right. \\ &\left. + \frac{n}{2n+1} \frac{\mathrm{d}A_n^{s/c}}{\mathrm{d}x} + \frac{n+3}{2n+5} \frac{\mathrm{d}A^{s/c}_{n+2}}{\mathrm{d}x} \right] \\ &+ x^2 \frac{\mathrm{d}^2 \omega_n^{s/c}}{\mathrm{d}x^2} + (4 - G)\, x \frac{\mathrm{d}\omega_n^{s/c}}{\mathrm{d}x} - n(n+3)\, \omega_n^{s/c} = -L^{Ts/c}_{n+1}, \quad n = 2, 4, \ldots \end{aligned} \tag{23}$$

where here

$$\sigma^* = R^2\sigma/\nu_T \approx 20. \tag{24}$$

Let us now attack the meridional components of the Reynolds equation (15). The Coriolis terms no longer contribute. By taking the "curl" of (15) we can even eliminate all the pressure terms, i.e. all the coefficients of the Kronecker tensor δ_{ij}:

$$\partial \nabla \times (\varrho\bar{\mathbf{u}})/\partial t - \nabla \times (\varrho\Omega^2\mathbf{s}) + \nabla \times \boldsymbol{I} = 0, \tag{25}$$

where

$$I_i = \partial(\varrho Q_{ik} - \mu^{-1}\bar{B}_i\bar{B}_k)/\partial x_k. \tag{26}$$

The longitudinal component of (25) is

$$\begin{aligned}
&\frac{\partial}{\partial t}\left[\frac{\partial}{\partial r}(r\varrho\bar{u}_\theta) - \frac{\partial(\varrho\bar{u}_r)}{\partial\theta}\right] - \frac{1}{r^2}\frac{\partial^2}{\partial r\,\partial\theta}(r^2\varrho Q_{rr}) + \frac{1}{\sin\theta}\frac{\partial^2}{\partial r\,\partial\theta}(\varrho\sin\theta Q_{\theta\theta}) \\
&- \cot\theta\frac{\partial}{\partial r}(\varrho Q_{\phi\phi}) + \frac{1}{r}\frac{\partial}{\partial\theta}[\varrho(Q_{\theta\theta} + Q_{\phi\phi})] \\
&- \frac{1}{r}\frac{\partial}{\partial\theta}\left[\frac{1}{\sin\theta}\frac{\partial}{\partial\theta}(\varrho\sin\theta Q_{r\theta})\right] + \frac{\partial}{\partial r}\left[\frac{1}{r^2}\frac{\partial}{\partial r}(r^3\varrho Q_{r\theta})\right] \\
&= -r\sin^2\theta\frac{\partial}{\partial\theta}(\varrho\Omega^2) + r^2\cos\theta\sin\theta\frac{\partial}{\partial r}(\varrho\Omega^2) \\
&+ \frac{1}{\mu}\left\{\frac{1}{r}\frac{\partial}{\partial\theta}(\bar{B}_\theta^2 + \bar{B}_\phi^2) - \cot\theta\frac{\partial\bar{B}_\phi^2}{\partial r} - \frac{1}{r^2}\frac{\partial^2}{\partial r\,\partial\theta}(r^2\bar{B}_r^2)\right. \\
&+ \frac{1}{\sin\theta}\frac{\partial^2}{\partial r\,\partial\theta}(\sin\theta\bar{B}_\theta^2) - \frac{1}{r}\frac{\partial}{\partial\theta}\left[\frac{1}{\sin\theta}\frac{\partial}{\partial\theta}(\bar{B}_r\bar{B}_\theta\sin\theta)\right] \\
&\left.+ \frac{\partial}{\partial r}\left[\frac{1}{r^2}\frac{\partial}{\partial r}(r^3\bar{B}_r\bar{B}_\theta)\right]\right\}.
\end{aligned} \tag{27}$$

This relation can readily be compared with (5.47). The correlations Q_{ij} are again related to the mean flow by (10.25).

For the stream function and angular velocity the representations (22) are used. The magnetic terms may be written as

$$\frac{1}{\mu}\left(\frac{1}{r}\frac{\partial\bar{B}_\phi^2}{\partial\theta} - \cot\theta\frac{\partial\bar{B}_\phi^2}{\partial r}\right) = \frac{\nu_T}{R^4}\sum_n [L_n^{P,s}\sin(\sigma t) + L_n^{P,c}\cos(\sigma t)]\,P_n^1, \tag{28}$$

where we have been able to neglect the (weak) poloidal components in comparison with the (strong) toroidal field. The result is

$$[+/-]\,\sigma^* \left(\frac{\mathrm{d}^2 A_n^{c/s}}{\mathrm{d}x^2} - \frac{n(n+1)}{x^2} A_n^{c/s}\right) + \mathcal{D} A_n^{s/c}$$
$$+ \frac{2x\Omega_\odot^{\;2} R^5 \varrho}{\nu_T} \left\{\frac{n-1}{2n-1}\left[(G+n-2)\,\omega_{n-2}^{s/c} - x\,\frac{\mathrm{d}\omega_{n-2}^{s/c}}{\mathrm{d}x}\right]\right.$$
$$\left. + \frac{n+2}{2n+3}\left[(G-n-3)\,\omega_n^{s/c} - x\,\frac{\mathrm{d}\omega_n^{s/c}}{\mathrm{d}x}\right]\right\} = L_n^{P,s/c}, \qquad (29)$$

the operator $\mathcal{D}$ being the same as in (5.61).

Boundary conditions must be applied to all components of the velocity vector. In order that the surfaces of the SCZ be stress-free, the correlations $Q_{r\theta}$ and $Q_{r\phi}$ must be set zero there. The resulting conditions at $x = x_i$ and $x = 1$ have already been derived in Chapter 5 and (10.53).

12.4. Preliminary Numerical Results

The appearance of the toroidal and meridional magnetic forces (Fig. 12.7 and 12.8) already suggests that the nature of the cyclic flows may depend on the dynamo model chosen. The flows can also be understood from the standpoint of Lenz's rule: their influence on active zones is to obstruct dynamo action.

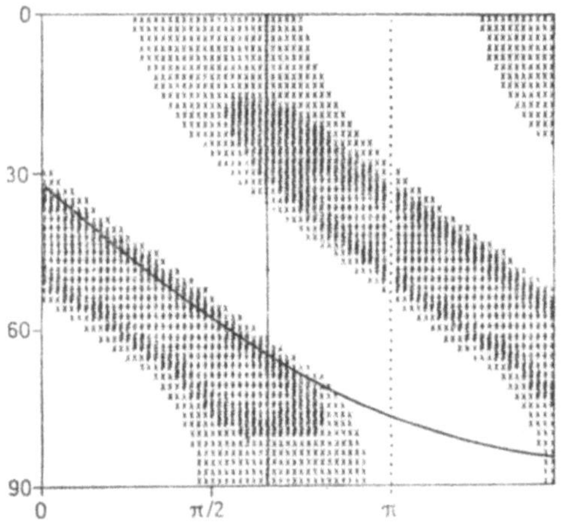

Fig. 12.7. The time-dependent part of the toroidal Lorentz force. The ordinate denotes the latitude and the abscissa the phase in the cycle. In the shaded areas the force is accelerating, while in the unshaded areas it is decelerating. At the boundaries of the differently shaded areas the force is 1/3 and 2/3 of its maximum value.

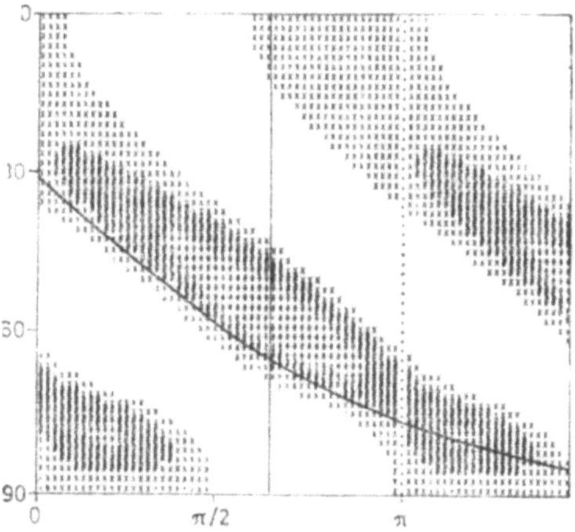

Fig. 12.8. The latitudinal Lorentz force. In the shaded areas the force is equatorwards, and in the unshaded areas it is polewards. Coordinates and markings are as in Fig. 12.7.

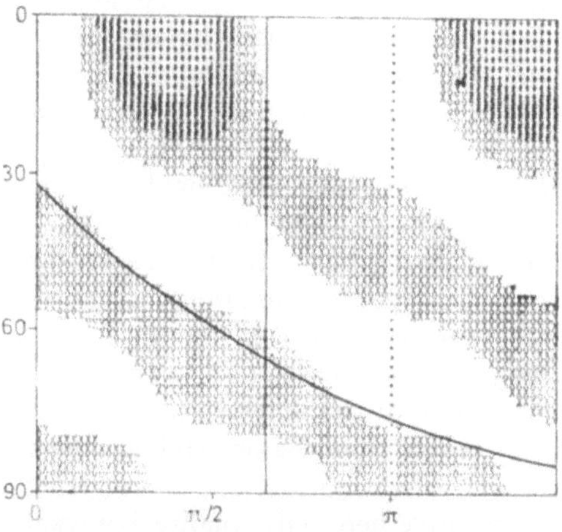

Fig. 12.9. The component of the variable angular velocity corresponding to the torsional oscillation. In the shaded areas the velocity is positive. The vertical continuous line defines the phase of the maximum toroidal field (the solar activity maximum) and the diagonal line in the center marks the active zone.

In order to study the flows, the mean-field equations (23) and (29) were solved. In order to compare directly the theoretical solutions with different components of the observed flows, it is helpful to transform the angular velocities to a common form, namely (1). As an example, some of the solutions for the thickzone model of Steenbeck and Krause are presented in Fig. 12.9 and 12.10. The solutions are only shown in a special case: both

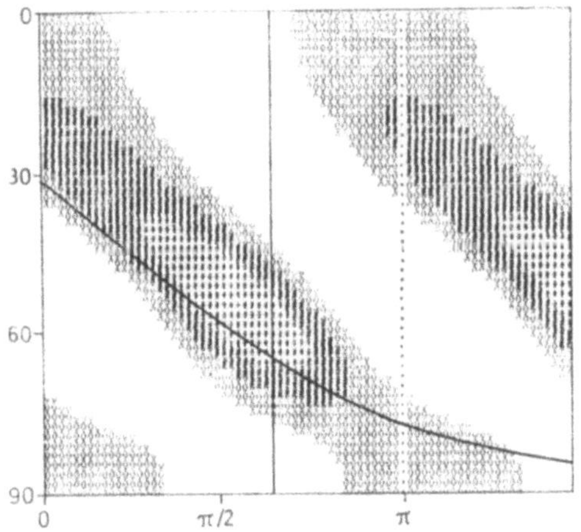

Fig. 12.10. The meridional circulation. In the shaded areas the motion is equatorwards, and in the unshaded areas it is polewards.

the meridional circulation, as a transporter of angular momentum in the toroidal equation, and the centrifugal term in the meridional equation are neglected (Rüdiger *et al.*, 1986). This means that only the viscosity and the magnetic force influence the flow and that there is no interaction between the two equations. Figure 12.9 displays the torsional wave as represented in (1): the meridional circulation is shown in Fig. 12.10. It may be seen that the observations of the torsional wave and of the meridional circulation are represented reasonably well in magnitude. The maximum of A is attained at the solar activity maximum, in conflict with the observations. The coefficients B and C are in antiphase, and B has a maximum at the solar maximum. Observation has, however, revealed that B is maximal at solar minimum (cf. Fig. 3.12). As the phase relations are sensitive to the dynamo model, one must be careful about the value of the eddy viscosity and the density distribution [which has been taken from Baker and Temesváry (1966)].

As for the magnitude of the flows, we give here only qualitative results, because the low radial resolution of the dynamo model makes more detailed inferences dangerous, particularly near the surface where the density changes rapidly. The magnitude of the meridional circulation obtained is a few m/s, and would be approximately 10 times larger were the assumed eddy viscosity 10 times smaller. The magnitude of the variations of the velocity of rotation (all coefficients) is smaller than is observed (the discrepancy depending on the viscosity). The relative magnitudes of A, B and C are much as have been observed, i.e. A changes only slightly, the changes in C being the greatest. As one regards the dependence of the variable flows on depth, it is evident that, because of the increasing density, the flows become smaller with increasing depth.

Part III
Microscopy

In this, the final part of the book, we will investigate whether the values of V, H, VV, HV, ν_T and χ_T needed to explain the observations of the Sun are consistent with turbulence theory. That we can find a set of values for these parameters that will correspond to the observed state of the solar surface is already known from earlier chapters. And we are therefore encouraged to believe in the existence in the SCZ of negative $V^{(0)} \cdot V^{(1)}$, positive $H^{(1)}$, and positive but rather small $VV^{(1)}$ and $HV^{(1)}$. But whether we can theoretically understand their magnitudes, and indeed other magnitudes such as the small Prandtl number, the positive eddy viscosity, etc. is quite another matter. The key question we wish to answer is this: Is it possible to construct turbulence models that, at the bottom of the convection zone (where the effect of rotation on the turbulence may be assumed to be sufficiently large), provide precisely the values required of $V_i^{(0)}$, $V_i^{(1)} \dots HV_i^{(1)}$.

In what follows we shall use the language of turbulence theory, and not merely because turbulence in nature is the rule rather than the exception. Although the principal alternative approach, in which ordered patterns of motion ("convection") are determined by large-scale numerical computations, undoubtedly has the ability to mimic the solar differential rotation, we believe that our approach is both simpler and more transparent. The physical insights it gives are therefore deeper. Our approach allows us to elucidate the basic physical mechanisms individually, since the overall process of solution can be divided into several separate steps.

At best our approach can, of course, perform no better than present day turbulence theory, but in general it is likely to do rather worse. Its weaknesses are the price we have to pay for the advantages mentioned above. First of all, there is the well-known difficulty of closure. The infinite system of equations, arising from the basic non-linearity of the governing equations, must be made tractable by a process of linearization. Here we use the Second Order Correlation Approximation (SOCA) in which the third and higher order statistical moments are neglected. Synonyms for this approach are FOSA (First Order Smoothing Approximation) or "Herring's Approximation". This procedure may be defended for "weak turbulence" with

$$\min(Re, S) \ll 1, \tag{1}$$

where, as usual, $Re = Lu'/\nu$ is the Reynolds number and $S = \tau_c u'/L$ is the Strouhal number. Even calculations based on SOCA provide insight into the physical process at work, since they do not trivially ignore the non-linearity. Only the auto-interaction terms, $(\boldsymbol{u}' \cdot \nabla)\,\boldsymbol{u}' - \overline{(\boldsymbol{u}' \cdot \nabla)\,\boldsymbol{u}'}$, which describe the flux of (energy and) momentum from one turbulent mode to another, are ignored. The interaction of the turbulence with the non-turbulent or "mean field" quantities remains untouched.

The approximation soon highlights a new difficulty: one must be able to distinguish between a fluctuation and the mean-field component of the flow, but how can this be done? This question concerns the applicability of the Reynolds rules. It might be argued that, within the framework of statistical physics, no such problem arises since averages are taken over a great number of identical excamples, the so-called "ensemble". Yet this convenient theoretical concept is not a practical reality. What we really require is a further constraint, that of "two-scaledness", i. e. the fluctuations must be on much smaller scales than the mean fields. The Reynolds rules are then satisfied. In considering averaging rules in their monograph, Krause and Rädler (1980) go so far as to state that this two-scale property is the basic constituent of mean-field physics.

Here we wish to emphasize another consequence of two-scaledness, namely that it implies that the relationships subsisting between correlations and mean quantities are local; the otherwise non-local *integral* expressions become local *differential* relations. For example, a general relation obtained using SOCA, such as

$$Q_{ij}(\boldsymbol{x}) = -\int \mathcal{N}_{ijp}(\boldsymbol{x} - \boldsymbol{x}')\,\bar{u}_p(\boldsymbol{x}')\,\mathrm{d}\boldsymbol{x}', \tag{2}$$

may be written as

$$Q_{ij}(\boldsymbol{x}) = -\mathcal{N}_{ijkl}\bar{u}_{k,l}, \tag{3}$$

when two-scaledness is assumed. Clearly this may be regarded principally as a matter of mathematical convenience and, indeed, some papers are able to deal with the behavior of small-scale mean fields using the full dispersion relations throughout, without at any stage invoking two-scaledness (Krause and Roberts, 1973; Krause and Rüdiger, 1974b).

All the deductions presented in Part II rested on the assumption that the mean fields have large scales, and the fluctuations small. In the remainder of this book all the coefficients in the two-scale relations will be determined on the basis of SOCA. For this reason the formulations which follow may be considered as the weakest part of our approach. Further progress in non-linear turbulence theory will show that the results we obtain are special cases of more general, and more exact, expressions.

Chapter 13. Turbulence Models

13.1. The Correlation Tensor

In this investigation we are concerned only with those properties of turbulence that depend on the two-point and two-time correlation tensor

$$Q_{ij}(\boldsymbol{x}, \boldsymbol{\xi}, t, \tau) = \overline{u_i'(\boldsymbol{x}, t)\, u_j'(\boldsymbol{x} + \boldsymbol{\xi}, t + \tau)}. \tag{1}$$

All computations of the Λ-effect, eddy viscosity and eddy conductivity presented in succeeding chapters are based on the structure of this tensor. We shall generally assume that these quantities arise through the action of mean fields, such as rotation and mean temperature, on an "original" or preexisting turbulence of relatively simple form. This original turbulence is regarded as being present even if the medium were non-rotating and isothermal, i.e. even if the mean-fields were absent. It is assumed to be maintained by a system of stochastic forces ("Rührwerk") and not by the mean flow. These forces take part in the basic rotation so that, in the co-rotating frame, they are independent of $\boldsymbol{\Omega}$. Rotation only modifies, and does not generate, the turbulence. All the fundamental structures of the correlation tensor are rotational modifications of the original turbulence, whose structure possesses only simple geometrical anisotropies, the most prominent direction being that of gravity (the radial direction). It is of fundamental importance here that the radial direction in a sphere varies from point to point, thus allowing the turbulence to be always globally inhomogeneous. The "Reynolds force" is the direct consequence of this.

We hope that, as a description of the stellar situation, this concept is not far from reality. The exact form of the original turbulence in a stellar atmosphere is not yet known. For certain physical groupings we can determine the general structure of the correlation tensor, but only in a heuristic way. ("Grouping" here refers to the form and number of inhomogeneities and/or anisotropies.) Although their general effects on the correlations can be stated readily enough, the precise structure of the corresponding scalar functions, of which we often know only the sign, remains open. This is one of the reasons why, in some instances, the determination of the turbulence parameters themselves is less accurate than we might wish. The main aim of these chapters is to demonstrate the results of the approximation we use ("SOCA"), and we will keep this restriction in mind when the results are discussed.

It is often advantageous to work with the Fourier transform of (1), i.e. its spectral tensor $\hat{Q}_{ij}$ defined by

$$Q_{ij}(\boldsymbol{x}, \boldsymbol{\xi}, t, \tau) = \iint_{-\infty}^{\infty} \hat{Q}_{ij}(\boldsymbol{x}, \boldsymbol{k}, t, \omega)\, \mathrm{e}^{i(\boldsymbol{k}\cdot\boldsymbol{\xi} - \omega\tau)}\, \mathrm{d}\boldsymbol{k}\, \mathrm{d}\omega . \tag{2}$$

The spectral tensor exhibits a number of quite important features. As we shall only deal with "original" turbulences that are homogeneous and steady, the dependencies on $\boldsymbol{x}$ and t in (2) can be omitted, to give

$$Q_{ij}(\boldsymbol{\xi}, \tau) = \iint \hat{Q}_{ij}(\boldsymbol{k}, \omega)\, \mathrm{e}^{i(\boldsymbol{k}\cdot\boldsymbol{\xi} - \omega\tau)}\, \mathrm{d}\boldsymbol{k}\, \mathrm{d}\omega . \tag{3}$$

Since Q_{ij} is a real quantity, the real part of $\hat{Q}_{ij}$ must be even in $\boldsymbol{k}$ and ω, and the imaginary part must be odd. Hence, the conjugate complex spectral tensor fulfills

$$\hat{Q}_{ij}^{*}(-\boldsymbol{k}, -\omega) = \hat{Q}_{ij}(\boldsymbol{k}, \omega) . \tag{4}$$

From the definition of Q_{ij} we know that

$$Q_{ij}(\boldsymbol{\xi}, \tau) = Q_{ji}(-\boldsymbol{\xi}, -\tau) , \tag{5}$$

so that only six independent correlations exist and generally

$$\hat{Q}_{ij}(\boldsymbol{k}, \omega) = \hat{Q}_{ji}(-\boldsymbol{k}, -\omega) . \tag{6}$$

Combining (4) and (6) we find the spectral tensor is Hermitian, i.e.

$$\hat{Q}_{ij} = \hat{Q}_{ji}^{*} . \tag{7}$$

Furthermore, according to Bochner's theorem (cf. Krause and Roberts, 1973), the spectral tensor Q_{ij} for steady homogeneous turbulence is always positive semi-definite,

$$\hat{Q}_{ij} X_i X_j^{*} \geqq 0 , \tag{8}$$

where $\boldsymbol{X}$ is an arbitrary vector quantity (Bochner, 1933). Quite generally, this relation implies

$$\hat{Q}_{ii} \geqq 0 , \tag{9}$$

which will enable us to fix the signs of certain spectral functions (see Krause and Rädler, 1980).

We shall refer to the turbulence as being anisotropic if one or more preferred directions exists in the turbulence, and if the correlations are not invariant with respect to an arbitrary rotation of the entire system. If the mean quantities are invariant under reflection in an arbitrary plane, then the turbulence is called "mirror-symmetric". Its spectral tensor for all homogeneous and isotropic turbulences satisfies

$$\hat{Q}_{ij}(-\boldsymbol{k}, \omega) = \hat{Q}_{ij}(\boldsymbol{k}, \omega), \tag{10}$$

hence it is a real quantity.

For more general situations we shall employ a similar argument. The term "mirror-symmetry" will be used to indicate that pseudo-tensors do not appear in the (polar) correlation tensor. Therefore, the coefficients of combinations that include the pseudo-tensor ε_{ijk}, will need to have pseudo-scalars as coefficients. Such scalars do not exist in normal situations, for example in a turbulence that is unaffected by a mean flow. Thus, ε-tensors will not normally appear in the correlation tensor. This argument does not, in general, imply that the spectral tensor is symmetric, as may be seen by quoting an example, namely the spectral tensor first derived by Rädler for an *inhomogeneous* turbulence. He assumed the turbulence to be anisotropic with preferred direction $\mathring{\boldsymbol{g}}$, to depend on $\mathring{\boldsymbol{g}} \cdot \boldsymbol{x}$ and to be linear in $\mathring{\boldsymbol{g}}$ and $\boldsymbol{x}$. The associated spectral tensor is (Krause and Rädler, 1980, p. 120)

$$\hat{Q}_{ij} = [Q_1 + (\mathring{\boldsymbol{g}} \cdot \boldsymbol{k})\, Q_2]\, (k^2\delta_{ij} - k_ik_j) - iQ_3k_i\mathring{g}_j + \cdots .$$

Those turbulence fields that are unaffected by the mean flow will be referred to as "original". They may even exist in nature, e.g. in non-rotating hydrostatic stellar atmospheres. Due to the gravitational force, they should be anisotropic but we see no reason to doubt that they are mirror-symmetric. In accordance with (10), we shall therefore assume that their spectral tensors are real and symmetric.

A special class of turbulence models are relevant to incompressible media, where

$$\nabla \cdot \boldsymbol{u}' = 0. \tag{11}$$

We shall adopt this simplification for most of our studies. Only one particular deduction (in § 14.4) will be depend on the more general relation

$$\nabla \cdot (\bar{\varrho}\boldsymbol{u}') = 0, \tag{12}$$

which is part of the so-called "anelastic approximation". Equation (11) implies

$$\partial Q_{ij}/\partial \xi_i = \partial Q_{ij}/\partial \xi_j = 0, \tag{13}$$

which in Fourier space becomes

$$k_i\hat{Q}_{ij} = k_j\hat{Q}_{ij} = 0. \tag{14}$$

13.2. The Spectral Tensor

The main objective of this chapter is to construct a spectral tensor for turbulence that has two preferred directions, which will be denoted by $\boldsymbol{e}$ ("east") and $\mathring{\boldsymbol{g}}$ ("gravity"). The theory will leave certain scalar functions free (i.e. unspecified). It would be possible to determine these were a properly founded physical turbulence theory available, which unfortunately is not yet the case.

Let us use an ansatz involving $\boldsymbol{k}$,

$$\hat{Q}_{ij} = k_m k_n Q_{ijmn}, \tag{15}$$

where Q_{ijmn} is to be constructed using only the vectors $\boldsymbol{e}$ and $\mathring{\boldsymbol{g}}$ together with the Kronecker delta. The tensor Q_{ijmn} must be symmetric with respect to i and j, as well as with respect to n and m. The simplest tensor combinations from which Q_{ijmn} can be formed are

$$\delta_{ij}\delta_{mn}, \qquad \delta_{im}\delta_{jn} + \delta_{in}\delta_{jm}, \tag{16}$$

thus giving

$$Q_{ijmn} = C_1\delta_{ij}\delta_{mn} + C_2(\delta_{im}\delta_{jn} + \delta_{in}\delta_{jm}). \tag{17}$$

The incompressibility condition,

$$k_i k_j k_m k_n Q_{ijmn} = 0, \tag{18}$$

yields

$$C_1 + 2C_2 = 0;$$

hence

$$Q_{ijmn} = C_1(2\delta_{ij}\delta_{mn} - \delta_{im}\delta_{jn} - \delta_{in}\delta_{jm})/2. \tag{19}$$

The spectral tensor of incompressible turbulence can thus be written as

$${}^{(3)}\hat{Q}_{ij} = {}^{(3)}q k^2 K_{ij}, \tag{20}$$

where

$$K_{ij} = \delta_{ij} - \mathring{k}_i\mathring{k}_j, \tag{21}$$

and $\mathring{\boldsymbol{k}} = \boldsymbol{k}/k$. In (20), ${}^{(3)}q$ is a free (spectral) function which, according to (9), can never be negative:

$${}^{(3)}q \geqq 0. \tag{22}$$

If in addition ${}^{(3)}q$ depends only on the magnitude of $\boldsymbol{k}$, the tensor (20) describes homogeneous, steady, isotropic and mirror-symmetric turbulence. Such motions are always free from helicity.

Next we shall take into account the preferred direction $\mathring{\boldsymbol{g}}$. One can again set up tensor products with which to form the required tensor. Assuming incompressibility, one obtains the spectral tensor

$$^{(2)}\hat{Q}_{ij} = {}^{(2)}q[k^2\delta_{ij} - k_ik_j - (\mathring{\boldsymbol{g}} \cdot \boldsymbol{k})^2\,\delta_{ij} - k^2\mathring{g}_i\mathring{g}_j + (\mathring{\boldsymbol{g}} \cdot \boldsymbol{k})\,(\mathring{g}_ik_j + \mathring{g}_jk_i)]. \quad (23)$$

Due to the special property,

$$\mathring{g}_i\mathring{g}_j{}^{(2)}\hat{Q}_{ij} = 0$$

of (23), motion parallel to $\mathring{\boldsymbol{g}}$ is absent. It is only in planes perpendicular to $\mathring{\boldsymbol{g}}$ that the material moves, and we shall therefore speak of horizontal turbulence. More generally, if half the horizontal intensity $\overline{\boldsymbol{u}_\perp{}^2}$ $(= Q_{ii} - \mathring{g}_i\mathring{g}_jQ_{ij})$ *exceeds* the vertical intensity, i.e. if

$$\delta I = (\overline{u'_\theta{}^2} + \overline{u'_\phi{}^2})/2 - \overline{u'_r{}^2} \equiv (Q_{ii} - 3\mathring{g}_i\mathring{g}_jQ_{ij})/2 \quad (24)$$

is positive, we shall speak of "horizontal-type turbulence". In the opposite case, $\delta I < 0$, the term "vertical-type turbulence" will be used. Purely vertical turbulence, in the strong sense, is impossible in incompressible media unless the vertical correlation length is infinite. If, moreover, the spectral function $^{(2)}q$ is proportional to the Dirac delta function $\delta(\mathring{\boldsymbol{g}} \cdot \boldsymbol{k})$, then the motion in every plane perpendicular to $\mathring{\boldsymbol{g}}$ is identical, being two-dimensional turbulence, which we have often expressed in the form

$$^{(2D)}\hat{Q}_{ij} = q_\perp(k_\perp, \omega)\,\delta(\mathring{\boldsymbol{g}} \cdot \boldsymbol{k})\,(k_\perp{}^2G_{ij} - k_{\perp i}k_{\perp j}), \quad (25)$$

where

$$G_{ij} = \delta_{ij} - \mathring{g}_i\mathring{g}_j, \qquad k_{\perp i} = G_{ij}k_j; \quad (26)$$

The two main properties of (25) should be recognized;

(i) The delta function prevents any turbulent quantity from depending on the coordinate in the direction of $\mathring{\boldsymbol{g}}$;

(ii) The structure of the tensor eliminates any motion in that direction.

What about the asymmetric parts of the spectral tensor? We know from (7) that they must be imaginary, and a term of the form $ip(g_ik_j - g_jk_i)$ with $p(-\boldsymbol{k}, \omega) = p(\boldsymbol{k}, \omega)$ appears to be possible. The spectral function p must, however, vanish if incompressibility is also required and (14) is applied.

With a second preferred direction, $\boldsymbol{e}$, one can form a tensor very similar to (23), i.e.

$$^{(1)}\hat{Q}_{ij} = {}^{(1)}q[k^2\delta_{ij} - k_ik_j - (\boldsymbol{e} \cdot \boldsymbol{k})^2\,\delta_{ij} - k^2e_ie_j + (\boldsymbol{e} \cdot \boldsymbol{k})\,(e_ik_j + e_jk_i)]: \quad (27)$$

This describes turbulent motions in planes perpendicular to $\boldsymbol{e}$,

$$e_ie_j{}^{(1)}\hat{Q}_{ij} = 0. \quad (28)$$

The ansatz

$$\hat{Q}_{ij} = {}^{(1)}\hat{Q}_{ij} + {}^{(2)}\hat{Q}_{ij} + {}^{(3)}\hat{Q}_{ij} \tag{29}$$

now expresses the general spectral tensor as the sum of three terms each of which vanishes when multiplied by the wave-number vector $\boldsymbol{k}$. It contains three free spectral functions, so there must be three different autocorrelations (literally "turbulent intensities"). Let $\mathring{g}$ be the radial direction and $\boldsymbol{e}$ the west-east one. If, as we have assumed, the spectral functions are even in every component of the wave-number vector, then in spherical coordinates we have for $\xi = 0$

$$Q_{ij} = \begin{pmatrix} \overline{u_r'^2} & 0 & 0 \\ 0 & \overline{u_\theta'^2} & 0 \\ 0 & 0 & \overline{u_\phi'^2} \end{pmatrix}, \tag{30}$$

with

$$\begin{aligned} \overline{u_r'^2} &= \iint [k_\theta^2 {}^{(1)}q + (k_\theta^2 + k_\phi^2)\, {}^{(3)}q] \, \mathrm{d}\boldsymbol{k} \, \mathrm{d}\omega, \\ \overline{u_\theta'^2} &= \iint [k_r^2 {}^{(1)}q + k_\phi^2 {}^{(2)}q + (k_r^2 + k_\phi^2)\, {}^{(3)}q] \, \mathrm{d}\boldsymbol{k} \, \mathrm{d}\omega, \\ \overline{u_\phi'^2} &= \iint [k_\theta^2 {}^{(2)}q + (k_r^2 + k_\theta^2)\, {}^{(3)}q] \, \mathrm{d}\boldsymbol{k} \, \mathrm{d}\omega. \end{aligned} \tag{31}$$

None of the cross-correlations $\overline{u_r'u_\theta'}$, $\overline{u_r'u_\phi'}$ or $\overline{u_\theta'u_\phi'}$ then appear.

There is an important exception, however: The existence of the two anisotropies, $\mathring{g}$ and $\boldsymbol{e}$, allows spectral tensors with cross-correlations. Indeed, by using terms that involve both $\boldsymbol{e}$ and $\mathring{g}$, one can construct a further tensor. Only four terms play a role:

$$\hat{Q}_{ij} = k^2 q(e_i \mathring{g}_j + e_j \mathring{g}_i) + q_1 \delta_{ij} + q_2(k_i \mathring{g}_j + k_j \mathring{g}_i) + q_3(k_i e_j + k_j e_i). \tag{32}$$

The coefficients $q, \ldots, q_3$ depend on all the components of the wave-number vector, i.e. on $\boldsymbol{e} \cdot \boldsymbol{k}$, $\mathring{g} \cdot \boldsymbol{k}$ and k^2. Incompressibility requires

$$q_1 = 2(\boldsymbol{e} \cdot \boldsymbol{k})\,(\mathring{g} \cdot \boldsymbol{k})\, q, \qquad q_2 = -(\boldsymbol{e} \cdot \boldsymbol{k})\, q, \qquad q_3 = -(\mathring{g} \cdot \boldsymbol{k})\, q, \tag{33}$$

so that

$$\begin{aligned} \hat{Q}_{ij} = q[&k^2(e_i \mathring{g}_j + e_j \mathring{g}_i) + 2(\boldsymbol{e} \cdot \boldsymbol{k})\,(\mathring{g} \cdot \boldsymbol{k})\, \delta_{ij} \\ &- (\boldsymbol{e} \cdot \boldsymbol{k})\,(k_i \mathring{g}_j + k_j \mathring{g}_i) - (\mathring{g} \cdot \boldsymbol{k})\,(k_i e_j + k_j e_i)]. \end{aligned} \tag{34}$$

This tensor has the properties that, if $\boldsymbol{e}$ and $\mathring{g}$ are perpendicular,

$$\hat{Q}_{ii} = 2(\boldsymbol{e} \cdot \boldsymbol{k})\,(\mathring{g} \cdot \boldsymbol{k})\, q, \qquad e_i e_j \hat{Q}_{ij} = \mathring{g}_i \mathring{g}_j \hat{Q}_{ij} = 0. \tag{35}$$

Hence, the turbulence (34) does not possess auto-correlations in the directions $\boldsymbol{e}$ and $\mathring{g}$.

Provided q is even,

$$e_i \mathring{g}_j \hat{Q}_{ij} = [k^2 - (\boldsymbol{e} \cdot \boldsymbol{k})^2 - (\mathring{\boldsymbol{g}} \cdot \boldsymbol{k})^2]\, q\,,$$

is nonzero, demonstrating yet again that a turbulence with two preferred directions can possess cross-correlations, unlike the tensors (23) and (27). Of course, this is precisely the crux of the matter for the TMT (see § 4.6): we are seeking nonzero cross-correlations $Q_{r\phi}$ and $Q_{\theta\phi}$. In this application the first direction is that of gravity, and the second is the direction of rotation, e.g. in the form

$$e_i = \varepsilon_{ipq} \mathring{\Omega}_p \mathring{g}_q\,. \tag{36}$$

If we take the form (36) then (34) becomes

$$\hat{Q}_{ij} = q\{\varepsilon_{qpk} k_p [2(\mathring{\boldsymbol{g}} \cdot \boldsymbol{k})\, \delta_{ij} - k_i \mathring{g}_j - k_j \mathring{g}_i] + k^2(\varepsilon_{qik} \mathring{g}_j + \varepsilon_{qjk} \mathring{g}_i) - (\mathring{\boldsymbol{g}} \cdot \boldsymbol{k})\,(k_i \varepsilon_{qjk} + k_j \varepsilon_{qik})\}\, \mathring{g}_q \mathring{\Omega}_k\,. \tag{37}$$

At the level where only modifications to anisotropic turbulence that are linear in the rotation are sought, this is the only spectral tensor that can arise, and it is symmetric. Using this construction, it is also possible to exhibit the cross-correlations $Q_{r\phi}$ and $Q_{\theta\phi}$ that result from the action of a uniform basic rotation. If again $\boldsymbol{e}$ is east and $\mathring{\boldsymbol{g}}$ is vertical we find

$$Q_{r\phi} = \mathring{g}_i e_j \iint \hat{Q}_{ij}\, \mathrm{d}\boldsymbol{k}\, \mathrm{d}\omega = \iint k_\theta^2 q\, \mathrm{d}\boldsymbol{k}\, \mathrm{d}\omega\,. \tag{38}$$

Note the special significance of the latitudinal wave-number.

It remains to discover the significance of "q". To this end we shall calculate in Chapter 14 the influence of rotation on some simple forms for the original turbulence which are themselves free of cross-correlations. They each have spectral tensors that contain three free functions representing the values of $\overline{u_r^2}$, $\overline{u_\theta^2}$ and $\overline{u_\phi^2}$, a possibility which has already been covered by (30). There the turbulence properties are not changed if $\boldsymbol{e}$ (or $\mathring{\boldsymbol{g}}$) is replaced by $-\boldsymbol{e}$ (or $-\mathring{\boldsymbol{g}}$). Thus, the words "top" and "bottom" have no meaning for such turbulences. In other words, only those parts of the turbulence fields that are invariant under the transformation $\mathring{\boldsymbol{g}} \to -\mathring{\boldsymbol{g}}$ contribute to the TMT. This is quite opposite to the situation that occurs in the theory of the α-effect in MHD. As we have already seen in § 4.6, all the relevant correlation tensors involving one-point cross-correlations are even in $\mathring{\boldsymbol{g}}$. It is therefore not easy to derive both the Λ-effect and the α-effect from one and the same original turbulence field.

13.3. Spectral Functions

Bochner's theorem for incompressible homogeneous turbulence requires the trace $\hat{Q}_{ii}$ to be non-negative. Thus, from (29), we find

$$k_r^2(2\,{}^{(3)}q + {}^{(1)}q) + k_\theta^2(2\,{}^{(3)}q + {}^{(2)}q) + k_\phi^2(2\,{}^{(3)}q + {}^{(2)}q + {}^{(1)}q) \geqq 0\,, \tag{39}$$

for all $\boldsymbol{k}$. For uncomplicated spectra this implies relations such as

$$2^{(3)}q + {}^{(1)}q \geqq 0, \qquad 2^{(3)}q + {}^{(2)}q \geqq 0, \qquad 2^{(3)}q + {}^{(2)}q + {}^{(1)}q \geqq 0.$$

The spectral functions are unrestricted, apart from such conditions and from the natural demand that $q \to 0$ as $\boldsymbol{k}$ or $\omega \to \infty$. It is through the specification of the functions, q, that we introduce various turbulence models. For example, models with ${}^{(1)}q = 0$ are isotropic in planes perpendicular to $\mathring{\boldsymbol{g}}$. If, in addition ${}^{(2)}q$ and ${}^{(3)}q$ depend only on k_r, $k_\perp$ and ω, then

$$q = q(k_r, k_\perp, \omega). \tag{40}$$

Such models whose anisotropy is confined to one direction, $\mathring{\boldsymbol{g}}$, will be called "axisymmetric". Their spectral functions must obey the Bochner conditions:

$$^{(3)}q \geqq 0, \qquad 2^{(3)}q + {}^{(2)}q \geqq 0. \tag{41}$$

When considering the frequency dependence of the spectra we should distinguish two possibilities. The first spectrum ("turbulence" type), does not vanish for $\omega = 0$ and decreases with increasing frequency, i.e. $\omega\, \partial q/\partial\omega \leqq 0$. Particular examples of this class are the spectra $\exp(-\omega^2)$ and $1/(1 + \omega^2)$ which are the Fourier transforms of the correlation functions $\exp(-\tau^2)$ and $\exp(-|\tau|)$. We generally expect "turbulence" type spectra to be the rule rather than the exception. It turns out that they represent fully-developed turbulences which, according to Fenstermacher *et al.* (1979) and Gollub and Benson (1980), do not show any periodic or quasiperiodic behavior.

The representation of turbulence that is most frequently employed, as a combination of "random waves", requires the spectrum to vanish at the origin. This can be readily verified by means of the definition

$$\hat{Q}_{ij}(\boldsymbol{k}, 0) = (2\pi)^{-4} \int\int \overline{u_i'(\boldsymbol{x}, t)\, u_j'(\boldsymbol{x} + \boldsymbol{\xi}, t + \tau)}\, e^{-i\boldsymbol{k}\cdot\boldsymbol{\xi}}\, \mathrm{d}\boldsymbol{\xi}\, \mathrm{d}\tau, \tag{42}$$

which vanishes for temporally-periodic correlations. If we suppose that factorization of the spectral functions is possible:

$$q(\boldsymbol{k}, \omega) = q_k(\boldsymbol{k})\, q_\omega(\omega), \tag{43}$$

then the simplest representation of "random waves" type is

$$q_\omega(\omega) = \delta(\omega - \tau_c^{-1})/\tau_c. \tag{44}$$

Here our standard normalization has been used, namely

$$W \equiv \int_{-\infty}^{\infty} q_\omega\, \mathrm{d}\omega = \tau_c^{-1}. \tag{45}$$

From this point onwards, τ_c will represent a "correlation" time and W the associated characteristic frequency. Later on, L will be used as a correlation length and K as the associated characteristic wave-number. The dimension of q_k is $cm^7 s^{-1}$.

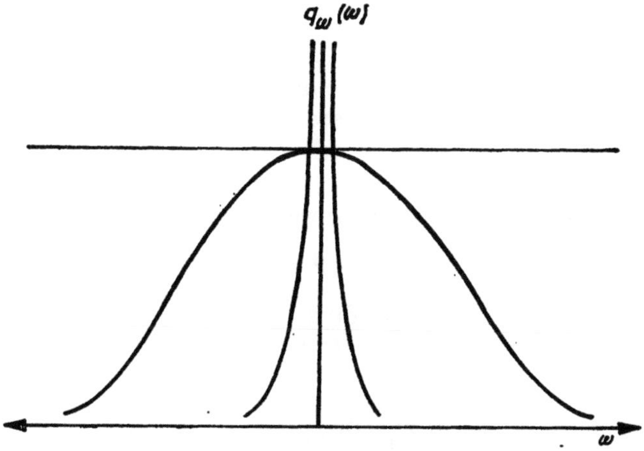

Fig. 13.1. The frequency spectrum of "turbulence" lies between the profiles for very short correlation times (q_ω = constant) and very large correlation times [$q_\omega \sim \delta(\omega)$]. We only use spectra that are even in the frequency ω.

In a certain sense a frequency spectrum of turbulence-type lies between two extreme cases: constant q (flat profiles) and a delta function at $\omega = 0$, i.e. $q \sim \delta(\omega)$ (steep profiles). Fig. 13.1 illustrates the principal configurations. Obviously, the first represents a correlation time dependence of delta function type. Vainshtein (1972) used such profiles to represent turbulence having a very short correlation time. This "white noise" particularly stresses the high frequencies. Insofar as the true frequency should be compared with the "viscosity frequency" νK^2,

$$\nu^* = \nu K^2/W ,$$

the constant spectrum tends to be appropriate for the case $\nu^* < 1$, which means "low viscosity" in the random-wave representation. At the other extreme, the spectra which are peaked at $\omega = 0$ imply infinite correlation times, $\tau_c \to \infty$. A particular correlation then no longer depends on the temporal interval between the measurements; it does not diminish with time. Since the high frequencies are filtered out, this model mimics high viscosities, $\nu^* > 1$, in the approximation of single modes. It is clear, therefore, that a turbulence-type representation with a monotonically decreasing spectrum lies somewhere between the two extremes, $\nu^* \ll 1$ and $\nu^* \gg 1$, of the random-wave representation. Detailed calculations are, however, needed before its exact position is established. We shall often use, as a frequency spectrum of "turbulence" type, the monotonic profile

$$q_\omega = \pi^{-1}(1 + \tau_c^2\omega^2)^{-1} \equiv \pi^{-1}W^2/(W^2 + \omega^2), \tag{46}$$

which corresponds to the correlation function $\exp(-|\tau|/\tau_c)$. Another example, which is more difficult to handle, is $q_\omega = \tau_c \exp(-\tau_c^2\omega^2)/\pi$, which when transformed corresponds to the correlation time law, $\exp(-\tau^2/\tau_c^2)$.

A similar scheme is also needed for the wave-number spectrum, and this is facilitated if the following different classes are introduced:

(a) *Few-mode spectra*

Once again the simplest spectrum is that for wave-like motions,

$$^{(i)}q_k(\boldsymbol{k}) = {}^{(i)}U\delta(\boldsymbol{k} - \boldsymbol{K}), \qquad i = 1, 2, 3, \tag{47}$$

where the dimension of U is cm^4/s. Depending on the choice of the U'. there are eight combinations of wave-numbers (see Durney and Spruit, 1979) Application of (47) to (31) yields

$$\begin{aligned} \tau_c\overline{u_r'^2} &= K_\theta^{2(1)}U + (K_\theta^2 + K_\phi^2)\,{}^{(3)}U, \\ \tau_c\overline{u_\theta'^2} &= K_r^{2(1)}U + K_\phi^{2(2)}U + (K_r^2 + K_\phi^2)\,{}^{(3)}U, \\ \tau_c\overline{u_\phi'^2} &= K_\theta^{2(2)}U + (K_r^2 + K_\theta^2)\,{}^{(3)}U, \end{aligned} \tag{48}$$

for the turbulent intensities, with $\boldsymbol{K}$ being that characteristic wave-number vector for which the spectrum is peaked. Inversion of (48) provides expressions for the constants $^{(i)}U$:

$$\begin{aligned} {}^{(1)}U/\tau_c &= [(K_\phi^2 - K_\theta^2)\,K_r^2\overline{u_r'^2} + K_\theta^2(K_\theta^2 + K_\phi^2)\,\overline{u_\theta'^2} \\ &\quad - K_\phi^2(K_\theta^2 + K_\phi^2)\,\overline{u_\phi'^2}]/2K_r^2K_\theta^2K_\phi^2, \\ {}^{(2)}U/\tau_c &= [-K_r^2(K_r^2 + K_\theta^2)\,\overline{u_r'^2} + K_\theta^2(K_r^2 + K_\theta^2)\,\overline{u_\theta'^2} \\ &\quad + K_\phi^2(K_r^2 - K_\theta^2)\,\overline{u_\phi'^2}]/2K_r^2K_\theta^2K_\phi^2, \\ {}^{(3)}U/\tau_c &= (K_r^2K_\theta^2\overline{u_r'^2} + K_\theta^2K_\phi^2\overline{u_\phi'^2} - K_\theta^4\overline{u_\theta'^2})/2K_r^2K_\theta^2K_\phi^2. \end{aligned} \tag{49}$$

Now we shall give some examples. Simple models lead to fixed $^{(i)}U$. The first illustration will be turbulence with axial symmetry, i.e. with complete isotropy in the horizontal plane. This requires

$$K_\phi = K_\theta, \qquad \overline{u_\phi'^2} = \overline{u_\theta'^2}, \tag{50}$$

so that

$$\begin{aligned} &{}^{(1)}U = 0, \qquad {}^{(2)}U/\tau_c = [2K_\phi^2\overline{u_\phi'^2} - (K_r^2 + K_\phi^2)\,\overline{u_r'^2}]/2K_\phi^4, \\ &{}^{(3)}U/\tau_c = \overline{u_r'^2}/2K_\phi^2. \end{aligned} \tag{51}$$

If, two-dimensionality is also demanded, i.e.

$$K_r = \overline{u_r'^2} = 0, \tag{52}$$

we find the very simple expressions

$$^{(1)}U = {}^{(3)}U = 0, \qquad {}^{(2)}U = \tau_c \overline{u_\phi'^2}/K_\phi^2. \tag{53}$$

On the other hand, purely radial motions, i.e. $\overline{u'_\phi{}^2} = \overline{u'_\theta{}^2} = 0$, are produced when we take

$$^{(2)}U = -{}^{(3)}U = -\tau_c \overline{u_r'^2}/2K_\phi^2. \tag{54}$$

Note that here $^{(2)}U$ has the opposite sign to that for 2D turbulence.

The second example concerns turbulence with horizontal anisotropy. For example, rolls with axes in the N-S direction occur when $K_\theta = 0$. If the latitudinal intensity is also made to vanish we find

$$K_r^{2\,(1)}U + K_\phi^{2\,(2)}U = -(1 + K_\phi^2/K_r^2)\,\overline{u_\phi'^2}, \qquad {}^{(3)}U = \tau_c \overline{u_r'^2}/K_\phi^2, \tag{55}$$

where we have used $K_r^2\overline{u_r'^2} = K_\phi^2\overline{u_\phi'^2}$, a necessary condition for convergence. Quite similar expressions may be derived for E-W rolls which are characterized by $K_\phi^2 = \overline{u_\phi'^2} = 0$ (Rüdiger, 1980).

Another rather tempting way of specifying the spectrum parameters $^{(1)}U$, $^{(2)}U$ and $^{(3)}U$ is to ignore $^{(1)}U$ and use a very simple relation between $^{(2)}U$ and $^{(3)}U$:

$$^{(2)}U = -\alpha\, {}^{(3)}U. \tag{56}$$

Then we have immediately

$$\begin{aligned} \overline{u_\theta'^2} &= [K_r^2 + (1-\alpha)\,K_\phi^2]\,\overline{u_r'^2}/K_\perp^2, \\ \overline{u_\phi'^2} &= [K_r^2 + (1-\alpha)\,K_\theta^2]\,\overline{u_r'^2}/K_\perp^2, \end{aligned} \tag{57}$$

and the appropriate Bochner condition

$$2K_r^2 + (2-\alpha)\,K_\perp^2 \geqq 0. \tag{58}$$

Obviously the constant quantity α must be less than 2. The simplest expressions result, however, if we write, as Durney and Spruit (1979) did,

$$\alpha = K^2/K_\perp^2, \tag{59}$$

and so obtain

$$\overline{u_\theta'^2} = (K_r^2K_\theta^2/K_\perp^4)\,\overline{u_r'^2}, \qquad \overline{u_\phi'^2} = (K_r^2K_\phi^2/K_\perp^4)\,\overline{u_r'^2}. \tag{60}$$

The implied ratio between the horizontal and vertical intensities is

$$\overline{u_\perp'^2}/\overline{u_r'^2} = K_r^2/K_\perp^2. \tag{61}$$

The particular significance of the conditions (56) and (59) will be clear later, when calculations of the influence of rotation on our turbulence models are presented. For the moment we will merely underscore the fact that, by using (59), we have satisfied the Bochner relation (58) in the most natural way: $2K_r^2 + (2 - \alpha) K_\perp^2 = K^2$, which is, of course, non-negative.

(b) Axisymmetric turbulence

Other models that have been discussed in detail are

$$^{(1)}q = 0, \qquad ^{(2)}q = -\alpha q, \qquad ^{(3)}q = q > 0, \qquad \alpha < 2. \tag{62}$$

The resulting spectral tensor is

$$\mathring{Q}_{ij} = q[k^2 K_{ij} - \alpha(k_\perp^2 G_{ij} - k_{\perp i} k_{\perp j})], \tag{63}$$

which describes axial symmetry if q depends only on the scalars k and $k_\perp$. For $\alpha \to -\infty$, only motions perpendicular to $\mathring{\boldsymbol{g}}$ occur (horizontal turbulence), whereas $\alpha = 0$ reproduces the isotropic case (20). The non-negativity of the spectral function q is ensured by Bochner's theorem. By specifying this function suitably one can establish different structures for the convective elements. For example, for $\alpha = 0$ the function $q = q(k, \omega)$ gives isotropic velocity distributions whereas $q = q(k, \omega)\, \delta(\mathring{\boldsymbol{g}} \cdot \boldsymbol{k})$ produces infinitely long cells (along $\mathring{\boldsymbol{g}}$). In contrast to this, spectral functions possessing a *minimum* at $\mathring{\boldsymbol{g}} \cdot \boldsymbol{k} = 0$ flatten the turbulent elements into "pancakes".

The accuracy of this claim may be demonstrated by considering the lines on which the auto-correlation perpendicular to $\mathring{\boldsymbol{g}}$ takes half its maximum value. To this end we calculate the correlation of the velocity $\boldsymbol{e} \cdot \boldsymbol{u}'(\boldsymbol{x})$ with $\boldsymbol{e} \cdot \boldsymbol{u}'(\boldsymbol{x} + \boldsymbol{\xi}_\perp)$, where $\boldsymbol{\xi}_\perp$ lies in the meridional plane: $\boldsymbol{e} \cdot \boldsymbol{\xi}_\perp = 0$. These transverse correlations decrease with increasing distance. To calculate the distance, $\xi_\perp$, at which the correlation considered is reduced by half, we must solve

$$\overline{u_\phi'(\boldsymbol{x})\, u_\phi'(\boldsymbol{x} + \boldsymbol{\xi}_\perp)}/\overline{u_\phi'^2(\boldsymbol{x})} = 0.5. \tag{64}$$

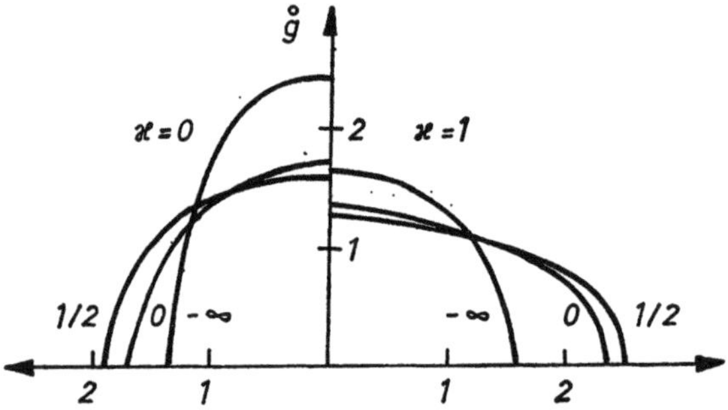

Fig. 13.2. Lines on which correlations take half their full values for the cell configuration of our models. The lines are labelled by their α-values. The most extreme cells are for $\varkappa = 0$, $\alpha = -\infty$ (oblong) and $\varkappa = 1$, $\alpha = 1/2$ (flat).

For isotropic turbulence this distance is independent of the angular position, whereas in all other cases it produces a line that is deformed to a lesser or a greater extent. We show these lines in Fig. 13.2 for the two models $\varkappa = 0$ and $\varkappa = 1$, where $\varkappa$ is the exponent in the spectral function:

$$q = q(k, \omega) \, (\mathring{\boldsymbol{g}} \cdot \mathring{\boldsymbol{k}})^{2\varkappa}. \tag{65}$$

In some applications $q(k, \omega)$ is replaced by delta functions. One can see in Fig. 13.2 that, for a given α, the cells for models with $\varkappa = 1$ are indeed flatter than those for models with $\varkappa = 0$.

As for the anisotropies of the auto-correlations, one immediately finds

$$\overline{u_r'^2}/\overline{u_\perp'^2} = \iint q k_\perp^2 \, \mathrm{d}\boldsymbol{k} \, \mathrm{d}\omega \Big/ \iint q[2k^2 - (1 + \alpha) \, k_\perp^2] \, \mathrm{d}\boldsymbol{k} \, \mathrm{d}\omega , \tag{66}$$

hence for $\varkappa = 0$

$$\overline{u_r'^2} \Big/ \frac{1}{2} \overline{u_\perp'^2} = 2/(2 - \alpha), \tag{67}$$

and for $\varkappa = 1$

$$\overline{u_r'^2} \Big/ \frac{1}{2} \overline{u_\perp'^2} = 2/(4 - \alpha): \tag{68}$$

Note that $\alpha < 0$ always provides horizontal-type turbulence, in which the horizontal intensity exceeds the vertical one. In contrast, vertical-type turbulence can only be formed from a model having $\varkappa = 0$ and a positive α (Fig. 13.3). In Table XIII.i. the main features of the models are listed for various values of the parameters.

Let $q = q_k \cdot q_\omega$, and let q_ω be positive and normalized according to (45). Then the intensity difference δI of the turbulence [cf. (24)] is in general

$$\delta I = [(2\varkappa - \alpha)/(2\varkappa + 1) \, (2\varkappa + 3) \, \tau_c] \int k^2 q_k \, \mathrm{d}\boldsymbol{k}. \tag{69}$$

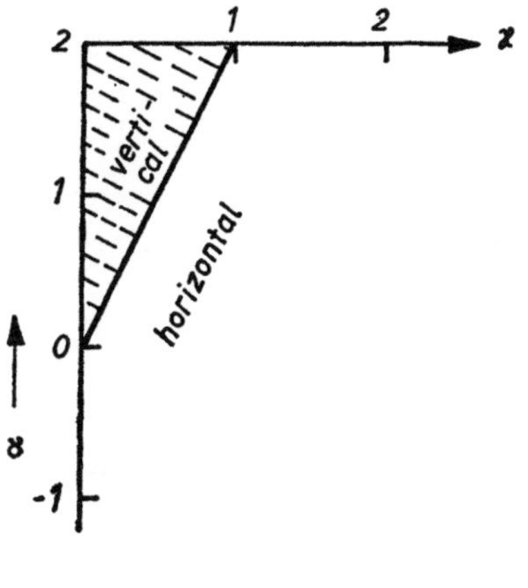

Fig. 13.3. Our model (65) prefers horizontal-type turbulences. Only for a few values in the parameter range ($\varkappa = 0$, $\alpha > 0$) does the vertical intensity exceed the horizontal intensity.

Table XIII.i. Intensities and cell configurations for various values of the parameters.

model	$\varkappa = 0$	$\varkappa = 1$
$\alpha < 0$	horizontal-type, oblong	horizontal-type, spherical
$\alpha = 0$	isotropic, spherical	horizontal-type, flat
$0 < \alpha < 2$	vertical-type, nearly spherical	horizontal-type, very flat

The simplest wave-number spectrum is that with a distinct peak at, for example, $K = L^{-1}$:

$$q_k = L\tau_c k^{-3} I \delta(k - L^{-1}), \qquad I > 0. \tag{70}$$

This gives the anisotropy parameter

$$\delta I = [(2\varkappa - \alpha)/(2\varkappa + 1)\,(2\varkappa + 3)]\, 4\pi I, \tag{71}$$

so that only an α in the range $2 > \alpha > 2\varkappa$ gives vertical-type turbulence (Fig. 13.3). Positive δI, i.e. $\alpha < 2\varkappa$, corresponds to turbulences having dominantly horizontal motions.

More extensive calculations are needed to clarify the cell structure. The results given in Fig. 13.4 demonstrate that the cells become flatter (along $\mathring{g}$) as the $\varkappa$ increases. For the special calculations of the following Chapters we shall use particular representations of the wave-number-part, q_k, of the spectral functions. These are all similar to one another, and have been chosen with a view to simplifying the integrations of the equations to which they are applied. Examples of such wave-number spectra are (15.13), (15.17) and (18.42), and using these the Λ-effect, eddy viscosity and turbulent conductivity were computed.

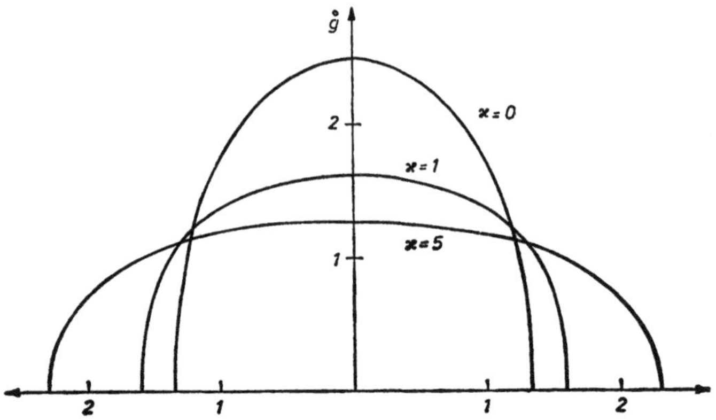

Fig. 13.4. The greater the value of $\varkappa$ in (65) the flatter the cell. The figure shows the half-value lines as defined by (64).

Chapter 14. Theory of the Λ-Effect

In this Chapter we develop a theory of the Λ-effect. To this end we will determine the influence of a rigid rotation on prescribed turbulences. The (stochastic) forces which maintain the stochastic motions are assumed to be independent of the basic rotation but to take part in the rotation. The prescribed "original" turbulence fields are also assumed to be so simple that the mirror-symmetric models of Chapter 13 can be adopted. For the preferred direction the radial direction ("gravity") $\mathring{\boldsymbol{g}}$ will, of course, be used. Since this varies from point to point, the resulting correlations are globally inhomogeneous and therefore give rise to nonzero Reynolds stresses. With this idea in mind, we hope to describe fairly accurately the real stellar situation.

14.1. The Spectral Tensor for Rotating Turbulence

It is convenient to start with the equation of motion in a co-rotating frame of reference:

$$\partial u_i/\partial t + (\boldsymbol{u} \cdot \nabla)\, u_i = -\varrho^{-1} \nabla_i p + \nu \Delta u_i - 2\varepsilon_{ijk}\Omega_j u_k + f_i, \qquad (1)$$

where $\boldsymbol{f}$ denotes the external forces per unit mass such as gravity and centrifugal force. The Reynolds equation, which results from (1) after averaging, is

$$\partial \bar{u}_i/\partial t + (\bar{\boldsymbol{u}} \cdot \nabla)\, \bar{u}_i + \overline{(\boldsymbol{u}' \cdot \nabla)\, u_i'} = -\overline{\varrho^{-1} \nabla_i p} + \nu \Delta \bar{u}_i - 2\varepsilon_{ijk}\Omega_j \bar{u}_k + \bar{f}_i. \quad (2)$$

If the density is assumed to be non-fluctuating, the result of subtracting (2) from (1) is

$$\begin{aligned} &\partial u_i'/\partial t - \nu \Delta u_i' + \varrho^{-1} \nabla_i p' + 2\varepsilon_{ijk}\Omega_j u_k' = f_i' + \overline{(\boldsymbol{u}' \cdot \nabla)\, u_i'} - (\boldsymbol{u}' \cdot \nabla)\, u_i', \\ &\nabla \cdot \boldsymbol{u}' = 0. \end{aligned} \qquad (3)$$

In these equations, which govern the velocity fluctuations, a fluctuating force $\boldsymbol{f}'$, per unit mass, appears. The nonlinear terms of the auto-interaction, namely $\overline{(\boldsymbol{u}' \cdot \nabla)\, \boldsymbol{u}'} - (\boldsymbol{u}' \cdot \nabla)\, \boldsymbol{u}'$, describe the energy flux between the different turbulent modes. They give rise to higher-order correlations unless some linearizing procedure is applied. Often they are neglected completely, which may be a valid step when the turbulence is weak:

$$\min\,(Lu'/\nu,\ \tau_c u'/L) \ll 1\,. \tag{4}$$

In our studies, a slight modification of this concept is preferred. It is assumed that a small-scale turbulence provides a background for the modes on which we focus. Thus, the large-scale fluctuations are attenuated by the eddy viscosity of the small-scale field rather than by the molecular viscosity. Indeed, turbulence that is sufficiently homogeneous and isotropic (i.e. small-scale) influences large-scale fields only through the eddy viscosity. In other words, a modified Boussinesq relation applies:

$$\overline{(\boldsymbol{u}' \cdot \nabla)\, u_i'} - (\boldsymbol{u}' \cdot \nabla)\, u_i' = \nu_T \Delta u_i'\,, \tag{5}$$

where the eddy viscosity ν_T is of order $5 \cdot 10^{12}$ cm^2/s for the Sun (Gierasch, 1974). This ansatz recognizes explicitly the only effect that is important, on all scales up to those of the granules. All numerical investigators have used exactly this ansatz to parameterize dissipation due to the small-scale part of the spectrum [see, for example, Marcus (1980)]. The eddy viscosity, which arises from a turbulence field that has no preferred direction (i.e. from an originally homogeneous and isotropic field, does not vanish when the molecular viscosity is zero (Krause and Rüdiger, 1974a; Kraichnan, 1976).

By the procedure we have just outlined, we overcome all the difficulties that would otherwise arise through the ineffectiveness of the molecular viscosity on the fluctuation scales with which we have to deal. The only new consequence of (5) is that the viscosity which must be used in any relations that neglect the fluctuating inertial terms is $\nu_e = \nu + \nu_T$ and not ν.

A similar procedure has been proposed by Vainshtein and Kichatinov (1983) who, on the basis of the closure scheme set out by Orszag (1970), suggest relaxation via

$$\partial u_i'/\partial t - \overline{(\boldsymbol{u}' \cdot \nabla)\, u_i'} + (\boldsymbol{u}' \cdot \nabla)\, u_i' = u_i'/\tau_c\,, \tag{6}$$

where τ_c is the relaxation time of the correlation. If this approach is followed it leads to

$$\partial u_i'/\partial t - \nu_e \Delta u_i' + \bar{\varrho}^{-1}\, \nabla_i p' + 2\varepsilon_{ijk}\Omega_j u_k' = f_i'\,, \tag{7}$$

where $\varrho \boldsymbol{f}'$ now represents the stochastic system of forces which maintains the turbulent motion, instead of (for example) a buoyancy force, as in the case of real convection. It is defined in the co-rotating system but it does not involve the angular velocity. This procedure could also be used in a non-rotating inertial system but then it would be necessary to include the overall rotation of the system in specifying the acceleration, $\boldsymbol{f}'$. Also the turbulence does not, in that case, arise from the global rotation.

The effect of rotation on the *onset* of turbulence has been ignored. We must, of course, ask whether the results to which we will then be led will dominate those associated with the effects of rotation at onset, which are really strong. "Actually, two distinct questions need to be considered. One is whether the condition for the onset of turbulence is altered by the presence of rotation; the second, whether turbulence, once established in a rotating star, is seriously different in its general properties from that in a non-rotating one" (Cowling, 1951). As far as the Sun is concerned, these two distinct aspects have to be assessed for a rotation that is not slow. Whereas the first question has been considered by Cowling (loc. cit.), by Tayler (1973) and by Durney and Spruit (1979), our work here is mainly concerned with the second.

The equation which must be considered is (7). No restriction need be placed on the strength of rotation in the following argument. In order to solve the governing equations, Fourier transforms

$$\boldsymbol{F}(\boldsymbol{x}, t) = \iint \hat{\boldsymbol{F}}(\boldsymbol{k}, \omega)\, \mathrm{e}^{\mathrm{i}(\boldsymbol{k}\cdot\boldsymbol{x} - \omega t)}\, \mathrm{d}\boldsymbol{k}\, \mathrm{d}\omega, \tag{8}$$

are introduced. This leads to a new form for (7), viz:

$$(-i\omega + \nu_e k^2)\, \hat{\boldsymbol{u}} + 2\boldsymbol{\Omega} \times \hat{\boldsymbol{u}} + i\varrho^{-1}\hat{p}\boldsymbol{k} = \hat{\boldsymbol{f}}, \qquad \boldsymbol{k}\cdot\hat{\boldsymbol{u}} = 0. \tag{9}$$

The pressure can be eliminated by multiplying $(9)_1$ by the tensor K_{ij} to give

$$(-i\omega + \nu_e k^2)\, \hat{u}_i + 2\varepsilon_{ipk}\Omega_p\hat{u}_k - 2[\mathring{\boldsymbol{k}}\cdot(\boldsymbol{\Omega}\times\hat{\boldsymbol{u}})]\, \mathring{k}_i = K_{ik}\hat{f}_k. \tag{10}$$

Since the vector product $\boldsymbol{\Omega} \times \boldsymbol{u}$ is difficult to handle we shall proceed as follows. Due to the incompressibility, the two vectors $\boldsymbol{k}$ and $\hat{\boldsymbol{u}}$ are orthogonal, i.e. $\boldsymbol{k}\cdot\hat{\boldsymbol{u}} = 0$. Hence, the three vectors $\boldsymbol{k}$, $\hat{\boldsymbol{u}}$ and $\boldsymbol{k}\times\hat{\boldsymbol{u}}$ form a Cartesian coordinate system to which we can refer the vector $\boldsymbol{\Omega}\times\boldsymbol{k}$. On applying the identity

$$\boldsymbol{\Omega}\times\hat{\boldsymbol{u}} = [\mathring{\boldsymbol{k}}\cdot(\boldsymbol{\Omega}\times\hat{\boldsymbol{u}})]\, \boldsymbol{k} + (\mathring{\boldsymbol{k}}\cdot\boldsymbol{\Omega})\, (\boldsymbol{k}\times\hat{\boldsymbol{u}}) \tag{11}$$

to (10), there results

$$(-i\omega + \nu_e k^2)\, \hat{u}_i + 2(\mathring{\boldsymbol{k}}\cdot\boldsymbol{\Omega})\, \varepsilon_{ijk}\mathring{k}_j\hat{u}_k = K_{ik}\hat{f}_k. \tag{12}$$

The general solution of these equations is

$$\hat{u}_i = D_{il}K_{lk}\hat{f}_k/(-i\omega + \nu_e k^2), \tag{13}$$

where the matrix D_{il} is given by

$$D_{il} = \frac{(-i\omega + \nu_e k^2)^2\, \delta_{il} + 2(\mathring{\boldsymbol{k}}\cdot\boldsymbol{\Omega})\, (-i\omega + \nu_e k^2)\, \varepsilon_{ilm}\mathring{k}_m}{(-i\omega + \nu_e k^2)^2 + (2\mathring{\boldsymbol{k}}\cdot\boldsymbol{\Omega})^2}. \tag{14}$$

This matrix represents the effect of the Coriolis forces on the original motion, and it collapses to the Kronecker delta if the system does not rotate. In that case, the stochastic system of forces $\boldsymbol{f}'$ would produce the velocity field $\boldsymbol{u}^{(0)}$ given by

$$\hat{u}_i^{(0)} = K_{ik}\hat{f}_k/(-i\omega + \nu_e k^2), \tag{15}$$

which represents the velocity field that, in Chapter 13, we called the "original" one.

Inserting this representation into (13) we obtain the more convenient formula

$$\hat{u}_i = D_{il}\hat{u}_l^{(0)}. \tag{16}$$

According to the Fourier transformation (8) we have

$$Q_{ij} = \int \overline{\hat{u}_i(\boldsymbol{k}, \omega)\, \hat{u}_j(\boldsymbol{k}', \omega')} \exp\{i[(\boldsymbol{k} + \boldsymbol{k}') \cdot \boldsymbol{x} - (\omega + \omega')\, t]\}\, d\boldsymbol{k}\, d\boldsymbol{k}'\, d\omega\, d\omega', \tag{17}$$

so that with (16)

$$Q_{ij} = \int D_{il}(\boldsymbol{k}, \omega)\, D_{jm}(\boldsymbol{k}', \omega')\, \overline{\hat{u}_l^{(0)}(\boldsymbol{k}, \omega)\, \hat{u}_m^{(0)}(\boldsymbol{k}', \omega')} \times \exp\{i[(\boldsymbol{k} + \boldsymbol{k}') \cdot \boldsymbol{x} - (\omega + \omega')\, t]\}\, d\boldsymbol{k}\, d\boldsymbol{k}'\, d\omega\, d\omega'. \tag{18}$$

Since $\hat{\boldsymbol{u}}^{(0)}$ belongs to a homogeneous turbulence, the well-known relation

$$\overline{\hat{u}_i^{(0)}(\boldsymbol{k}, \omega)\, \hat{u}_j^{(0)}(\boldsymbol{k}', \omega')} = \hat{Q}_{ij}^{(0)}(\boldsymbol{k}', \omega')\, \delta(\boldsymbol{k} + \boldsymbol{k}', \omega + \omega') \tag{19}$$

can be applied and we find

$$\hat{Q}_{ij} = D_{il}^* D_{jm} \hat{Q}_{lm}^{(0)}. \tag{20}$$

Written out in detail, this is

$$\hat{Q}_{ij} = [(\omega^2 + \nu_e^2 k^4)/NN^*]\, \{[\omega^2 + \nu_e^2 k^4 - (2\boldsymbol{k} \cdot \boldsymbol{\Omega})^2]\, \hat{Q}_{ij}^{(0)} + (2\hat{\boldsymbol{k}} \cdot \boldsymbol{\Omega})^2 K_{ij} \hat{Q}_{ll}^{(0)} + (2\boldsymbol{k} \cdot \boldsymbol{\Omega})\, \hat{k}_m[\nu_e k^2(\varepsilon_{ilm}\hat{Q}_{lj}^{(0)} + \varepsilon_{jlm}\hat{Q}_{il}^{(0)}) + i\omega(\varepsilon_{jlm}\hat{Q}_{il}^{(0)} - \varepsilon_{ilm}\hat{Q}_{lj}^{(0)})]\}, \tag{21}$$

where

$$N = (-i\omega + \nu_e k^2)^2 + (2\hat{\boldsymbol{k}} \cdot \boldsymbol{\Omega})^2.$$

Expression (21) has already been derived elsewhere (Rüdiger, 1977a). It demonstrates either the weak or the strong influence of slow or rapid rotation on turbulence generated by a homogeneous system of forces that is independent of the rotation. In inviscid media the denominator of (14) may vanish for

$$\omega = \pm 2\hat{\boldsymbol{k}} \cdot \boldsymbol{\Omega}, \tag{22}$$

which is the dispersion relation for free "inertial waves".

The one-point correlations for rotating turbulence can now be established. We could even determine two-point correlations but this is not necessary in the Reynolds stress theory of differential rotation. It is clear at a glance that turbulence which is two-dimensional with respect to the z-axis [its spectral tensor containing the delta function $\delta(\boldsymbol{k} \cdot \boldsymbol{\Omega})$] is not influenced by rotation. This is one consequence of the Taylor-Proudman theorem. The correlation length parallel to $\boldsymbol{\Omega}$ is infinite, and the motion therefore does not depend on the coordinate z. Rotation cannot influence such fields since the Coriolis force becomes conservative and can be absorbed into the pressure.

14.2. The Two Parts of the Correlation Tensor

The spectral tensor of the original turbulence, $\hat{Q}_{ij}^{(0)}$, is assumed to be real, i.e. symmetric in i and j. Thus we postulate that only rather simple fields exist in a non-rotating star. The main examples of such fields were introduced in Chapter 13, where we characterized them as "homogeneous and without helicity". For example, it is not possible to define "above" and "below" in such a field from properties of its correlations. We are mainly interested in the one-point correlation tensor (21) of rotating turbulence, which appears as the Reynolds stress in mean-field hydrodynamics. It may be divided into two parts depending on its behavior under the transformation $\Omega \to -\Omega$:

$$Q_{ij}^{(r)} = Q_{ij}^{\text{even}} + Q_{ij}^{\text{odd}}, \tag{23}$$

where Q^{even} is the part even in Ω, and Q^{odd} is the part odd in Ω. The contributions can be separated immediately by using (21):

$$\begin{aligned} Q_{ij}^{\text{even}} &= \iint \frac{\omega^2 + \nu_e^2 k^4}{NN^*} \{[\omega^2 + \nu_e^2 k^4 - (2\boldsymbol{k} \cdot \boldsymbol{\Omega})^2]\, \hat{Q}_{ij}^{(0)} \\ &\quad + K_{ij} (2\boldsymbol{k} \cdot \boldsymbol{\Omega})^2\, \hat{Q}_{ll}^{(0)}\}\, d\boldsymbol{k}\, d\omega, \\ Q_{ij}^{\text{odd}} &= \iint \frac{\omega^2 + \nu_e^2 k^4}{NN^*} (2\boldsymbol{k} \cdot \boldsymbol{\Omega})\, \nu_e k_m (\varepsilon_{ilm} \hat{Q}_{lj}^{(0)} + \varepsilon_{jlm} \hat{Q}_{il}^{(0)})\, d\boldsymbol{k}\, d\omega. \end{aligned} \tag{24}$$

The denominator is

$$NN^* = (\omega^2 + \nu_e^2 k^4)^2 + 2(\nu_e^2 k^4 - \omega^2)\,(2\boldsymbol{k} \cdot \boldsymbol{\Omega})^2 + (2\boldsymbol{k} \cdot \boldsymbol{\Omega})^4. \tag{25}$$

The trace of the even part is

$$Q_{ll}^{\text{even}} = \iint [(\omega^2 + \nu_e^2 k^4)/NN^*]\, [\omega^2 + \nu_e^2 k^4 + (2\boldsymbol{k} \cdot \boldsymbol{\Omega})^2]\, \hat{Q}_{ll}^{(0)}\, d\boldsymbol{k}\, d\omega, \tag{26}$$

while that of the odd part vanishes. Hence, the energy of the turbulence is contained in the even part, whereas the cross-correlations with ϕ, i.e. the

Λ-effect, are contained in Q^{odd}, because they, too, are by definition odd in Ω. In contrast to this situation, the auto-correlations and the cross-correlation $Q_{r\theta}$ are independent of the sign of the rotation. Further differences can easily be discerned. It is an almost trivial assertion that, for sufficiently slow rotations, the even tensor possesses the symbolic form

$$Q^{\text{even}} \sim \cdots 1 + \cdots + \Omega^2 + \cdots, \tag{27}$$

where only the second term appears in the correlation $Q_{r\theta}$. In contrast, the odd tensorial part arises only because of the basic rotation, so that its expansion is

$$Q^{\text{odd}} \sim \cdots \Omega + \cdots + \Omega^3 + \cdots. \tag{28}$$

The first term gives the fundamental mode of the Λ-effect, while the subsequent terms give rise to the higher modes. But we have defined the latter in powers of $\sin\theta$ rather than Ω. The coefficient of (say) Ω^3 may therefore also contribute to the fundamental mode.

The above series expansion of (24) can be carried out quickly. For example, for the even tensor we find

$$Q_{ij}^{\text{even}} = Q_{ij}^{(0)} + \iint \left(K_{ij} \hat{Q}_{ll}^{(0)} + \frac{\omega^2 - 3\nu_e^2 k^4}{\omega^2 + \nu_e^2 k^4} \hat{Q}_{ij}^{(0)} \right) \frac{(2\hat{\boldsymbol{k}} \cdot \boldsymbol{\Omega})^2}{\omega^2 + \nu_e^2 k^4} \, \mathrm{d}\boldsymbol{k} \, \mathrm{d}\omega + \dots \tag{29}$$

with the expression for the odd tensor being similar [cf. (15.1)]. We can even make statements about the case of rapid rotation. To this end we make the substitutions

$$\hat{\nu} = (\nu_e^2 k^4 - \omega^2)/(\nu_e^2 k^4 + \omega^2), \qquad \hat{\boldsymbol{k}} \cdot \hat{\boldsymbol{\Omega}} = \cos\theta, \qquad 2\Omega \cos\theta = t, \tag{30}$$

which, when applied to the even tensor, yields

$$Q^{\text{even}} \sim \int_0^\pi \frac{(a_1 + b_1 \cos^2\theta + \cdots) + t^2(a_2 + b_2 \cos^2\theta + \cdots)}{1 + 2\hat{\nu} t^2 + t^4} \sin\theta \, \mathrm{d}\theta, \tag{31}$$

or, what is the same thing,

$$Q^{\text{even}} \sim \frac{1}{\Omega} \int_0^\Omega \frac{(a_1 + b_1 t^2/\Omega^2 + \cdots) + t^2(a_2 + b_2 t^2/\Omega^2 + \cdots)}{1 + 2\hat{\nu} t^2 + t^4} \, \mathrm{d}t. \tag{32}$$

Thus in the limit as $\Omega \to \infty$ we have

$$\lim_{\Omega \to \infty} \Omega Q^{\text{even}} = \int_0^\infty [(a_1 + a_2 t^2)/(1 + 2\hat{\nu} t^2 + t^4)] \, \mathrm{d}t, \tag{33}$$

which is normally finite. Hence, for very rapid rotations, the even part of the correlation tensor vanishes as it is $O(1/\Omega)$.

A similar procedure for the odd part of the tensor starts from

$$Q^{\text{odd}} \sim \int\limits_0^\pi t(1 + 2\mathring{\vartheta} t^2 + t^4)^{-1} (a \cos\theta + b \cos^3\theta + \cdots) \sin\theta \, d\theta , \tag{34}$$

and leads to $Q^{\text{odd}} = O(\Omega^{-2})$. In this general discussion, the possibility that sometimes some a's may "accidentally" vanish, and that the results must in consequence be slightly modified, has been ignored.

It is noteworthy that differences between the two parts of the tensor not only exist for slow rotations but also for rapid rotation: as $\Omega \to \infty$, the TMT vanish more rapidly than the auto-correlations. One is tempted to believe that this phenomenon arises from a rotationally-created reorganization of the turbulence at the expense of the ϕ-correlations. Indeed, the odd tensor does vanish identically for some simple turbulence models, one of which (namely two-dimensional turbulence) is appropriate for strong rotations.

First, however, we shall present the results for the simplest field, i.e. homogeneous, isotropic turbulence. Substituting its spectral tensor (13.20) into (21), we find

$$Q_{ij} = \iint \frac{\omega^2 + \nu_e^2 k^4}{NN^*} \{[\omega^2 + \nu_e^2 k^4 + (2\mathring{\boldsymbol{k}} \cdot \boldsymbol{\Omega})^2] \, k^2 K_{ij} - 2i\omega(2\boldsymbol{k} \cdot \boldsymbol{\Omega}) \, \varepsilon_{ijm} k_m\} \, {}^{(3)}q \, d\boldsymbol{k} \, d\omega . \tag{35}$$

Clearly $Q^{\text{odd}} = 0$ provided, as we always assume, q is even in the frequency. Rotationally-disturbed isotropic turbulence, therefore, never transports angular momentum and cannot by itself maintain the non-uniform rotation.

The consequences of (35) for the auto-correlations have been presented in detail in a separate publication (Rüdiger, 1978a). It need only be reported here that, for very rapid rotation, the turbulence intensity $\overline{u_z^2}$ is equal to the sum of both equatorial auto-correlations:

$$\lim_{\Omega \to \infty} Q_{ii} / \mathring{\Omega}_i \mathring{\Omega}_j Q_{ij} = 2 . \tag{36}$$

Hence, the rotation favors motions in the direction of $\boldsymbol{\Omega}$ at the expense of the perpendicular flows. We found exactly the same behavior for the effect of a strong magnetic field on an originally isotropic turbulence.

Interestingly enough, the simplest task does not yield the simplest model. We can see immediately that all spectral tensors having the delta function $\delta(\boldsymbol{k} \cdot \boldsymbol{\Omega})$ as a factor are uninfluenced by rotation. Such fields have infinite correlations along $\boldsymbol{\Omega}$, so that the motion does not depend on the coordinate z. In such cases the Coriolis force is conservative and can be balanced by the pressure gradient. Rotation therefore has no effect, in agreement with the Taylor-Proudman theorem, which states that the slow steady motion of a rotating inviscid fluid is two-dimensional. When Ω is sufficiently large, as measured by $\nu/\Omega L^2$ and $v'/\Omega L$, one may expect the theorem to be roughly obeyed.

Rotation enlarges the correlation length in the direction of $\boldsymbol{\Omega}$, and thus reduces the cross-correlations. This is the reason why $Q^{\rm odd}$ vanishes more rapidly than $Q^{\rm even}$ as $\Omega \to \infty$. The part $Q^{\rm odd}$, and hence the Λ-effect, only exists when the turbulence is neither isotropic nor two-dimensional (with respect to the rotation axis). This means that neither spherical nor cylindrical (with respect to $\boldsymbol{\Omega}$) configurations of cells can carry angular momentum. But the angular momentum flux vanishes in any case when the rotation is too slow or too rapid. It appears that to create a Λ-effect is not an easy undertaking!

Finally we shall present the complete spectral tensor for the horizontal turbulence (13.23) which, for the specification ${}^{(2)}q \sim \delta(\boldsymbol{g} \cdot \boldsymbol{k})$, describes a two-dimensional flow. Substituting this tensor into (21) we obtain the rather complex expression

$$\begin{aligned}\hat{Q}_{ij} = {} & [(\omega^2 + \nu_e^2 k^4)/NN^*] \big\{(\omega^2 + \nu_e^2 k^4)\,(k_\perp^2 G_{ij} - k_{\perp i} k_{\perp j}) \\ & + (2\boldsymbol{k} \cdot \boldsymbol{\Omega})^2\,[k^2 \mathring{g}_i \mathring{g}_j + (\mathring{\boldsymbol{g}} \cdot \boldsymbol{k})^2\, k_i k_j - (\mathring{\boldsymbol{g}} \cdot \boldsymbol{k})\, \mathring{g}_i k_j - (\mathring{\boldsymbol{g}} \cdot \boldsymbol{k})\, \mathring{g}_j k_i] \\ & - 2\nu_e (\boldsymbol{k} \cdot \boldsymbol{\Omega})\, k_n \mathring{g}_l (\varepsilon_{iln}[k^2 \mathring{g}_j - (\mathring{\boldsymbol{g}} \cdot \boldsymbol{k})\, k_j] + \varepsilon_{jln}[k^2 \mathring{g}_i - (\mathring{\boldsymbol{g}} \cdot \boldsymbol{k})\, k_i]) \\ & - 2i\omega (\hat{\boldsymbol{k}} \cdot \boldsymbol{\Omega})\, \hat{k}_m (2[k^2 - (\mathring{\boldsymbol{g}} \cdot \boldsymbol{k})^2]\, \varepsilon_{ijm} + k^2 \varepsilon_{jlm} \mathring{g}_l \mathring{g}_i - k^2 \varepsilon_{ilm} \mathring{g}_l \mathring{g}_j \\ & + (\mathring{\boldsymbol{g}} \cdot \boldsymbol{k})\, \varepsilon_{ilm} \mathring{g}_l k_j - (\mathring{\boldsymbol{g}} \cdot \boldsymbol{k})\, \varepsilon_{jlm} \mathring{g}_l k_i)\big\}\, {}^{(2)}q(k, \omega). \end{aligned} \tag{37}$$

From this relation we can read off the Λ-effect for the general spectral tensor (13.29) with ${}^{(1)}Q = 0$. It is contained in the expression

$$\begin{aligned}\hat{Q}_{ij}^{\rm odd} = {} & -2[(\omega^2 + \nu_e^2 k^4)/NN^*]\, \nu_e k_n (\boldsymbol{k} \cdot \boldsymbol{\Omega})\, \mathring{g}_l \{\varepsilon_{iln}[k^2 \mathring{g}_j - (\mathring{\boldsymbol{g}} \cdot \boldsymbol{k})\, k_j] \\ & + \varepsilon_{jln}[k^2 \mathring{g}_i - (\mathring{\boldsymbol{g}} \cdot \boldsymbol{k})\, k_i]\}\, {}^{(2)}q(k, \omega). \end{aligned} \tag{38}$$

From $(24)_1$ we obtain the even part of the correlations, leading directly to the radial (i.e. vertical) intensity:

$$\begin{aligned}\mathring{g}_i \mathring{g}_j \hat{Q}_{ij} = {} & [(\omega^2 + \nu_e^2 k^4)/NN^*]\, [(\omega^2 + \nu_e^2 k^4)\, k_\perp^{2\,(3)}q \\ & + k_\perp^2 k^{-2} (2\hat{\boldsymbol{k}} \cdot \boldsymbol{\Omega})^2\, (k_\perp^{2\,(2)}q + k^{2\,(3)}q)]. \end{aligned} \tag{39}$$

Note that, for

$$k_\perp^{2\;(2)}q + k^{2\;(3)}q = 0, \tag{40}$$

a simple relation subsists between $\overline{u_r'^2}$ and the ratio ${}^{(2)}q/NN^*$:

$$\begin{aligned}\hat{Q}_{ij}^{\rm odd} = {} & \hat{Q}_{rr} \frac{2\nu_e k^2 k_n \mathring{g}_l}{k_\perp^4 (\omega^2 + \nu_e^2 k^4)}\, \{\varepsilon_{iln}[k^2 \mathring{g}_j - (\mathring{\boldsymbol{g}} \cdot \boldsymbol{k})\, k_j] \\ & + \varepsilon_{jln}[k^2 \mathring{g}_i - (\mathring{\boldsymbol{g}} \cdot \boldsymbol{k})\, k_i]\}\, (\boldsymbol{k} \cdot \boldsymbol{\Omega}). \end{aligned} \tag{41}$$

Turbulence that satisfies (40), a relation identical to (13.59), behaves exceptionally in that all its correlations can be expressed in terms of the vertical intensity $\overline{u_r'^2}$. Durney and Spruit (1979) consider that the dependence of $\overline{u_r'^2}$ on position $\boldsymbol{x}$ and angular velocity $\boldsymbol{\Omega}$ is already known.

The vanishing of Λ for high rotation rates seems to be a rather general result [see also Hathaway (1984)]. It allows a prediction to be made: a rapidly rotating convection zone should be Λ-free; if, therefore, the Λ-effect plays the dominant role in determining differential rotation, the convection zone should be free of differential rotation also. The rotation can be considered as rapid if its period τ_{rot} is substantially shorter than the correlation time τ_c of the turbulence:

$$\tau_{\text{rot}} \ll 2\tau_c/(1 + 4\pi\, \nu_e^2 L^{-4} \tau_c^2)^{1/2}. \tag{42}$$

This condition is satisfied in the inner convective zones of early-type main sequence stars, for which characteristic scales are $\tau_{\text{rot}} \approx 3\text{d}$ and $\tau_c \approx 30\text{d}$ (see Tayler, 1973). According to the present arguments, Λ should be very small in such zones, and the rotation within them should be moderately uniform. This is confirmed by the observations of Ap-stars which have magnetic fields that are essentially asymmetric with respect to the rotation axis. Extensive calculations by Rädler (1979) show such fields are incompatible with non-uniform rotation. It is, therefore, very tempting to use asymmetric magnetic fields as indicators for an underlying solid-body rotation. If this is accepted, Ap-stars may serve as illustrations of our arguments about the effects of rapid rotation.

Magnetic fields can, however, also serve as effective indicators of non-uniform rotation when that magnetism is maintained by an $\alpha\Omega$-dynamo. As is well-known, this type of dynamo generates large-scale, cyclic magnetic fields by a combination of α-effect and differential rotation (Parker, 1955; Steenbeck and Krause, 1969). If the latter becomes weak because the rotation is too rapid, dynamo action is harder to achieve. Hence, the rapid rotators among the late-type stars ($\tau_{\text{rot}} < 20\text{d}$, so that the Coriolis frequency 2Ω exceeds the eddy frequency) should exhibit a different magnetic behavior to the slow rotators. There is some observational evidence that this is the case, as reported in the work of Vaughan *et al.* (1981). These authors were able to measure the rotational modulation of CaII emission from the chromospheres of late-type stars, and they found "almost exclusively" solar-like cycles" in stars having rotation periods in excess of about 20 days". The interpretation by Durney *et al.* (1981) also shows that, for a given stellar structure, magnetic "cycles" and "possible cycles" occur only for modest rotation rates (see their Fig. 2). Perhaps the Vaughan-Preston (1980) gap indicates the upper limit of rotation for each spectral class for which an influential differential rotation exists. This would imply that any future searches for differential rotation will be successful only for those rotators that are neither too slow nor too fast. Because of their large timescales, the solar giant cells should also be treated theoretically as modes of a very rapidly rotating turbulence, and therefore of minor relevance to the theory of differential rotation.

14.3. The Fundamental Mode of Λ

In order to determine the required ϕ-components of the correlation tensor, we have to take the scalar product of $(24)_2$ with the eastward unit vector, $\boldsymbol{e}$. We also adopt the highly anisotropic spectral tensor (13.29), which can exist even in a non-rotating star, as the correlation function of the original turbulence $Q_{ij}^{(0)}$. The result is

$$Q_{i\phi}^{(r)} = \nu_e \iint [(\omega^2 + \nu_e^2 k^4)/NN^*] \, (2\boldsymbol{k} \cdot \boldsymbol{\Omega}) \, M_i \, \mathrm{d}\boldsymbol{k} \, \mathrm{d}\omega, \tag{43}$$

where the vector $\boldsymbol{M}$ is

$$\boldsymbol{M} = [k^2 - (\boldsymbol{e} \cdot \boldsymbol{k})^2] \, (\boldsymbol{k} \times \boldsymbol{e}) \, {}^{(1)}q + [k^2 - (\mathring{\boldsymbol{g}} \cdot \boldsymbol{k})^2] \, (\boldsymbol{e} \times \boldsymbol{k}) \, {}^{(2)}q + 2(\mathring{\boldsymbol{g}} \cdot \boldsymbol{k}) \, (\boldsymbol{e} \cdot \boldsymbol{k}) \, (\mathring{\boldsymbol{g}} \times \boldsymbol{k}) \, {}^{(2)}q. \tag{44}$$

It possesses the meridional components

$$\begin{aligned} M_r &= [\mathring{\boldsymbol{g}} \cdot (\boldsymbol{e} \times \boldsymbol{k})] \, \{[k^2 - (\mathring{\boldsymbol{g}} \cdot \boldsymbol{k})^2] \, {}^{(2)}q - [k^2 - (\boldsymbol{e} \cdot \boldsymbol{k})^2] \, {}^{(1)}q\}, \\ M_\theta &= (\mathring{\boldsymbol{g}} \cdot \boldsymbol{k}) \, \{[k^2 - (\mathring{\boldsymbol{g}} \cdot \boldsymbol{k})^2 - 2(\boldsymbol{e} \cdot \boldsymbol{k})^2] \, {}^{(2)}q - [k^2 - (\boldsymbol{e} \cdot \boldsymbol{k})^2] \, {}^{(1)}q\}. \end{aligned} \tag{45}$$

Again, it should be noted that ${}^{(3)}q$, which corresponds to the isotropic part of the field, does not appear.

Let us discuss these expressions for various turbulence models. The simplest models are those that describe turbulence having statistically isotropic motions in a plane perpendicular to $\mathring{\boldsymbol{g}}$, i.e. with axial symmetry. To examine these, we set ${}^{(1)}q = 0$ and immediately find

$$V = (2/\sin\theta) \iint [(\omega^2 + \nu_e^2 k^4)/NN^*] \, [\mathring{\boldsymbol{g}} \cdot (\boldsymbol{e} \times \boldsymbol{k})] \, k_\perp{}^{2(2)}q \cdot (\boldsymbol{k} \cdot \mathring{\boldsymbol{\Omega}}) \, \mathrm{d}\boldsymbol{k} \, \mathrm{d}\omega, \tag{46}$$

where we have to use the relation

$$\boldsymbol{k} \cdot \mathring{\boldsymbol{\Omega}} = [\mathring{\boldsymbol{g}} \cdot (\boldsymbol{e} \times \boldsymbol{k})] \sin\theta + (\mathring{\boldsymbol{g}} \cdot \boldsymbol{k}) \, (\mathring{\boldsymbol{g}} \cdot \mathring{\boldsymbol{\Omega}}). \tag{47}$$

Equation (46) demonstrates that the shape of ${}^{(2)}q$ determines the sign of the vertical Λ. Positive ${}^{(2)}q$ leads to positive V, and vice versa (or so we may suppose).

The equivalent result for the latitudinal TMT is

$$H = (2/\cos\theta) \iint [(\omega^2 + \nu_e^2 k^4)/NN^*] \, (\mathring{\boldsymbol{g}} \cdot \boldsymbol{k}) \times [k_\perp{}^2 - 2(\boldsymbol{e} \cdot \boldsymbol{k})^2] \, {}^{(2)}q \, (\boldsymbol{k} \cdot \mathring{\boldsymbol{\Omega}}) \, \mathrm{d}\boldsymbol{k} \, \mathrm{d}\omega, \tag{48}$$

which, due to the difference term in the numerator, proves to be more sensitive than (46) to the choice of spectral function. In contrast to V, it vanishes for 2D turbulence because of the delta function, $\delta(\mathring{\boldsymbol{g}} \cdot \boldsymbol{k})$, in the spectrum. The vertical flux (46) for such a turbulence is simply

$$V = 2 \int \frac{(\omega^2 + \nu_e^2 k_\perp^4)\, k_\perp^5 \cos^2 \phi\, q_\perp(k_\perp, \omega)\, \mathrm{d}\phi\, \mathrm{d}k_\perp\, \mathrm{d}\omega}{(\omega^2 + \nu_e^2 k_\perp^4)^2 + 2(\nu_e^2 k_\perp^4 - \omega^2)(2\Omega_\perp \cos\phi)^2 + (2\Omega_\perp \cos\phi)^4}, \tag{49}$$

which is positive definite. Observe that for any rotational rate this 2D turbulence always transports the angular momentum radially outwards. It never transports angular momentum latitudinally. We shall meet these statements again when we focus our attention on the case of *slow* rotation. There has been mediocre success in analysing relations such as (43) because of the generality of their dependence on the angular velocity *and* the turbulence model. It is advisable therefore to suppose either a very simple turbulence or a very simple mean flow, i.e. slow rotation. Only the latter possibility will be treated in the remainder of this Section. It can be derived from (24) simply by setting $\Omega = 0$ in the denominator. The resulting Λ-quantities are $V^{(0)}$ and $H^{0)}$, where the latter must vanish unless singularities occur at the poles. Calculations of this term must therefore be regarded as having only a formal significance.

In the case of "slow rotation", when Coriolis forces have little influence, the rotational frequency may be supposed to be small compared with the characteristic frequencies of the turbulence:

$$\Omega^2 \ll \omega^2 + \nu_e^2 k^4, \tag{50}$$

or, using a less precise formulation, $\Omega^* < 1$. The tensor D_{ij} is now simplified by the disappearance of the inertial-wave resonance. The resulting expression

$$D_{ij} = \delta_{ij} + \varepsilon_{ijm}(2\mathring{\boldsymbol{k}} \cdot \boldsymbol{\Omega})\, \mathring{k}_m / (-i\omega + \nu_e k^2), \tag{51}$$

merely represents the solution of the frequently discussed first order equation

$$\partial u_i'/\partial t - \nu \Delta u_i' + \bar{\varrho}^{-1} \nabla_i p' + 2\varepsilon_{ijk} \Omega_j u_k^{(0)} = 0, \tag{52}$$

which was used by Durney and Spruit. As a result, the spectral tensor is linear in Ω. The desired rotationally-created one-point correlation tensor is then given by the general Λ-tensor

$$\Lambda_{ijk} = 2\nu_e \iint (\omega^2 + \nu_e^2 k^4)^{-1} (\varepsilon_{iln} \hat{Q}_{lj}^{(0)} + \varepsilon_{jln} \hat{Q}_{il}^{(0)})\, k_n k_k\, \mathrm{d}\boldsymbol{k}\, \mathrm{d}\omega. \tag{53}$$

With the formulae

$$\mathring{\boldsymbol{g}} \cdot (\boldsymbol{e} \times \boldsymbol{k}) = -k_\theta, \qquad \boldsymbol{k} \cdot \mathring{\boldsymbol{\Omega}} = k_r \cos\theta - k_\theta \sin\theta, \tag{54}$$

we deduce the relations

$$\begin{aligned} Q^{(r)}_{r\phi} &= 2\nu_e\Omega \iint (k_\theta \sin\theta - k_r \cos\theta)\, k_\theta \\ &\quad\times \left[(k_\theta^2 + k_\phi^2)\,{}^{(2)}q - (k_r^2 + k_\theta^2)\,{}^{(1)}q\right] \frac{\mathrm{d}\boldsymbol{k}\,\mathrm{d}\omega}{\omega^2 + \nu_e^2 k^4}, \\ Q^{(r)}_{\theta\phi} &= 2\nu_e\Omega \iint (k_r \cos\theta - k_\theta \sin\theta)\, k_r \\ &\quad\times \left[(k_\theta^2 - k_\phi^2)\,{}^{(2)}q - (k_r^2 + k_\theta^2)\,{}^{(1)}q\right] \frac{\mathrm{d}\boldsymbol{k}\,\mathrm{d}\omega}{\omega^2 + \nu_e^2 k^4}. \end{aligned} \tag{55}$$

Finally, we shall suppose that the remaining spectral functions are even in k_r and k_θ; this seems to be the normal state of affairs. In this way we arrive at the expressions

$$\begin{aligned} V^{(0)} &= 2 \iint (\omega^2 + \nu_e^2 k^4)^{-1} \left[(k_\phi^2 + k_\theta^2)\,{}^{(2)}q - (k_r^2 + k_\theta^2)\,{}^{(1)}q\right] k_\theta^2\, \mathrm{d}\boldsymbol{k}\,\mathrm{d}\omega, \\ H^{(0)} &= 2 \iint (\omega^2 + \nu_e^2 k^4)^{-1} \left[(k_\theta^2 - k_\phi^2)\,{}^{(2)}q - (k_r^2 + k_\theta^2)\,{}^{(1)}q\right] k_r^2\, \mathrm{d}\boldsymbol{k}\,\mathrm{d}\omega. \end{aligned} \tag{56}$$

A first general consequence is that rotation produces no radial TMT when the turbulence has extremely strong correlations in the latitudinal direction, because in this case the spectral functions are proportional to the Dirac function $\delta(k_\theta)$, and therefore V vanishes. The same holds true for the latitudinally directed flux when the correlations are strong in the radial direction.

Turbulence models under conditions of axial symmetry [cf. (13.40)] have the quite elementary properties

$$\begin{aligned} V^{(0)} &= 2 \iint \frac{(k_\theta^2 + k_\phi^2)\, k_\theta^2}{\omega^2 + \nu_e^2 k^4}\,{}^{(2)}q\, \mathrm{d}\boldsymbol{k}\,\mathrm{d}\omega = \iint \frac{k_\perp^{4}\,{}^{(2)}q}{\omega^2 + \nu_e^2 k^4}\, \mathrm{d}\boldsymbol{k}\,\mathrm{d}\omega, \\ H^{(0)} &= 0, \end{aligned} \tag{57}$$

so that the sign of $V^{(0)}$ is closely related to the sign of the spectral function ${}^{(2)}q$. The flux is directed outwards for models with positive ${}^{(2)}q$, and inwards for models with negative ${}^{(2)}q$.

Let us turn to the widely accepted and greatly simplifying approximation in which the turbulence spectrum is replaced by a single mode. Plausibly, this may be an adequate representation of a spectrum having one predominating peak. (In the opposite case of a nearly flat spectrum, the following statements cannot be justified at all.) We shall assume therefore that the spatial correlations of the original turbulence field are harmonic with only one characteristic wave-number vector, $\boldsymbol{K}$. Thus, the spectral functions degenerate into Dirac functions (13.47), in which the ${}^{(i)}U$ represent the distinct power in each of the three parts of the spectral tensor.

With the abbreviation

$$Q_0 = \nu_e \int_{-\infty}^{\infty} (\omega^2 + \nu_e^2 k^4)^{-1}\, k^2 q_\omega\, \mathrm{d}\omega, \tag{58}$$

and using the one-mode spectra (13.47), the basic expressions (55) reduce to the simpler formulae

$$\begin{aligned} \Lambda_V &= 2(K_\theta/K)^2\, Q_0[(K_\theta^2 + K_\phi^2)\, {}^{(2)}U - (K_r^2 + K_\theta^2)\, {}^{(1)}U], \\ \Lambda_H &= 2(K_r/K)^2\, Q_0[(K_\theta^2 - K_\phi^2)\, {}^{(2)}U - (K_r^2 + K_\theta^2)\, {}^{(1)}U]. \end{aligned} \tag{59}$$

We see at once that this Λ_V vanishes for all modes for which $K_\theta = 0$; these are commonly regarded as rolls with N-S axes. Such rolls never produce a radial TMT, a point made in detail by Hathaway *et al.* (1979). A similar result holds for the latitudinal flux with respect to the radial wave number component K_r. If a mode shows any form of radial two-dimensionality, the rotation does not generate a latitudinal Λ.

Both the quantities under consideration are proportional to the integral Q_0 in which dissipation is an essential ingredient. It is positive for a positive-semidefinite frequency spectrum q_ω and in the limit of vanishing viscosity satisfies

$$\lim_{\nu_c \to 0} Q_0 = \pi q_\omega(0). \tag{60}$$

Thus, rotation is unable to produce Λ in non-dissipative media if the frequency spectrum vanishes at the origin, $\omega = 0$: This happens if, for example, the temporal behavior of the original turbulence is periodic. Oscillating inviscid media must be considered, therefore, to be free from the Λ-effect and, consequently, free from differential rotation. This conclusion is independent of special assumptions concerning the spatial behavior of the correlations, and seems to be quite general. Provided the spectral functions do not vanish at $\omega = 0$, the right-hand side of (60) may be a good approximation to the integral Q_0 provided that the viscosity is relatively small. It should be noted, however, that the approximation of vanishing dissipation does not play a fundamental role. Without loss of generality, Q_0 may be assumed to be of order unity and to include, for example, all the unstable (growing) turbulent modes as well as the normal temporal correlation functions.

As far as the intensity parameters ${}^{(1)}U$ and ${}^{(2)}U$ are concerned, we may use the intensities implied by (13.49) and arrive at the expressions

$$\begin{aligned} \Lambda_V &= 2\tau_c K^{-2}[(K_\theta^2 + K_\phi^2)\, \overline{u_\phi^{(0)^2}} - (K_r^2 + K_\theta^2)\, \overline{u_r^{(0)^2}}], \\ \Lambda_H &= 2\tau_c K^{-2}[(K_r^2 + K_\phi^2)\, \overline{u_\phi^{(0)^2}} - (K_r^2 + K_\theta^2)\, \overline{u_\theta^{(0)^2}}], \end{aligned} \tag{61}$$

which give the rotationally produced TMT for any mode of the original stochastic field. In order to test (61) and to demonstrate its general character, we shall now specify the wave-numbers and intensities so that they probe a variety of models. First we will explore turbulence models having axial symmetry.

(a) Axisymmetric turbulence

Because of the extreme simplicity of conditions (13.50) for axial symmetry, the latitudinal flux again vanishes identically: $\Lambda_H = 0$. A small manipulation of the expression $(61)_1$ for the vertical flux gives

$$\Lambda_V = 2\tau_c K^{-2}[(K_\theta^2 + K_\phi^2)\,(\overline{u_\phi^{(0)^2}} - \overline{u_r^{(0)^2}}) + (K_\phi^2 - K_r^2)\,\overline{u_r^{(0)^2}}], \tag{62}$$

which takes nonzero values when the intensities $\overline{u_\phi^{(0)^2}}$ and $\overline{u_r^{(0)^2}}$ are different and/or the wave-number components K_ϕ and K_r are different. We must, in consequence distinguish between two contributions to the radial TMT: anisotropy in the intensities and/or anisotropy in the wave-number components. If the characteristic wave-numbers K_r and K_ϕ are nearly equal, the horizontal motion provides upward fluxes and the radial motion provides downward fluxes:

$$\Lambda_V \approx 2\tau_c(\overline{u_\phi^{(0)^2}} - \overline{u_r^{(0)^2}}). \tag{63}$$

As was to be expected, this result strongly reminds us of the early finding of Biermann.

Next, the quite simple but rather telling model of strictly two-dimensional motion is treated. In the one-mode approximation, this is defined by (13.52) and leads to an upwardly directed vertical Λ:

$$\Lambda_V = 2\tau_c\overline{u_\phi^{(0)^2}} > 0. \tag{64}$$

Here, the correlation $Q_{r\phi}$ arises because the action of the Coriolis force on the given azimuthal $u_\phi^{(0)}$ produces a radial velocity fluctuation u_r' which is, of course, strictly correlated with the original $u_\phi^{(0)}$. Obviously, the neglect of radial motion in the original field does not in general imply the absence of all radial motions. According to (62), the flux is also directed upwards when only the first of the conditions (13.52) holds but not the second. If the Coriolis force acts on a field of purely radial motion $[\overline{u_\phi^{(0)^2}} = 0]$ a downwardly directed TMT is created (Rüdiger, 1977b):

$$\Lambda_V = -\tau_c\overline{u_r^{(0)^2}} < 0. \tag{65}$$

The existence of this Λ depends crucially on whether the "original" turbulence field shows sufficiently strong anisotropy.

Gough (1978) also considered a radially anisotropic but horizontally isotropic motion. His final result depended on a special choice of mode:

$$K_r^2 = 2(\Phi - 1)\,K_\phi^2, \qquad \overline{u_\phi^{(0)^2}} = \frac{1}{2}\,(\Phi - 1)\,\overline{u_r^{(0)^2}}, \tag{66}$$

from which (62) provides a downwardly directed TMT, independent of the free anisotropy parameter Φ:

$$\Lambda_V = -\tau_c \overline{u_r^{(0)^2}} < 0. \tag{67}$$

We find a good agreement between Gough's original expression and our more general result.

b) Non-axisymmetric turbulence.

As we pointed out above, in contrast to turbulence having horizontal isotropy, most of those that lack this symmetry are able to provide a latitudinal Λ. The degree of arbitrariness, however, is somewhat greater than for the simpler models. Several proposals have been made and most give finite values of Λ_H. In this way Ruzmaikin and Vainshtein (1978) discussed the action of the Coriolis force on convective elements possessing only an azimuthal component $u_\phi^{(0)}$. By continuity, the azimuthal wave-number must vanish. If we apply our expressions (61) to this situation we obtain

$$\Lambda_V = 2\tau_c (K_\theta/K)^2 \overline{u_\phi^{(0)^2}} > 0, \qquad \Lambda_H = 2\tau_c (K_r/K)^2 \overline{u_\phi^{(0)^2}} > 0. \tag{68}$$

Hathaway *et al.* (1979) calculated the effect of (unrestricted) rotation on N-S orientated rolls, as well as on rolls with axes in the E-W direction. The first case arises for vanishing K_θ corresponding to vanishing Λ_V. If the latitudinal intensity $\overline{u_\theta^{(0)^2}} = 0$, we obtain for this type of turbulence

$$\Lambda_H = 2\tau_c \overline{u_\phi^{(0)2}} > 0. \tag{69}$$

For the E-W orientated rolls, $K_\phi = u_\phi^{(0)} = 0$, it follows

$$\Lambda_V = -2\tau_c \overline{u_r^{(0)^2}} < 0, \qquad \Lambda_H = -2\tau_c \overline{u_\theta^{(0)^2}} < 0. \tag{70}$$

If these results are to be compared with those of Hathaway *et al.* (1979), one must, of course, expand the latter for slow rotation. Both TMTs have negative signs, in contrast to those arising from the non-axisymmetric modes (68) or (69). This is the main difference between the axisymmetric and non-axisymmetric cases.

Durney and Spruit (1979) considered a special model with $\nu = {}^{(1)}q = 0$ and ${}^{(2)}q = -k^{2(3)}q/k_\perp^2$ [cf. (13.59)]. Although they stress that their work is applicable to rapid rotations, we can understand their basic findings for the Reynolds stress even from the approximation of hydrodynamically slow rotation. First of all, (56) leads to

$$\Lambda_V = -2\pi \int k_\theta^{2(3)} q(\boldsymbol{k}, 0)\, \mathrm{d}\boldsymbol{k}, \quad \Lambda_H = 2\pi \int k_r^2 k_\perp^{-2} (k_\phi^2 - k_\theta^2)\, {}^{(3)}q(\boldsymbol{k}, 0)\, \mathrm{d}\boldsymbol{k}. \tag{71}$$

The second of these can also be written as

$$\Lambda_H = 2\tau_c(\overline{u_\phi^{(0)^2}} - \overline{u_\theta^{(0)^2}}), \tag{72}$$

which is strongly reminiscent of (3.22) in Iroshnikov's paper of 1969.

No matter what spectral function is chosen, the latitudinal TMT is directly proportional to the difference between the horizontal intensities. In order to verify their results, we need to express these intensities in terms of the radial intensity. This is practicable only for the extremely peaked spectrum (13.47). If (13.60) is substituted into (71), we arrive at the final results

$$\Lambda_V = -2\tau_c K_\theta^2 K_\perp^{-2}\overline{u_r^{(0)^2}} < 0, \qquad \Lambda_H = 2\tau_c K_r^2 K_\perp^{-4}(K_\phi^2 - K_\theta^2)\,\overline{u_r^{(0)^2}}. \tag{73}$$

As was to be expected, these results, which are valid only for small rotation, are identical with the findings of Durney and Spruit.

14.4. Stratified Medium with Isotropic Forces

Quite another approach has been suggested by Gailitis and Rüdiger (1982) in order to make use of several different sets of observations of the solar surface. What we see on the Sun is a stochastic motion which does not transport energy and which quite definitely corresponds to modes for which the solar rotation can be considered to be slow: the supergranulation. Its horizontal velocities clearly exceed vertical ones. Nevertheless, we can observe radial gradients of density and angular velocity, and we know the latitude dependence of Ω. When the latter is ignored (because its sources lie deeper in the SCZ where modes exist for which the rotation must be considered to be fast), we may describe the supergranulation layer by

$$\delta I > 0, \quad G \simeq 10^3, \quad V^{(0)} \simeq -1, \quad V^{(1)}, H^{(1)}, \dots \approx 0, \tag{74}$$

and this is what has to be explained. Instead of resorting to numerical methods (cf. Gilman and Foukal, 1979), we tried to establish a reasonable turbulence model on the basis of "slow rotation" and "strong density stratification". As no definite ideas about the wave-numbers in the layer are known, expressions (61) are not directly applicable.

Gailitis and Rüdiger (1982) formulated the question in the following way: Is it possible to assume isotropy in the stochastic forces which, because of the density stratification, simultaneously provides positive δI and negative $V^{(0)}$? The answer was "Yes", although horizontal turbulence expressions (61) yield positive $V^{(0)}$, not negative. If horizontal-type turbulence is postulated, a necessary condition for the negativity of $V^{(0)}$ would be $K_r \gg K_\phi$, but we do not know yet whether this would be physically realistic.

In order to establish a model we must take into account four main features of the supergranulation:

(i) the time scale τ_c is much shorter than the rotation period τ_{rot};

(ii) the Strouhal-number does not exceed unity;
(iii) no heat is transported;
(iv) the density gradient is very large.

As a result of the first three of these features, we may neglect:

(i) the non-linear influence of Ω;
(ii) correlations higher than second order;
(iii) density fluctuations.

We must, however, include the density stratification. Thus, the hydrodynamic equations are:

$$\begin{aligned} \partial \boldsymbol{u}'/\partial t - \nu_e \Delta \boldsymbol{u}' &= -\varrho^{-1}\nabla p' + 2\boldsymbol{u}' \times \boldsymbol{\Omega} + \boldsymbol{f}', \\ \nabla \cdot (\varrho \boldsymbol{u}') = 0, \qquad \varrho &= \varrho_0 \exp(-\boldsymbol{G}\cdot\boldsymbol{x}), \end{aligned} \tag{75}$$

where the actual non-linearities maintaining the motion are replaced by a stochastic field $\boldsymbol{f}'$. Here $\boldsymbol{G}$ is a radial vector of constant magnitude G. Clearly, due to the inclusion of the density stratification, the system (75) is more complex than that of the Boussinesq approximation. It may be interpreted as a simplified, and therefore analytically tractable, version of the anelastic approximation.

In order to illustrate the effect of stratification at its simplest, we restrict ourselves to forces $\boldsymbol{f}'$ that are completely isotropic. We shall demonstrate that in this case (75) is equivalent to negative $V^{(0)}$. It will turn out that the magnitude of $\boldsymbol{G}$ determines whether or not the horizontal intensity dominates, but this does not influence the sign of $V^{(0)}$.

Let us discuss the consequences of the above system by means of the correlation tensor

$$Q_{ij} = \iint (\hat{Q}_{ij}^{(0)} + \hat{Q}_{ij}^{(r)})\, \mathrm{d}\boldsymbol{k}\, \mathrm{d}\omega, \tag{76}$$

where $Q_{ij}^{(r)}$ contains the effects of rotation, but only as far as the terms linear in Ω. First, we shall consider the lowest-order tensor $Q_{ij}^{(0)}$, for which the system (75) yields

$$\hat{Q}_{ij}^{(0)} = \varkappa_{im}^{*}\varkappa_{jn}\hat{F}_{mn}/(\omega^2 + \nu_e^2 k^4), \tag{77}$$

where

$$\varkappa_{im} = \delta_{im} - \hat{\varkappa}_i \hat{\varkappa}_m, \qquad \boldsymbol{\varkappa} = \boldsymbol{k} + i\boldsymbol{G}. \tag{78}$$

For the spectral tensor $\hat{F}_{ij}$ of the driving force field, we use the simplest possible form, i.e. a tensor corresponding to a stationary, homogeneous and isotropic field:

$$\hat{F}_{mn} = f_1(k, \omega)\, K_{mn} + f_2(k, \omega) \hat{k}_m \hat{k}_n. \tag{79}$$

All the resulting anisotropies are direct consequences of terms in (75) and, according to Bochner's theorem, both the functions $f_1(k, \omega)$ and $f_2(k, \omega)$ are positive semi-definite. The zeroth-order tensor $Q_{ij}^{(0)}$ provides only auto-correlations. After some short manipulations we obtain for small G

$$\delta I \simeq -\frac{1}{30} \iint [G^2(6f_1 + 7f_2)/k^2(\omega^2 + \nu_e^2 k^4)] \, \mathrm{d}\boldsymbol{k} \, \mathrm{d}\omega , \tag{80}$$

and for large G

$$\delta I \simeq \frac{1}{3} \iint [(2f_1 + f_2)/(\omega^2 + \nu_e^2 k^4)] \, \mathrm{d}\boldsymbol{k} \, \mathrm{d}\omega . \tag{81}$$

Clearly δI is negative for small G and positive for large G. Hence, a weakly stratified turbulence should be mainly of the vertical type, whereas a strongly stratified turbulence should be mainly of the horizontal type. In view of the observations of the Sun, our attention is focussed on the latter case.

As the next step, we need to evaluate the first-order spectral tensor $\hat{Q}_{ij}^{(r)}$ which contains the effects of the basic rotation. After lengthy calculations the system (75) proves to be compatible with

$$\hat{Q}_{ij}^{(r)} = 2 \left(\frac{\varkappa_{im}^* \varepsilon_{jnq} \mathring{\varkappa}_q \mathring{\varkappa}_l}{-i\omega + \nu_e k^2} + \frac{\varkappa_{jn} \varepsilon_{imq} \mathring{\varkappa}_q^* \mathring{\varkappa}_l^*}{i\omega + \nu_e k^2} \right) \frac{\hat{F}_{mn} \Omega_l}{\omega^2 + \nu_e^2 k^4} . \tag{82}$$

For the single non-vanishing component of this tensor, we find

$$V^{(0)} = -2G^2 \iint \frac{(k^2 + k_r^2) \, f_1 + (k^2 - k_r^2) \, f_2}{(k^2 - G^2)^2 + 4G^2 k_r^2} \, \frac{k_\perp^2 \, \mathrm{d}\boldsymbol{k} \, \mathrm{d}\omega}{(\omega^2 + \nu_e^2 k^4)^2} < 0 , \tag{83}$$

which is always negative, no matter what parameters are chosen! Within the framework of our assumptions, the density stratification allows only inwardly directed TMT to occur. In particular, for strongly stratified media we may therefore expect the simultaneous existence of a dominating horizontal motion, which can be directly observed on the solar surface, and a negative $V^{(0)}$, which is inferred indirectly by the form of the sub-rotation in the supergranulation layer. Unfortunately, further work on this model, i.e. the determination of $H^{(1)}$, $V^{(1)}$, has not yet been undertaken.

We have shown how the predominantly negative sign of the fundamental mode of the vertical Λ-effect can come about. In fact, the turbulence models of Gough, Hathaway *et al.*, Durney and Spruit, and Gailitis and Rüdiger have generally yielded negative signs. Adopting the usual relation $\nu_T \approx \overline{u'^2} \tau_c$, the corresponding $V^{(0)}$ is always of order unity, i.e.

$$V^{(0)} \approx -1 . \tag{84}$$

No depth-dependence remains. From (62) we see that negative $V^{(0)}$ follow for dominating $\overline{u_r'^2}$ and/or K_r. Hence vertical types of turbulence, as well as horizontal types with very flat cells (along $\mathring{\boldsymbol{g}}$), lead to the result (84).

Chapter 15. The Higher Modes of Λ

The fluid elements envisaged by mixing-length theory have, in the lower part of the SCZ, lifetimes of about 10—20 days, so that Ω^* is of order unity. We are therefore encouraged to express the Λ's for the Sun as series in powers of Ω. We shall now present calculations up to the terms of order Ω^3, and we will concentrate on the case of moderate rotation, i.e. neither too slow nor too rapid. With this understanding, (14.43) may be written as (Rüdiger, 1983b)

$$Q_{i\phi}^{(r)} = \iint \frac{\nu_e M_i(\boldsymbol{k})\,(2\boldsymbol{k}\cdot\boldsymbol{\Omega})}{\omega^2 + \nu_e^2 k^4} \left[1 + 2\,\frac{\omega^2 - \nu_e^2 k^4}{(\omega^2 + \nu_e^2 k^4)^2}\,(2\mathring{\boldsymbol{k}}\cdot\boldsymbol{\Omega})^2\right] \mathrm{d}\boldsymbol{k}\,\mathrm{d}\omega\,. \tag{1}$$

Using the notation of our representation (2.20)—(2.21), we shall deduce the Λ-modes $V^{(1)}$ and $H^{(1)}$ for some special turbulence models. For the first model we shall assume axial symmetry, i.e. isotropy with respect to the horizontal plane, whereas the second model makes use of the one-mode approximation.

15.1. Axisymmetric Turbulence

As has already been stressed in Chapter 13, the spectral functions of turbulence models with symmetry in the horizontal plane depend only on k_r and $k_\perp$. We then find from (1), for moderate rotation,

$$\begin{aligned} Q_{r\phi}^{(r)} &= 2\int k_\perp^2 k^{-2}[\mathring{\boldsymbol{g}}\cdot(\boldsymbol{e}\times\boldsymbol{k})]\,(\boldsymbol{k}_\perp\cdot\boldsymbol{\Omega}) \\ &\quad\times \{Q_0 + 8Q_2 k^{-2}[3k_r^2(\mathring{\boldsymbol{g}}\cdot\boldsymbol{\Omega})^2 + (\boldsymbol{k}_\perp\cdot\boldsymbol{\Omega})^2]\}\,\mathrm{d}\boldsymbol{k}, \\ Q_{\theta\phi}^{(r)} &= 2(\mathring{\boldsymbol{g}}\cdot\boldsymbol{\Omega})\int k_r^2 k^{-2}[k_\perp^2 - 2(\boldsymbol{e}\cdot\boldsymbol{k}_\perp)^2] \\ &\quad\times \{Q_0 + 8Q_2 k^{-2}[k_r^2(\mathring{\boldsymbol{g}}\cdot\boldsymbol{\Omega})^2 + 3(\boldsymbol{k}_\perp\cdot\boldsymbol{\Omega})^2]\}\,\mathrm{d}\boldsymbol{k}, \end{aligned} \tag{2}$$

where only those terms possessing the symmetries of (14.55) have been retained, and

$$Q_0 = \int \frac{\nu_e k^{2\,(2)}q\,\mathrm{d}\omega}{\omega^2 + \nu_e^2 k^4}, \qquad Q_2 = \int \frac{\nu_e k^2(\omega^2 - \nu_e^2 k^4)\,{}^{(2)}q}{(\omega^2 + \nu_e^2 k^4)^3}\,\mathrm{d}\omega\,. \tag{3}$$

After integrating over azimuth in the horizontal plane, one arrives at the general results

$$\begin{aligned} V^{(1)} &= 6\Omega^2 \nu_T^{-1} \int k_\perp^4 (k_\perp^2 - 4k_r^2)\, k^{-4} Q_2 \,\mathrm{d}\boldsymbol{k}, \\ H^{(1)} &= 12\Omega^2 \nu_T^{-1} \int k_\perp^4 k_r^2 k^{-4} Q_2 \,\mathrm{d}\boldsymbol{k}. \end{aligned} \tag{4}$$

Expressions (4) are correct for turbulence having axial symmetry, i.e. for which only the radial direction, $\mathring{\boldsymbol{g}}$, is preferred. Of course, the extremely simplified model of two-dimensional turbulence also belongs to this class; its spectral function involves the Dirac delta function: $q(\boldsymbol{k}, \omega) = q_\perp(k_\perp, \omega) \times \delta(\boldsymbol{k}_r)$. In this case (4) becomes

$$V^{(1)} = 6\Omega^2 \iint (\omega^2 + \nu_e^2 k_\perp^4)^{-3}\, k_\perp^4 (\omega^2 - \nu_e^2 k_\perp^4)\, q_\perp \,\mathrm{d}\boldsymbol{k}_\perp \,\mathrm{d}\omega,$$

which is similar to (14.49). This extraordinarily simple turbulence provided the first application of the general theory designed to add the influence of rotation to an "original" turbulence (Rüdiger, 1977a). This paper established the existence of the mode $V^{(1)}$ in stars rotating relatively rapidly. A more realistic model, unrestricted by the assumption of two-dimensionality, was investigated very soon afterwards (Rüdiger, 1977b).

It will be noted that, in contrast to the expressions of § 14.3 for slow rotation, the characteristic combination $\omega^2 - \nu_e^2 k^4$ occurs in the integrals in (4), and it is this that complicates the analysis. Nevertheless, let us look at its implications. A cursory inspection clearly indicates that $Q_2 > 0$ implies $H^{(1)} > 0$.

In § 15.4 we shall discuss in detail the constraints on Λ that follow from the results of Chapters 6 and 11. The condition $Q_2 > 0$ certainly means the viscosity ν_e must in some sense be small. Indeed, in what follows these conlectures will be confirmed by a model that is very close to (13.65), namely

$${}^{(2)}q = q_k q_\omega (\mathring{\boldsymbol{g}} \cdot \mathring{\boldsymbol{k}})^{2\varkappa}, \qquad {}^{(3)}q = q(k, \omega). \tag{5}$$

Substituting (5) into (4) gives us

$$\begin{aligned} V^{(1)} &= \frac{384\pi(1 - 4\varkappa)\, \Omega^2 \nu_T^{-1}}{(2\varkappa + 1)(2\varkappa + 3)(2\varkappa + 5)(2\varkappa + 7)} \int\limits_0^\infty k^4 Q_2' q_k \,\mathrm{d}k, \\ H^{(1)} &= \frac{384\pi \Omega^2 \nu_T^{-1}}{(2\varkappa + 3)(2\varkappa + 5)(2\varkappa + 7)} \int\limits_0^\infty k^4 Q_2' q_k \,\mathrm{d}k, \end{aligned} \tag{6}$$

with the kernel

$$Q_2' = \int\limits_{-\infty}^{\infty} \nu_e k^2 (\omega^2 - \nu_e^2 k^4)(\omega^2 + \nu_e^2 k^4)^{-3} q_\omega \,\mathrm{d}\omega. \tag{7}$$

First it will be noted that

$$V^{(1)}/H^{(1)} = (1 - 4\varkappa)/(1 + 2\varkappa), \tag{8}$$

which runs from 1 to -2 for $\varkappa = 0, 1, 2, \ldots$ Notice also that the sign of $V^{(1)}$ reverses between small and large $\varkappa$. Clearly, $V^{(1)}$ and $H^{(1)}$ take opposite signs for $\varkappa \geqq 1$, i.e. for relatively flat cells. Since the interpretation of the observations in Chapter 11 required that $V^{(1)}$ and $H^{(1)}$ should have the same sign, we infer that cells in the SCZ cannot be flat in structure. But the question remains as to whether the turbulence is of a horizontal or vertical type, i.e. whether its horizontal intensity exceeds its vertical one, or vice versa.

To clarify this point we may appeal to the positivity of $H^{(1)}$. Clearly, $H^{(1)} > 0$ requires that the integral appearing in $(6)_2$ be positive, but its complicated structure does not permit any general conclusions to be drawn. We must therefore undertake an analysis to find out which features of turbulence models lead to positive kernels Q_2' or, what is nearly the same thing, which predict positive $H^{(1)}$. There are, in fact, some quite enlightening trends. The kernel Q_2', whose sign is already heavily dependent on the frequency profile of the spectral function under consideration, plays a crucial role. Since

$$Q_2' = \int \frac{\nu_e k^2 \omega}{\omega^2 + \nu_e^2 k^4} \frac{\partial}{\partial \omega} \left(\frac{q_\omega}{\omega^2 + \nu_e^2 k^4} \right) \mathrm{d}\omega , \tag{9}$$

the kernel Q_2' proves to be negative-definite for all spectra with monotonically decreasing profiles, $\omega\, \partial q_\omega / \partial \omega \leqq 0$. Also both for constant q_ω, and for very steep profiles $[q_\omega \approx \delta(\omega)]$, Q_2' is negative. According to $(6)_2$, the Λ-coefficient, $H^{(1)}$, is positive for negative Q_2' provided q_k is mainly negative. A "turbulence" which produces positive $H^{(1)}$ *must*, therefore, have a dominatingly vertical motion. If also $V^{(1)}$ is required to be positive, then (8) requires $\varkappa$ to be small, namely $\varkappa < 0.25$ (see § 15.4).

In contrast, ω-profiles which possess a distinct peak at any nonvanishing ω lead to non-definite kernels. The sign of these kernels depends strongly on the choice of the input parameters: a sufficiently small viscosity generally gives rise to positive kernels and vice versa. It is probably best to demonstrate the two cases separately:

(a) Monotonic frequency spectra

The simplest example of this class, i.e. (13.46), produces the negative kernel

$$Q_2' = -(1 + 3\tau_c \nu_e k^2 + 4\tau_c^2 \nu_e^2 k^4)/[4\nu_e^2 k^4 (1 + \tau_c \nu_e k^2)^3]. \tag{10}$$

Some recent experiments seem to indicate that such profiles, i.e. those with $\omega\, \partial q / \partial \omega < 0$, are normal for well-established turbulence. Such monotonically decreasing spectra always produce negative kernels Q_2', so that $q_k < 0$ is necessary in order to produce positive $H^{(1)}$. We also know that negative q_k is appropriate for turbulence of vertical-type.

First, the kernel (10) will be applied to a simple wave-number spectrum with a distinct, delta function-like peak, namely (13.70), from which we obtain

$$H^{(1)} = -\frac{6(2\varkappa+1)(1+3\nu^*+4\nu^{*2})}{(2\varkappa+5)(2\varkappa+7)\nu^{*3}(1+\nu^*)^3}\frac{\tau_c^2\delta I}{L^2}(2\Omega\tau_c)^2, \tag{11}$$

where we have introduced the normalized eddy viscosity

$$\nu^* = \nu_e\tau_c/L^2 = \nu_e/L^2W. \tag{12}$$

Alternatively, the less peaked, but possibly more realistic, profile

$$q_k = 8k\tau_c L^6 I/(1 + L^4k^4)^2 \tag{13}$$

may be chosen. Its predictions hardly differ from the earlier expression (11):

$$H^{(1)} = -\frac{6\pi(2\varkappa+1)(1+5\nu^{*2}-17\nu^{*4}+27\nu^{*6})}{(2\varkappa+5)(2\varkappa+7)\nu^{*3}(1+\nu^{*2})^4}\frac{\tau_c^2\delta I}{L^2}(2\Omega\tau_c)^2. \tag{14}$$

It is clear that in both cases the inviscid limit, $\nu^* \to 0$, cannot be taken. This may be attributed to the presence of inertial waves, which principally appear in those cases that are not limited to the linear approximation in Ω.

In summary, positive $H^{(1)}$ results from negative δI, i.e. from dominating vertical intensities.

(b) Peaked frequency spectra

Another, quite different, situation arises from profiles that have maxima at non-zero frequencies. In order to avoid lengthy calculations, only their simplest representation (13.44), i.e.

$$q_\omega = W\delta(\omega - W), \qquad W \equiv \tau_c^{-1},$$

will be considered. Clearly, as a result of (7), the kernel Q_2' is given by

$$Q_2' = W\nu_e k^2(W^2 - \nu_e^2k^4)(W^2 + \nu_e^2k^4)^{-3}, \tag{15}$$

which is to be substituted into (6). For the k-spectrum, we again use the delta function-example (13.70), and this leads to

$$H^{(1)} = \frac{24(2\varkappa+1)}{(2\varkappa+5)(2\varkappa+7)}\frac{1-\nu^{*2}}{(1+\nu^{*2})^3}\frac{\tau_c^2\delta I}{L^2}(2\Omega\tau_c)^2: \tag{16}$$

As a second example, the monotonic profile

$$q_k = 4L^4\tau_c k^{-1}(1 + L^4k^4)^{-2}I \tag{17}$$

was used, and this led to

$$H^{(1)} = \frac{12(2\varkappa + 1)(2 - \nu^*)}{(2\varkappa + 5)(2\varkappa + 7)(1 + \nu^*)^4} \frac{\tau_c^2 \delta I}{L^2} (2\Omega\tau_c)^2 . \tag{18}$$

Again, the two results hardly differ from each other, but strikingly, their signs differ from those obtained in (a). These signs depend on the product of two differences, namely δI and $1 - \nu^*$. Positive values of $H^{(1)}$ arise if relations such as

$$\delta I \cdot (1 - \nu^*) > 0 \tag{19}$$

are fulfilled. It now depends on the actual value of the viscosity whether turbulence of vertical or horizontal type produces positive $H^{(1)}$. Relatively small viscosities, in conjunction with turbulence of horizontal-type ($\delta I > 0$), imply positive $H^{(1)}$. Any large viscosity, however, requires dominantly vertical motions in order to reproduce the observations. But this possibility may be excluded, since $H^{(1)}$ vanishes rapidly with increasing ν.

It is worth noting that there is some observational evidence for a "small" eddy viscosity. If one combines the scales of the supergranulation, as well as for the giant cells, with an eddy viscosity of about 10^{13} cm^2/s, he finds that the parameter ν^* is less than unity. The required turbulence must then be of horizontal type. The key question to be answered by every theory that explains the observed solar equatorial acceleration is whether that acceleration should be associated with a radially increasing or a radially decreasing angular velocity (sub-rotation or super-rotation). It will become clear in § 15.4 that in our theory the answer depends essentially on the form assumed for the (frequency) spectrum of the rotating turbulence.

We have also used the spectral tensor (13.63) which models a variety of different turbulences (see Table XIII.i.). The required fluxes of the angular momentum, $Q_{r\phi}$ and $Q_{\theta\phi}$, are proportional to α since, for $\alpha = 0$, the tensor collapses to K_{ij} which does not contribute to the correlations. By numerical computation we were able, in our original paper (Rüdiger, 1978b), to proceed without placing any restriction on the rotation rate. Substituting (13.63) into (14.46) and (14.48), we find

$$\begin{aligned} V &= -\frac{2\alpha}{\sin^2\theta} \iint \frac{\omega^2 + \nu_e^2 k^4}{NN^*} q (\mathring{\boldsymbol{g}} \cdot \mathring{\boldsymbol{k}})^{2\varkappa} k_\perp^2 (\boldsymbol{k} \cdot \mathring{\boldsymbol{\Omega}}) (\boldsymbol{k}_\perp \cdot \mathring{\boldsymbol{\Omega}}) \, d\boldsymbol{k} \, d\omega , \\ H &= -\frac{2\alpha}{\Omega^2 \cos\theta \sin^2\theta} \iint \frac{\omega^2 + \nu_e^2 k^4}{NN^*} q (\mathring{\boldsymbol{g}} \cdot \mathring{\boldsymbol{k}})^{2\varkappa} (\mathring{\boldsymbol{g}} \cdot \boldsymbol{k}) (\boldsymbol{k} \cdot \mathring{\boldsymbol{\Omega}}) \\ &\quad \times [2(\boldsymbol{k}_\perp \cdot \boldsymbol{\Omega})^2 - k_\perp^2 \Omega_\perp^2] \, d\boldsymbol{k} \, d\omega . \end{aligned} \tag{20}$$

We will analyze these integrals in a polar coordinate system in which

$$\mathring{\boldsymbol{k}} \cdot \mathring{\boldsymbol{\Omega}} = \cos \Theta . \tag{21}$$

The unit vector $\mathring{\Omega}$ defines the z-axis and the local vector $\mathring{g}$ lies in the yz-plane. Expressed in terms of x, y, z components, we then have

$$\mathring{g} = \begin{pmatrix} 0 \\ \sin\theta \\ \cos\theta \end{pmatrix}, \quad e = \begin{pmatrix} -1 \\ 0 \\ 0 \end{pmatrix}, \quad \mathring{k} = \begin{pmatrix} \sin\Theta\cos\Phi \\ \sin\Theta\sin\Phi \\ \cos\Theta \end{pmatrix}, \tag{22}$$

where θ is the polar angle of the radius vector $\mathring{g}$. We obtain the expressions

$$\begin{aligned} V &= -\frac{2\alpha}{\sin^2\theta} \int \frac{\eta^{2\varkappa}(1-\eta^2)\cos\Theta(\cos\Theta - \eta\cos\theta)(\omega^2 + \nu_e{}^2k^4)\,k^6q}{(\omega^2+\nu_e{}^2k^4)^2 + 8(\nu_e{}^2k^4 - \omega^2)\,\Omega^2\cos^2\Theta + 16\Omega^4\cos^4\Theta} \\ &\quad \times \sin\Theta\,\mathrm{d}\Theta\,\mathrm{d}\Phi\,\mathrm{d}k\,\mathrm{d}\omega, \\ H &= -\frac{2\alpha}{\cos\theta\sin^2\theta} \\ &\quad \times \int \frac{\eta^{2\varkappa+1}\cos\Theta[2\cos^2\Theta - \sin^2\theta - 4\eta\cos\Theta\cos\theta + \eta^2(2-\sin^2\theta)]}{(\omega^2+\nu_e{}^2k^4)^2 + 8(\nu_e{}^2k^4-\omega^2)\,\Omega^2\cos^2\Theta + 16\Omega^4\cos^4\Theta} \\ &\quad \times (\omega^2+\nu_e{}^2k^4)\,k^6q\sin\Theta\,\mathrm{d}\Theta\,\mathrm{d}\Phi\,\mathrm{d}k\,\mathrm{d}\omega, \end{aligned} \tag{23}$$

where we have used the abbreviation $\eta = \mathring{g}\cdot\mathring{k}$. Since the integration variable ϕ only occurs in the numerators, we can easily carry out the ϕ-integrations. Elementary calculations and comparisons of the $\sin\theta$ terms lead to the final result:

$$\begin{pmatrix} V^{(1)} \\ H^{(1)} \end{pmatrix} = \iint \begin{pmatrix} \mathcal{V} \\ \mathcal{H} \end{pmatrix} \frac{qk^4}{\omega^2 + \nu_e{}^2k^4}\,\mathrm{d}k\,\mathrm{d}\omega, \tag{24}$$

which is given here in the form of integrands. It is interesting to examine the values of $\mathcal{V}$ and $\mathcal{H}$ for $\varkappa = 0$ and $\varkappa = 1$. We find that, for $\varkappa = 0$,

$$\mathcal{V} = \mathcal{H} = -\alpha\int_0^1 (5x^2 - 3)\,x^2(1 + 2\hat{\nu}\Omega^{*2}x^2 + \Omega^{*4}x^4)^{-1}\,\mathrm{d}x, \tag{25}$$

while, for $\varkappa = 1$, we have

$$\begin{aligned} \mathcal{V} &= -\alpha\int_0^1 (6 - 25x^2 + 21x^4)\,x^2(1 + 2\hat{\nu}\Omega^{*2}x^2 + \Omega^{*4}x^4)^{-1}\,\mathrm{d}x, \\ \mathcal{H} &= -\alpha\int_0^1 (3 - 20x^2 + 21x^4)\,x^2(1 + 2\hat{\nu}\Omega^{*2}x^2 + \Omega^{*4}x^4)^{-1}\,\mathrm{d}x. \end{aligned} \tag{26}$$

Here we have introduced the dimensionless parameters $\Omega^* = 2\Omega/(\omega^2 + \nu_e{}^2k^4)^{1/2}$ and $\hat{\nu} = (\nu_e{}^2k^4 - \omega^2)/(\nu_e{}^2k^4 + \omega^2)$, which basically determine the integrals. If delta function-like peak spectra are chosen for q, the integrations (24) over $\boldsymbol{k}$ and ω may be regarded as having been carried out. For the Sun, Ω^* is small for the granulation but large for the giant cells, $\hat{\nu}$ is negative for small viscosity ($\nu^* < 1$) and positive when the viscosity is large ($\nu^* > 1$).

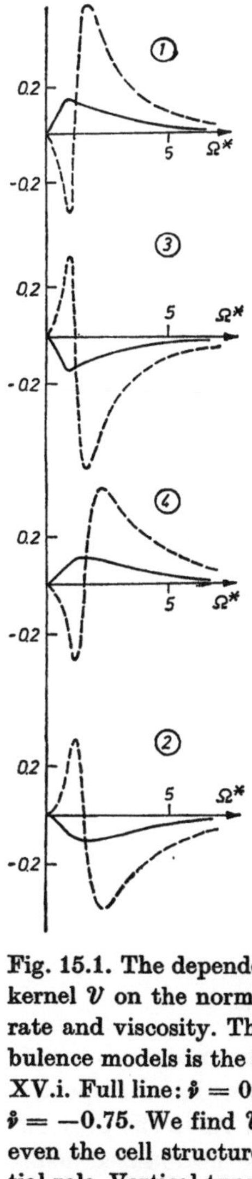

Fig. 15.1. The dependence of the integral kernel $\mathcal{V}$ on the normalized rotation rate and viscosity. The sequence of turbulence models is the same as in Table XV.i. Full line: $\hat{\nu} = 0.75$; broken line: $\hat{\nu} = -0.75$. We find $\mathcal{V}$ varies greatly, even the cell structure plays an influential role. Vertical-type turbulence ① and horizontal-type turbulence of spherical structure ④ behave similarly, but in quite an opposite way to the horizontal models that have flat configurations. For low viscosity ($\hat{\nu} < 0$) the inertial wa ve resonance appears at $\Omega^* = 1$.

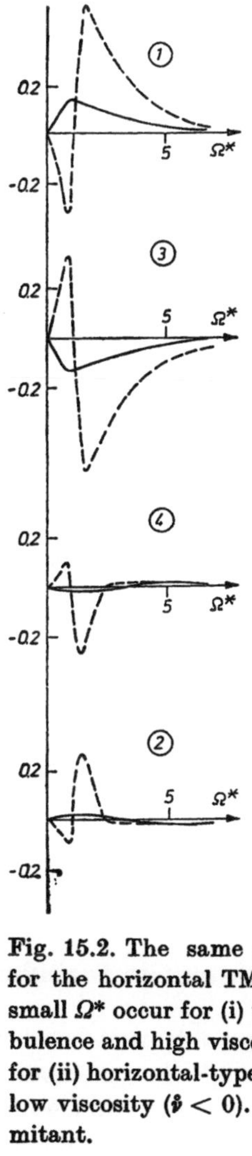

Fig. 15.2. The same as in Fig. 15.1 but for the horizontal TMT. Positive $\mathcal{H}$ fo small Ω^* occur for (i) vertical-type turbulence and high viscosity ($\hat{\nu} > 0$) and for (ii) horizontal-type turbulence and low viscosity ($\hat{\nu} < 0$). Model ② is intermitant.

Table XV.i. Cell-characteristics for various turbulence models

No	turbulence type, cell shape	α	$\varkappa$
1	vertical, spherical	1/2	0
2	horizontal, very flat	1/2	1
3	horizontal, oblong	$-\infty$	0
4	horizontal, spherical	$-\infty$	1

All the known solar turbulence modes have $\nu^* < 1$, i.e. $\hat{\nu}$ is negative. Due to the inertial wave resonance in inviscid media, the resulting profiles are always peaked at $\Omega^* = 1$ for negative ν. The kernels $\mathcal{V}$ and $\mathcal{H}$ for several models are presented in Fig. 15.1 and 15.2, from which it becomes obvious why no simple theory of differential rotation exists. All turbulent quantities depend strongly on several parameters. Even their signs vary as various dependencies change. For small Ω^* the signs differ for positive and negative $\hat{\nu}$, i.e. the quantities vary with $\hat{\nu}$. For fast rotation, however, $\hat{\nu}$ loses its importance and the quantities vary as an even power of $\hat{\nu}$. When $\hat{\nu}$ is close to unity, $\mathcal{V}$ and $\mathcal{H}$ do not change sign. Table XV.i lists four distinct turbulence models. The resulting $\mathcal{V}$ are similar for models 1 and 4, as well as for models 2 and 3. In contrast, the $\mathcal{H}$ are similar for models 1 and 2, and also for models 3 and 4. Finally, we see that the structure of the turbulence has an important bearing on the calculated TMT. Even the form of the cells plays a fundamental role. It can be seen that positive $\mathcal{H}$ result from $\Omega^* \leqq 1$ and negative $\hat{\nu}$ for horizontal turbulence. This is in clear contrast to the result $(6)_2$ where, even for very flat cells, positive $H^{(1)}$ are implied by horizontal-type turbulence.

A final discussion of the implications of the arguments of this Section on the differential rotation is postponed to § 15.4.

15.2. Non-axisymmetric Turbulence

Up to the present time it has been possible to treat non-axisymmetric models only through the one-mode approximation, for which the spectral functions degenerate into Dirac δ-functions, as in (13.47). As in the previous Sections, the common frequency spectrum, q_ω, can be normalized using (13.45). After substituting (13.47) into the general result (1), we may deduce the first Λ-modes by lengthy but straightforward calculations:

$$V^{(1)} = 4(2\tau_c\Omega)^2 (K_\theta^2 - 3K_r^2) [(K_\theta^2 + K_\phi^2) \overline{u_\phi^2} - (K_r^2 + K_\theta^2) \overline{u_r^2}] Q_2/\nu_T\tau_c K^4, \tag{27}$$

$$H^{(1)} = 4(2\tau_c\Omega)^2 (3K_\theta^2 - K_r^2) [(K_r^2 + K_\phi^2) \overline{u_\phi^2} - (K_r^2 + K_\theta^2) \overline{u_\theta^2}] Q_2/\nu_T\tau_c K^4,$$

where again the quantity

$$Q_2 = \int \nu K^2(\omega^2 - \nu^2 K^4)\,(\omega^2 + \nu^2 K^4)^{-3}\, q_\omega\, \mathrm{d}\omega\,, \tag{28}$$

which has the dimension of (time)^2, has been introduced. Several terms compete as differences in these expressions, so that the determination of the sign and magnitude of the quantities is a complicated matter. The situation is improved slightly if the turbulence model proposed by Durney and Spruit (1979) is adopted. The fluid is assumed to be inviscid, and the horizontal velocities are expressed in terms of the vertical ones by means of (13.60). These relations can be substituted into (27). Another simplification concerns the quantity Q_2 which, for vanishing viscosity, can now be expressed as the square of the correlation time. As with familiar kernels such as

$$\nu k^2/\pi(\omega^2 + \nu^2 k^4)\,, \quad 2\nu k^2\omega^2/\pi(\omega^2 + \nu^2 k^4)^2\,, \quad 2\nu^3 k^6/\pi(\omega^2 + \nu^2 k^4)^2\,,$$

the form

$$4\omega^2 \nu k^2(\omega^2 - \nu^2 k^4)/\pi(\omega^2 + \nu^2 k^4)^3$$

also becomes a Dirac δ-function for $\nu \to 0$. Hence

$$\lim_{\nu \to 0} Q_2 = \frac{1}{4}\,\pi \int \omega^{-2} q_\omega \delta(\omega)\, \mathrm{d}\omega\,, \tag{29}$$

and thus, if it is to converge, the frequency spectrum q_ω must vanish at $\omega = 0$. Unless the spectrum varies like ω^2 near the origin, we will necessarily obtain infinite values of Q_2. The assumption $\nu = 0$ turns out to be an assumption about the slow-mode frequency behavior of the turbulence, and Q_2 reduces to $Q_2 = \frac{1}{2}\,\tau_c^2$, apart from a factor close to unity that depends on the precise profile assumed. To sum up, the inviscid turbulence of Durney and Spruit should generate, for a moderate basic rotation,

$$\begin{aligned} V^{(1)} &= 2(2\tau_c\Omega)^2\,(3K_r^2 - K_\theta^2)\,K_\theta^2\tau_c\overline{u_r^2}/\nu_T K^2 K_\perp^2\,, \\ H^{(1)} &= 2(2\tau_c\Omega)^2\,(K_r^2 - 3K_\theta^2)\,(K_\theta^2 - K_\phi^2)\,K_r^2\tau_c\overline{u_r^2}/\nu_T K^2 K_\perp^4\,. \end{aligned} \tag{30}$$

These expressions are rather similar to those given by the authors (see § 15.3).

15.3. Another Approach

Let us apply the linearized Navier-Stokes equation (14.52) to an axisymmetric turbulence whose spectral function already possesses a suitable latitudinal profile. In order to demonstrate this mechanism we simply set

${}^{(2)}q = {}^{(2)}q_0 + {}^{(2)}q_2 \sin^2\theta$ and assume further that ${}^{(1)}q = 0$ (horizontal isotropy). From (14.56) we find immediately that

$$V = 2 \iint k_\perp^2 k_\theta^2 (\omega^2 + \nu^2 k^4)^{-1}\, {}^{(2)}q \, \mathrm{d}\boldsymbol{k}\, \mathrm{d}\omega, \tag{31}$$

so that

$$V^{(1)} = 2 \iint k_\perp^2 k_\theta^2 (\omega^2 + \nu^2 k^4)^{-1}\, {}^{(2)}q_2 \, \mathrm{d}\boldsymbol{k}\, \mathrm{d}\omega. \tag{32}$$

The corresponding overall differential rotation $\delta\Omega$ is then, according to (6.59),

$$\delta\Omega/\mathsf{d} = -\iint k_\perp^2 k_\theta^2 (\omega^2 + \nu^2 k^4)^{-1}\, {}^{(2)}q_2 \, \mathrm{d}\boldsymbol{k}\, \mathrm{d}\omega, \tag{33}$$

and the observed negativity of (33) therefore requires positive ${}^{(2)}q_2$. Thus, for this particular situation, equatorial acceleration would require spectral functions that increase from poles to equator: ${}^{(2)}q$ (pole) $<$ ${}^{(2)}q$ (equator).

There is, however, a striking inconsistency. When the latitudinal dependence of ${}^{(2)}q_2$ is due to the influence of rotation on the onset of turbulent convection, as suggested by Durney and Spruit, we have to adopt the form for the Λ-effect that applies when the rotation is not slow, rather than the simple form (31). Generally speaking, unless the condition of slow rotation ($\Omega^* \ll 1$) is satisfied, we must apply the complete Navier-Stokes equation (14.7) rather than its approximate form (14.52) that is linear in Ω, and which led to (14.53), analysed in Chapter 14. Durney and Spruit argue that the actual value of Ω^* is indeed close to unity in the lower SCZ. In this case, the influence of the basic rotation on the onset of convection generates latitudinal profiles of $\overline{u_r'^2}$ such that [cf. Durney and Spruit (1979), see their (29)]

$$\overline{u_r'^2} = K^{-2} L^2 \{[gT^{-1} \Delta \nabla T K_\perp^2 - 4\Omega^2 K_r^2] + 4\Omega^2 (K_r^2 - K_\theta^2) \sin^2\theta\}. \tag{34}$$

Here T is temperature, L mixing length, $\Delta \nabla T$ superadiabatic temperature gradient and g local gravity [see Schmidt and Stix (1983) for a numerical discussion].

If we now substitute (34) into the expressions (14.73) for the Λ's (which, however, are not relevant to rapid rotation), this "inhomogeneous intensity" leads to the modes

$$\begin{aligned} V^{(1)} &= 8(\Omega\tau_c)^2 (K_\theta^2 - K_r^2)\, K_\theta^2 L^2/\tau_c \nu_T K^2 K_\perp^2, \\ H^{(1)} &= 8(\Omega\tau_c)^2 (K_\theta^2 - K_r^2)(K_\theta^2 - K_\phi^2)\, K_r^2 L^2/\tau_c \nu_T K^2 K_\perp^4. \end{aligned} \tag{35}$$

These terms are of order unity if

$$\Omega^* \equiv 2\Omega\tau_c \approx 1, \qquad \tau_c \nu_T K^2 \approx 1, \tag{36}$$

which may be true for the Sun. Of course, we still have to ask whether the terms (35) dominate those which arise because of the really strong influence of rotation on the turbulent fluctuations assumed.

Another question involving (35), as well as our relations (30), concerns the components of the dominating wave-number vector, $\boldsymbol{K}$. If they are unknown, the usefulness of relations deduced from the one-mode approximation is rather diminished. Nevertheless, one cannot too cavalierly specify them.

Hathaway (1984) has taken the work of Durney and Spruit further. He included, in particular, the case of fast rotation, so that his eddy fluxes for the momentum, $Q_{r\phi}$ and $Q_{\theta\phi}$, tend to zero for $\Omega \to \infty$. For example, his approach gives, for the vertical TMT,

$$Q_{r\phi} = -2L^2\Omega \sin\theta \cdot K_\theta^2 K_\perp^{-2} \sigma^3 (\sigma^2 + 4\Omega^2 \cos^2\theta)^{-1}, \tag{37}$$

where σ is the growth rate, which satisfies a certain quartic equation. It is interesting to expand (37) in powers of $\sin^2\theta$. For σ we use the same normalization as for the angular velocity of rotation, viz.:

$$\Omega^* = 2\Omega(g\Delta\, \nabla T/T)^{-1/2}, \tag{38}$$

and write

$$\sigma = \sigma_0 + \sigma_1 \sin^2\theta\,. \tag{39}$$

This means that a pole-equator difference in growth rates may exist, reflecting the rotational influence of the *onset* of convection. For positive σ_1, the equatorial modes grow faster than the polar ones. The result of the expansion is

$$V = -\frac{4L^2K_\theta^2}{\nu K_\perp^2} \frac{\sigma_0^3}{\sigma_0^2 + \Omega^{*2}} \left(1 + \frac{\sigma_1\sigma_0 + \Omega^{*2}}{\sigma_0^2 + \Omega^{*2}} \sin^2\theta\right) \frac{\Omega}{\Omega^*}. \tag{40}$$

We see that $V^{(0)}$ is again negative as, indeed, is $V^{(1)}$, provided σ_1 is not too negative. Relation (40) demonstrates clearly that $V^{(0)}$ is generated by two effects: the direct action of the Coriolis force, and the effect of rotation on the onset of convection ($\sigma_0\sigma_1$).

From the work by Hathaway, Toomre and Gilman (1980) we expect the equatorial modes to be the faster growing. Hence, σ_1 should be positive, and $V^{(1)}$ will then necessarily be negative. The growth rate equation given by Hathaway gives, however, a small but negative σ_1, if one accepts his concept of a "dominant eddy". Hathaway argued that one can model the behavior of the characteristic wave-numbers in the presence of rotation by simple expressions. Based on numerical computations, he suggested that the effect of rotation on the shape of the cells is to be seen in:

(i) a decrease in K_θ compared with K_ϕ for low latitudes;
(ii) an increase in K_ϕ compared with K_r at high latitudes;

(equatorial cells are elongated in the N-S direction, polar cells are oblong in the vertical direction). Two heuristic relations said to describe this conjecture are

$$\frac{K_\theta^2}{K_\phi^2} = \frac{1 + 3\Omega^{*2} - 3\Omega^{*2} \sin^2 \theta}{1 + 3\Omega^{*2}}, \qquad \frac{K_\phi^2}{K_r^2} = \frac{1 + \cos^2 \theta + 4\Omega^{*2} \cos^2 \theta}{4(1 + \cos^2 \theta)}. \tag{41}$$

The radial wave-number is, quite simply, $K_r = \pi/L$. With the ansatz (41) one can:

(i) find an expression for the factor $K_\theta^2/K_\perp^2$ in (37); and
(ii) solve the growth rate equation.

From the latter one obtains

$$\sigma = 3^{-1/2}[1 - (5/12)\,\Omega^{*2} \sin^2 \theta], \tag{42}$$

which is quite the reverse of the behavior anticipated. The dominant cell has the fastest motions at the poles. The magnitude of the associated term in (37) is, however, too small to affect the negativity of $V^{(1)}$.

From Fig. 5 of Hathaway's paper, one can also remove the negativity of $V^{(0)}$ and $V^{(1)}$. $H^{(1)}$ is then positive.

15.4. The TMT Modes Required

After the investigations of Part II we are confronted with the following scenario. In order to explain the observations, the horizontal momentum flux mode $H^{(1)}$ must be positive in all cases. Also the product $V^{(0)}V^{(1)}$ proves to be negative definite. For one class of solutions (say P), however, $V^{(0)}$ is positive and $V^{(1)}$ negative; for the other class (say N), the reverse is true. Those models with uniform $V^{(1)}, \ldots, HV^{(1)}$, as well as the models which artificially ignored the meridional flow, belong to class P. Class N arises as the result of calculations with a finite exponent λ; the higher order TMT modes $V^{(1)}, \ldots, HV^{(1)}$ then decrease outward (and meridional flow is present). Belonging to this class also are the two-shell models (with $x_i = 0.95$), after we have allowed a necessary, but small, temperature difference to be present.

We cannot conceal our surprise about the existence of such a wide disparity in the explanations. We believe, intuitively, that mixing-length theory lies behind models of class N. The correlation time increases inwards, and the rotationally created eddy fluxes should do likewise. This might be a convincing standpoint if the observations for $Q_{\theta\phi}$ (1) did not tell a different story. These reveal directly the presence of significant $H^{(1)}$ close to the surface, as otherwise ν_T would be negative. Since, however, the sunspots might not be anchored in very deep layers, the coefficient $H^{(1)}$ must exist even in the surface layers. This appears to be incompatible with mixing length concepts. The alternative models with depth-independent coefficients have more to do with the "global convection" simulations of Gilman and Glatzmaier. There is no reason to postulate any depth dependence in the correlation time of the giant granulations; uniform eddy fluxes should therefore be a useful approximation.

The two-shell model with $x_j = 0.95$ clearly lies between the two extremes. There *is* an outward decrease, but only very close to the surface. In this, it resembles the $\lambda = 0$ model. Indeed, the required Λ's for the pure two-shell model ["pure" will be taken to mean $t_2(1) = 0$] lie between the two classes P and N. Both $V^{(0)}$ and $V^{(1)}$ are negative; from P we obtain negative $V^{(1)}$, and from N negative $V^{(0)}$. We shall not, however, discuss this case further in view of the intrinsic uncertainties of these models.

Of course, we do not know the real situation in the SCZ, so the two approaches will be treated separately. Nevertheless, the conditions

$$V^{(0)}V^{(1)} < 0, \qquad H^{(1)} > 0, \tag{43}$$

hold for both classes. The second relation, in particular, proves to be revealing. In addition, the implications of the two sets of Λ-modes,

$$\begin{aligned} &P\,(\lambda = 0): \qquad V^{(0)} > 0 \qquad V^{(1)} < 0, \\ &N\,(\lambda \gtrsim 2): \qquad V^{(0)} < 0 \qquad V^{(1)} > 0, \end{aligned} \tag{44}$$

will be discussed. For P, the eddy fluxes were assumed to be uniform, in contrast to N where they had a remarkably steep gradient inwards. We should stress that, in both cases, the observed boundary condition on the radial rotation law, i.e. that $\partial\Omega/\partial r < 0$ at the surface, is satisfied. Only in case N does it immediately lead to a negative $V^{(0)}$. As we know from the foregoing Chapter, negative $V^{(0)}$ are favored in turbulence theory. We shall now see that class N of mean-field models does indeed lead to a consistent picture. We must be careful, however, in applying the results obtained from SOCA in the present Chapter to the mean-field calculations of Chapters 6 and 11. SOCA may be too simple, all the more so since its treatment of the density gradient is incorrect. Nevertheless, we shall risk a discussion in order to illuminate all possibilities. One may be sure that final answers are still lacking, but much can be achieved if at least the right questions are asked.

Models with different λ's also require different orders of magnitude for the eddy coefficients that must be assumed. The magnitudes of $V^{(1)}$ and $H^{(1)}$ are the same for both classes, but in class N we have

$$|V^{(0)}| < |V^{(1)}| \tag{45}$$

as opposed to class P where these two quantities are nearly equal. Let us examine first the positivity of $H^{(1)}$. As has already been stressed on the basis of (4) ("axisymmetric turbulence"), Q_2 must be positive, and from (6)

$Q_2' q_k > 0$. The kernel Q_2' contains the frequency spectrum and is negative if that spectrum decreases monotonically with increasing frequency. If this is the case, q_k must be negative. Negative q_k describes turbulence of vertical-type, i.e. where vertical motions dominate horizontal ones.

According to (14.57), vertical-type turbulences have negative $V^{(0)}$. This is consistent with case N. Also the turbulence models of § 14.4 gave negative $V^{(0)}$ in both cases. For $V^{(1)}$ to be positive for positive Q_2 (as is required for N), the expression $(4)_1$ for $V^{(1)}$ demands that the wave-number k_r be small in comparison with $k_\perp$. One possibility is the delta function-like spectrum $q \approx \delta(k_r)$, i.e. two-dimensionality. But as $(6)_1$ demonstrates, even the spectrum with $\varkappa = 0$ leads to a positive $V^{(1)}$, which is then equal to the resulting $H^{(1)}$. So the sequence of cell configurations that are permitted ranges from those that are very long (parallel to $\mathring{\boldsymbol{g}}$) to those that are nearly spherical.

To sum up, one of the simplest examples which produces the constraints N in (44) is the turbulence defined by

$$^{(2)}q = -^{(3)}q = -W^2 q_k/\pi(W^2 + \omega^2), \qquad q_k > 0, \tag{46}$$

i.e. $\alpha = 1$ or

$$\hat{Q}_{ij} = q_k[(\mathring{\boldsymbol{g}} \cdot \boldsymbol{k})^2\, \delta_{ij} + k^2 \mathring{g}_i \mathring{g}_j - (\mathring{\boldsymbol{g}} \cdot \boldsymbol{k})\, (\mathring{g}_i k_j + \mathring{g}_j k_i)]\; W^2/\pi(W^2 + \omega^2). \tag{47}$$

This is a vertical-type turbulence having cells that are not flat and a frequency spectrum which decreases monotonically as $\omega \to \infty$. The associated TMT modes are

$$\begin{aligned} V^{(0)} &= -(8W/15\nu_e) \int k^2 q_k (W + \nu_e k^2)^{-1}\, \mathrm{d}\boldsymbol{k} < 0, \\ V^{(1)} = H^{(1)} &= \frac{8\Omega^2 W}{35\nu_e^3} \int \frac{q_k}{k^2}\, \frac{W^2 + 3W\nu_e k^2 + 4\nu_e^2 k^4}{(W + \nu_e k^2)^3}\, \mathrm{d}\boldsymbol{k} > 0. \end{aligned} \tag{48}$$

Two additional remarks should be made:

(i) A term of order Ω^2, which hitherto has normally been neglected, must be added to the given $V^{(0)}$. It can be determined using

$$V^{(0)} = 24\Omega^2 \nu_e^{-1} \iint (\mathring{\boldsymbol{g}} \cdot \boldsymbol{k})^2\, k_\perp^4 k^{-4} Q_2\, \mathrm{d}\boldsymbol{k}, \tag{49}$$

from which it is clear that it will be positive for positive Q_2. It will therefore reduce the given $V^{(0)}$ so that even condition (45) is satisfied. The expression (49) is of the same order as $V^{(1)}$;

(ii) The equality of $V^{(1)}$ and $H^{(1)}$ holds only for the chosen values of $\varkappa$ and α. It is not hard to vary them in such a way that $H^{(1)}$ does indeed exceed $V^{(1)}$.

Our results so far have tended to favor case N. We could, however, have argued in quite another way. Positive $H^{(1)}$ arise from a delta functionlike

Table XV.ii. Main features of the two classes of solution

	class P	class N
$V^{(0)}$	> 0	< 0
λ	0,1	$\geqq 2$
model	global convection	mixing length
$V^{(1)}$	< 0	> 0
$H^{(1)}$	> 0	> 0
q_ω	random waves	turbulence
q_k	> 0	< 0
turbulence-type	horizontal	vertical
$\varkappa$	> 0.25	< 0.25
cell structure	flat	not flat

frequency spectrum only when the viscosity is so small that $\nu^* < 1$. Then $q_k > 0$ follows from $(4)_2$, i.e. the turbulence is of horizontal-type. The associated fundamental mode $V^{(0)}$ is positive, as is required in class P. So the two types of frequency spectrum q_ω ("turbulence" and "random-waves") are closely related to the two classes of mean-field models for the differential rotation. *"Turbulence" spectra correspond to solutions for which the modes (class N) decrease in the outward direction; "random-waves" imply results that hold for constant Λ-modes (class P).* We imagine that the latter case is related to the global convection simulations, while the case of large λ provides a better way of realizing the mixing length concept.

Both these possibilities also arise if, at the start, relation $(43)_1$ is utilized: Through (14.57) and $(6)_1$ this means, in a symbolic manner,

$$Q_2'(1 - 4\varkappa) < 0. \tag{50}$$

"Turbulence" spectra yield negative Q_2'; "random-wave" spectra with $\nu^* < 1$ yield positive Q_2'. So "turbulence" *cannot* have $\varkappa > 0.25$, while random waves *must* have $\varkappa > 0.25$. The relation $(6)_1$, together with (4), yields $q_k > 0$ for class P, and $q_k < 0$ for class N. The turbulence in the former case is of the horizontal-type, while in the latter it is of vertical-type. Table XV.ii. summarizes the results. A decision between these two possibilities can be hoped for from:

(i) refinements in SOCA, and/or an appeal to the positivity of $VV^{(1)}$ and $HV^{(1)}$;
(ii) "observation" of the internal rotation and flow pattern;
(iii) dynamo theory.

Chapter 16. The Theory of Eddy Viscosity

Having dealt with the influence of rigid rotation on relatively simple turbulences, we now have to assess the effect of rotational shearing, i.e. the tensor $Q_{ij}^{(d)}$. Certainly, this is also responsible for the generation of spatial inhomogeneities which, through a divergence, produce Reynolds stresses. Let us suppose that these stresses depend only linearly on the rates of strain that create them. Then, in specifying the correlation tensor, we need consider only the gradients of $\bar{\boldsymbol{u}}$, i.e. $\bar{u}_{i,j}$. If, in addition, we take the symmetry in i and j into account, we may write

$$Q_{ij} = a\delta_{ij} + b(\bar{u}_{i,j} + \bar{u}_{j,i}), \tag{1}$$

provided that there is no additional preferred direction in the turbulence. Equation (1) implies that, in the presence of a mean flow, the turbulent region behaves like a medium of a different viscosity. We therefore write $b = -\nu_T$, where the correlation coefficient ν_T formally acts as a viscosity. Since ν_T also appears as a coefficient in the one-point correlation tensor, it is sufficient, when determining it, to calculate the effect of the shear flow on the prescribed turbulence. When we refer to 'the eddy viscosity', we invariably refer to the role played by ν_T in determining the correlation.

From a physical point of view there is little difference between eddy viscosity and molecular viscosity: molecules, as well as eddies, carry momentum. If a turbulence element moves from a rapidly moving region to another having a lower mean velocity, it supplies momentum from its origins to its new surroundings. Also, when motion in the reverse direction takes place, differences in mean velocity tend to be reduced. The momentum flux which is carried through surface II in Fig. 16.1 is clearly $\varrho[v'\bar{u}^{III} + (-v')\,\bar{u}^{I}]$, which in the limit of small separations is $-\varrho v'L\,\partial\bar{u}/\partial y$, where L is a correlation length and v' is the root-mean-square (rms) velocity. In the case of non-uniformity in the y-direction, viscous forces result, with a viscosity of the order of $v'L$. Here L considerably exceeds the characteristic microscopic length (e.g. the mean free-path between collisions), though the eddy velocity does not attain the speed of molecules ($\approx$ speed of sound).

Otherwise e.g. a maelstrom with a diameter of 100 km would last for about 10^8 years ($\nu_{\text{water}} \approx 0.01\ \text{cm}^2/\text{s}$)!

Let us construct a more detailed illustration. In Fig. 16.2 a turbulent element E moves to E' with a velocity consisting of a fluctuating velocity, u', superimposed on a mean velocity $\bar{u} = y(\partial\bar{u}/\partial y)_{y=0}$. (We omit the suffix

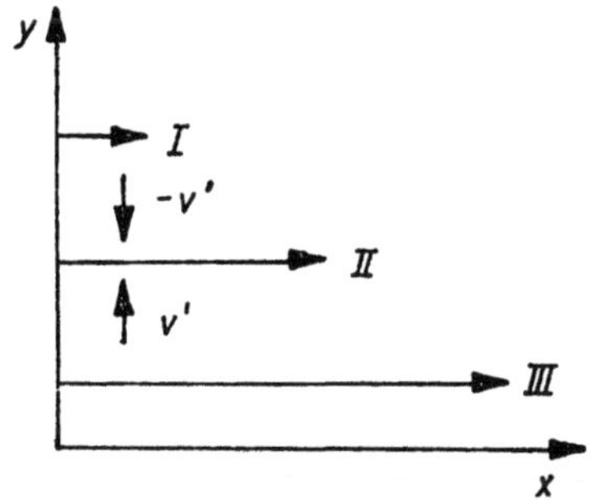

Fig. 16.1. On the nature of the viscous momentum exchange in a shear flow (see text).

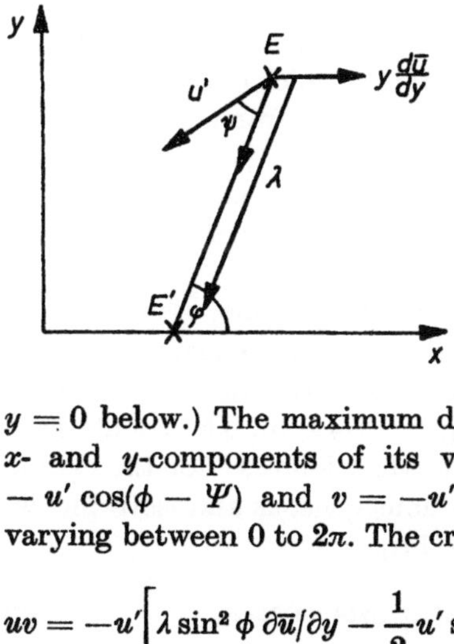

Fig. 16.2. The context of the geometrical estimate of the eddy viscosity as a coefficient in the Boussinesq stress-strain relation.

$y = 0$ below.) The maximum distance between E and E' is L. At E', the x- and y-components of its velocity are respectively $u = \lambda \sin\phi\ \partial u/\partial y - u' \cos(\phi - \Psi)$ and $v = -u' \sin(\phi - \Psi)$, the fluctuation parameter ϕ varying between 0 to 2π. The cross-product at E' proves to be, for small Ψ,

$$uv = -u'\left[\lambda \sin^2\phi\ \partial\bar{u}/\partial y - \frac{1}{2}u' \sin^2 2\phi + \Psi\left(u' \cos^2 2\phi - \frac{1}{2}\lambda \sin^2 2\phi\ \partial\bar{u}/\partial y\right)\right]. \tag{2}$$

For geometrical reasons we have $\tan\phi = u/v$, so that for small Ψ

$$\Psi \simeq (\lambda/u') \sin^2\phi\ \partial\bar{u}/\partial y. \tag{3}$$

Using this relation and averaging, we have

$$\overline{uv} = -(u'/2\pi L) \int_0^L \int_0^{2\pi} (\lambda \sin^2\phi + \lambda \cos^2 2\phi \sin^2\phi)\, d\phi\, d\lambda\ \partial\bar{u}/\partial y, \tag{4}$$

so that from (2) we obtain the result

$$\overline{uv} = -\frac{1}{4} Lu' \, \partial\bar{u}/\partial y . \tag{5}$$

Equation (5) may be taken to be a particular realization of (1), where the eddy viscosity is $u'L/4$, which is always positive. Our calculations demonstrate that positivity is not automatic in all cases. Sometimes the pressure gradient may strongly influence the results, and this has been ignored in the illustration we have just given. Nevertheless, it does provide a good insight into the nature of eddy viscosity.

16.1. The Viscosity Terms

In order to include non-uniform rotation, one can use the ansatz

$$u' = u^{(r)} + u^{(d)}, \qquad u^{(d)}/_{\Omega=\mathrm{constant}} = 0, \tag{6}$$

where $u^{(r)}$ is the fluctuating flow that would be present were the rotation rigid, and $u^{(d)}$ is an additional part of the fluctuation which owes its existence to the gradient of the angular velocity; the co-rotating frame of reference is used. Here we are only able to work to terms linear in the deviation from the uniform state, and for this reason the idea of an eddy viscosity is appropriate. For weak differential rotation this procedure can certainly be justified.

One could, of course, also work in the non-rotating frame, assuming that the force fields which maintain the turbulence are statistically steady in the inertial frame. The corresponding eddy viscosities have been determined in two papers by Krause and Rüdiger (1974a, b). The results have a strikingly simple structure compared with what follows. For the case in which the original turbulence is completely homogeneous and isotropic, we found that [see also Roberts and Soward (1975)]

$$^{(3)}\nu_T = (4/15) \int\!\!\int \nu^3 k^6 \hat{Q}_{ll}^{(0)}(k, \omega)\, (\omega^2 + \nu^2 k^4)^{-2}\, \mathrm{d}\boldsymbol{k}\, \mathrm{d}\omega . \tag{7}$$

For two-dimensional turbulence the result was

$$\begin{aligned} ^{(2)}\nu_T &= \frac{1}{2} \int\!\!\int \nu k_\perp^4 (\nu^2 k_\perp^4 - \omega^2)\, (\omega^2 + \nu^2 k_\perp^4)^{-2}\, q_\perp(k_\perp, \omega)\, \mathrm{d}\boldsymbol{k}_\perp\, \mathrm{d}\omega \\ &= -\frac{1}{2} \int\!\!\int \nu k_\perp^4 \omega (\omega^2 + \nu^2 k_\perp^4)^{-1}\, \partial q_\perp(k_\perp, \omega)/\partial\omega\, \mathrm{d}\boldsymbol{k}_\perp\, \mathrm{d}\omega , \end{aligned} \tag{8}$$

which is clearly indefinite in sign. It is always positive for spectra of turbulence-type ($\omega\, \partial q/\partial\omega < 0$), but it might be negative in other cases. Krause and Rüdiger (1974b) have discussed this point in great detail. They showed that, as the scales of the mean quantities are decreased, ${}^{(2)}\nu_T$ always becomes positive; were it otherwise, the smallest scales would grow fastest, a physical absurdity.

In the limit of vanishing viscosity ($\nu \to 0$), expression (8) vanishes identically, whereas (7) remains finite:

$$\lim_{\nu \to 0} {}^{(3)}\nu_T = (2\pi/15) \int \hat{Q}_{ii}(k, 0)\, \mathrm{d}\boldsymbol{k}. \tag{9a}$$

The three-dimensional eddy viscosity therefore remains positive-definite even in the limit $\nu \to 0$ and we can therefore write

$${}^{(3)}\nu_T = (1/15) \int_{-\infty}^{\infty} \overline{\boldsymbol{u}'(\boldsymbol{x}, t)\, \boldsymbol{u}'(\boldsymbol{x}, t + \tau)}\, \mathrm{d}\tau, \tag{9b}$$

a formula which might well be tested by measurements. As must be the case, calculations made in the inertial frame with rotating forces lead to identically the same results as calculations made in the co-rotating frame with stationary forces.

When the turbulence is anisotropic, the deformation tensor will not be the only one that appears in the correlation tensor. In particular, the vertical direction, $\mathring{\boldsymbol{g}}$, is preferred because of gravity, and will introduce anisotropy to the turbulence. The eddy viscosity is then no longer a scalar quantity. In general, a tensor relation of the form

$$Q_{ij} = \cdots - \mathcal{N}_{ijkl} \bar{u}_{k,l} \tag{10}$$

must be expected. For two-dimensional turbulence, we have even extended this ansatz to the nonlinear case (Rüdiger, 1976), but in the present context such studies are irrelevant. The eddy viscosity tensor, $\mathcal{N}_{ijkl}$, is formed in a complicated way using the tensors δ_{ij} and ε_{ijk}, as well as the vectors $\boldsymbol{g}$ and $\boldsymbol{\Omega}$. Its structure is simple only in the cases "$\boldsymbol{g} = 0$" or "$\boldsymbol{\Omega} = 0$". For turbulence that is both rotating and anisotropic, we have restricted ourselves to determining only those components of Q_{ij} that are relevant to the rotational problem, i.e. $Q_{r\phi}$ and $Q_{\theta\phi}$.

With regard to the rate of strain, we suggest the following scenario. For non-uniform rotation the velocity of the flow at a point $\boldsymbol{x}$ in the neighborhood of an arbitrary point $\boldsymbol{x}_0$ is assumed to be

$$\bar{u}_k(\boldsymbol{x}) = [\Omega(\boldsymbol{x}) - \Omega(\boldsymbol{x}_0)]\, \varepsilon_{kpq} \mathring{\Omega}_p x_q, \tag{11a}$$

where as usual $\mathring{\boldsymbol{\Omega}}$ is the unit vector in the z-direction. The rate of strain at $\boldsymbol{x}_0$ is therefore

$$\bar{u}_{k,l}(\boldsymbol{x}_0) = r \varepsilon_{kpq} \Omega_{,l} \mathring{\Omega}_p \mathring{g}_q. \tag{11b}$$

Substituting this expression in (10), we find an expression for the required correlation tensor, $Q_{ij}^{(d)}$, the only components of which that interest us being

$$\begin{aligned} Q_{\phi r}^{(d)} &= -(r/\sin\theta)\, \varepsilon_{jmn}\varepsilon_{kpq}\mathring{g}_i\mathring{g}_n\mathring{g}_q\mathring{\Omega}_m\mathring{\Omega}_p\mathcal{N}_{ijkl}\Omega_{,l}, \\ Q_{\theta\phi}^{(d)} &= (r/\sin\theta)\, \varepsilon_{jmn}\varepsilon_{kpq}\mathring{g}_n\mathring{g}_q\mathring{\Omega}_{\perp i}\mathring{\Omega}_m\mathring{\Omega}_p\mathcal{N}_{ijkl}\Omega_{,l}. \end{aligned} \tag{12}$$

Again the original tensor is free of cross-correlations. In accordance with (7.38), we now introduce the definitions

$$Q_{r\phi}^{(d)} = -r\begin{pmatrix} \nu_{vv}\sin\theta \\ \nu_{vh}\cos\theta \\ 0 \end{pmatrix}\nabla\Omega, \quad Q_{\theta\phi}^{(d)} = -r\begin{pmatrix} \nu_{hv}\cos\theta \\ \nu_{hh}\sin\theta \\ 0 \end{pmatrix}\nabla\Omega, \tag{13}$$

to which we may immediately add

$$\begin{aligned} \nu_{vv} &= (\sin^2\theta)^{-1}\, \varepsilon_{jmn}\varepsilon_{kpq}\mathring{g}_i\mathring{g}_n\mathring{g}_q\mathring{g}_l\mathring{\Omega}_m\mathring{\Omega}_p\mathcal{N}_{ijkl}, \\ \nu_{vh} &= -(\sin\theta\cos\theta)^{-1}\, \varepsilon_{jmn}\varepsilon_{kpq}\mathring{g}_i\mathring{g}_n\mathring{g}_q\mathring{\Omega}_{\perp l}\mathring{\Omega}_m\mathring{\Omega}_p\mathcal{N}_{ijkl}, \\ \nu_{hv} &= -(\sin\theta\cos\theta)^{-1}\varepsilon_{jmn}\varepsilon_{kpq}\mathring{g}_n\mathring{g}_q\mathring{g}_l\mathring{\Omega}_{\perp i}\mathring{\Omega}_m\mathring{\Omega}_p\mathcal{N}_{ijkl}, \\ \nu_{hh} &= (\sin^2\theta)^{-1}\, \varepsilon_{jmn}\varepsilon_{kpq}\mathring{g}_n\mathring{g}_q\mathring{\Omega}_{\perp l}\mathring{\Omega}_{\perp i}\mathring{\Omega}_m\mathring{\Omega}_p\mathcal{N}_{ijkl}. \end{aligned} \tag{14}$$

These scalars are components of the fourth rank tensor $\mathcal{N}_{ijkl}$. We shall call them "viscosity terms". For purely isotropic viscosity we have $\nu_{vv} = \nu_{hh} = \nu$, $\nu_{hv} = \nu_{vh} = 0$.

We can derive an important property of the scalars from (14). As only two vectors (namely $\mathring{g}$ and $\mathring{\Omega}$) are involved, no scalars can be formed using the ε-tensor. Hence the number of ε-tensors in (14) must be even. Therefore we need only those terms in $\mathcal{N}_{ijkl}$ that are even in ε_{ijk}.

16.2. Determination of the Viscosity Tensor

Let us develop a theory for the viscosity tensor $\mathcal{N}_{ijkl}$. The equation governing the required part, $\boldsymbol{u}^{(d)}$, of the turbulent flow is

$$\partial u_i^{(d)}/\partial t - \nu\Delta u_i^{(d)} + 2\varepsilon_{ijk}\Omega_j u_k^{(d)} = -\varrho^{-1}\nabla_i p^{(d)} - \bar{u}_{l,m}x_m u_{i,l}^{(r)} - u_l^{(r)}\bar{u}_{i,l}, \tag{15}$$

where formally $\bar{u}_i(x) = \bar{u}_i + x_m\bar{u}_{i,m}$ has been used. Hereafter we shall revert to a simple viscosity ν (and later to a simple thermal conductivity, χ). In contrast to the Λ-effect considered earlier, a fully isotropic original turbulence is associated with both a turbulent viscosity and a turbulent conductivity. It is still an open question whether a molecular or a turbulent value should be inserted for "ν" in (15). For simplicity we will leave (15) in its original, unmodified state. Whether the molecular or some other value of ν is used depends on the philosophy adopted. In this Chapter we shall call ν "the microscopic diffusivity". (In the next chapter, we shall use the same name for χ.)

Because of the inhomogeneous terms on the right-hand side of (15), we must slightly modify the Fourier transform, by writing

$$u_i^{(d)}(\boldsymbol{x}, t) = \iint (\hat{v}_i + ik_i\hat{u}_{ll}k^{-2} + x_m\hat{u}_{im})\, e^{i(\boldsymbol{k}\cdot\boldsymbol{x} - \omega t)}\, d\boldsymbol{k}\, d\omega\,. \qquad (16)$$

Here the quantities $\hat{v}_i$ and $\hat{u}_{im}$ represent functionals of homogeneous stochastic fields. We have chosen this representation because the functionals satisfy exactly the usual incompressibility requirement:

$$k_i\hat{v}_i = k_i\hat{u}_{ij} = 0\,. \qquad (17)$$

Of course, our ansatz (16), which is linear in $\boldsymbol{x}$, can be used only for weak inhomogeneities. On substituting it into (15), we arrive at the two equations:

$$(-i\omega + \nu k^2)\,(\hat{v}_i + ik_i\hat{u}_{ll}k^{-2}) - 2\nu ik_l\hat{u}_{il} + 2\varepsilon_{ijk}\Omega_j\hat{v}_k + 2\varepsilon_{ijk}\Omega_j ik_k\hat{u}_{ll}k^{-2}$$
$$= -\varrho^{-1}(ik_i\hat{p} + \hat{p}_i) - \hat{u}_l^{(r)}\overline{u}_{i,l}\,, \qquad (18)$$
$$(-i\omega + \nu k^2)\,\hat{u}_{im} + 2\varepsilon_{ijk}\Omega_j\hat{u}_{km} = -i\varrho^{-1}k_i\hat{p}_m - ik_l\hat{u}_i^{(r)}\overline{u}_{l,m}\,.$$

First we shall solve $(18)_2$ by multiplying it through by K_{ij} to obtain

$$\hat{u}_{im} = -iD_{ij}(-i\omega + \nu k^2)^{-1}\, k_l\hat{u}_j^{(r)}\overline{u}_{l,m}\,. \qquad (19)$$

An expression for $\hat{p}_i$, the gradient of the pressure p, can now be determined and substituted in (18). Here also the pressure is eliminated by using K_{ij}, and we obtain

$$\hat{v}_i = D_{it}(-i\omega + \nu k^2)^{-1}(2i\nu K_{jt}k_l\hat{u}_{jl} - 2i\varepsilon_{tjk}\Omega_j k_k k^{-2}\hat{u}_{ll}$$
$$-2iK_{jt}\varepsilon_{nmk}\Omega_m k_n k^{-2}\hat{u}_{kj} - K_{jt}\hat{u}_l^{(r)}\overline{u}_{j,l})\,. \qquad (20)$$

Solutions (19) and (20) may now be used to evaluate the correlation tensor. We shall proceed as follows.

Up to first order in the velocity gradient, we have

$$Q_{ij}^{(d)} = q_{ij}(\boldsymbol{x}, \boldsymbol{\xi}, t, \tau) + q_{ji}(\boldsymbol{x} + \boldsymbol{\xi}, -\boldsymbol{\xi}, t + \tau, -\tau), \qquad (21)$$

where

$$q_{ij} = \overline{u_i^{(r)}(\boldsymbol{x}, t)\, u_j^{(d)}(\boldsymbol{x} + \boldsymbol{\xi}, t + \tau)}\,. \qquad (22)$$

For the homogeneous part of the correlations this means, in terms of Fourier transforms,

$$\hat{Q}_{ij}^{(d)} = \hat{q}_{ij} + \hat{q}_{ji}^*, \qquad (23)$$

where $\hat{q}_{ij}$ is the Fourier transform of q_{ij} which can be written as

$$\hat{q}_{ij} = (2\pi)^{-4} \int \exp \{i[(\boldsymbol{k}'' - \boldsymbol{k})\,\boldsymbol{\xi} + (\boldsymbol{k}' + \boldsymbol{k}'')\,\boldsymbol{x} - (\omega'' - \omega)\,\tau - (\omega' + \omega'')\,t]\} \times \overline{\hat{u}_i^{(r)}(\boldsymbol{k}', \omega')\, u_j^{(d)}(\boldsymbol{k}'', \omega'')}\, \mathrm{d}\boldsymbol{\xi}\, \mathrm{d}\boldsymbol{k}'\, \mathrm{d}\boldsymbol{k}''\, \mathrm{d}\tau\, \mathrm{d}\omega'\, \mathrm{d}\omega''. \tag{24}$$

On using (16), we reduce (24) to

$$\hat{q}_{ij} = \int\int \exp \{i[(\boldsymbol{k} + \boldsymbol{k}')\,\boldsymbol{x} - (\omega + \omega')\,t]\} \times \overline{\hat{u}_i^{(r)}(\boldsymbol{k}', \omega')\,[\hat{v}_j(\boldsymbol{k}, \omega) + ik^{-2}k_j \hat{u}_{ll}(\boldsymbol{k}, \omega)]}\, \mathrm{d}\boldsymbol{k}'\, \mathrm{d}\omega'. \tag{25}$$

After substituting solutions (19) and (20) in this expression, we see that $\hat{q}_{ij}$ may be thought of as a sum of integrals over the correlations $\overline{u_i^{(r)}(\boldsymbol{k}', \omega')} \times \overline{u_j^{(r)}(\boldsymbol{k}, \omega)}$. These correlations refer to the homogeneous field, and they can therefore be expressed through relations such as (14.19). Consequently, the integration over $\boldsymbol{k}'$ and ω' can be carried out, and we find

$$\hat{q}_{ij} = -\nu_{ijkl} \overline{u}_{k,l}, \tag{26}$$

with

$$\nu_{ijkl} = \frac{-\hat{Q}_{qj}^{(r)}}{-i\omega + \nu k^2} \left[\frac{2\nu k_l k_k D_{it} K_{pt} D_{pq}}{(-i\omega + \nu k^2)} - \frac{2\varepsilon_{tmn} D_{it} D_{lq} \mathring{k}_n \mathring{k}_k \Omega_m}{(-i\omega + \nu k^2)} - \frac{2\Omega_m D_{it} K_{lt} D_{pq} \varepsilon_{nmp} \mathring{k}_n \mathring{k}_k}{(-i\omega + \nu k^2)} - D_{it} K_{kt} \delta_{ql} + \mathring{k}_i \mathring{k}_k D_{lq} \right]. \tag{27}$$

These manipulations lead to the rather complex result

$$\begin{aligned} \nu_{ijkl} = {} & N^{-1}\{(-i\omega + \nu k^2)\,(\delta_{jk} - 2\mathring{k}_j \mathring{k}_k)\, \hat{Q}_{il}^{(r)} \\ & + 2\mathring{k}_n(\boldsymbol{k} \cdot \boldsymbol{\Omega})\,(\varepsilon_{lnq} \mathring{k}_j \mathring{k}_k \hat{Q}_{iq}^{(r)} + \varepsilon_{jnk} \hat{Q}_{il}^{(r)})\} \\ & + 2N^{-2}\{([(2\mathring{\boldsymbol{k}} \cdot \boldsymbol{\Omega})^2\,(-i\omega + 2\nu k^2) - \nu k^2(-i\omega + \nu k^2)^2]\, \mathring{k}_l \mathring{k}_k \\ & - 4\mathring{\boldsymbol{k}} \cdot \boldsymbol{\Omega}(-i\omega + \nu k^2)\, \mathring{k}_k \Omega_l)\, \hat{Q}_{ij}^{(r)} + \\ & [(-i\omega + \nu k^2)^2 - (2\mathring{\boldsymbol{k}} \cdot \boldsymbol{\Omega})^2]\,(\varepsilon_{jmn} \hat{Q}_{il}^{(r)} - \varepsilon_{qmn} \hat{Q}_{iq}^{(r)} K_{jl})\, \mathring{k}_n \mathring{k}_k \Omega_m \\ & - 4\nu k^2(-i\omega + \nu k^2)\,(\mathring{\boldsymbol{k}} \cdot \boldsymbol{\Omega})\, \varepsilon_{jqn} \mathring{k}_n \mathring{k}_k \mathring{k}_l \hat{Q}_{iq}^{(r)}\}, \end{aligned} \tag{28}$$

where again

$$N = (-i\omega + \nu k^2)^2 + (2\mathring{\boldsymbol{k}} \cdot \boldsymbol{\Omega})^2.$$

The tensor $\hat{Q}_{ij}^{(r)}$ which appears here is the well-known combination (14.21). When this is substituted into (28), we obtain the final result by using

$$\mathcal{N}_{ijkl} = \int\int (\nu_{ijkl} + \nu_{jikl}^*)\, \mathrm{d}\boldsymbol{k}\, \mathrm{d}\omega. \tag{29}$$

16.3. Eddy Viscosities of Simple Turbulences

In this Section we shall apply (28) to a variety of simple forms of turbulence:

(*a*) *Three-dimensional Isotropic Turbulence*

This turbulence does not have any characteristic vectors of its own that can be used when forming the viscosity tensor, and $\mathcal{N}_{ijkl}$ must therefore have the form

$$\begin{aligned}\mathcal{N}_{ijkl} = {} & a(\delta_{ik}\delta_{jl} + \delta_{jk}\delta_{il}) + a_1\delta_{ij}\delta_{kl} \\ & + b(\Omega_i\Omega_l\delta_{jk} + \Omega_j\Omega_l\delta_{ik}) + b_1(\Omega_i\Omega_k\delta_{jl} + \Omega_j\Omega_k\delta_{il}) \\ & + b_2\Omega_i\Omega_j\delta_{kl} + b_3\Omega_k\Omega_l\delta_{ij} + c\Omega_i\Omega_j\Omega_k\Omega_l . \end{aligned} \tag{30}$$

On using this expression in definitions (14), one obtains

$$\begin{aligned} \nu_{vv} &= a + b\Omega^2\cos^2\theta , \qquad \nu_{hh} = a + b\Omega^2 \sin^2\theta , \\ \nu_{vh} &= \nu_{hv} = -b\Omega^2 \sin^2\theta . \end{aligned} \tag{31}$$

As is required, the off-diagonal values, ν_{vh} and ν_{hv}, vanish as $\Omega \to 0$, and the remaining two diagonal values become equal. These conditions hold under conditions of complete isotropy, a state normally realized in laminar flows. A sufficiently high rotation rate introduces anisotropy. In order to determine the viscosity tensor $\mathcal{N}_{ijkl}$ we have to substitute the complete spectral tensor (14.21) into (28). The only terms required are those which are even in the ε-tensors, i.e. the real part of (14.21) is used in the first four lines of (28) and the imaginary part is used in the remaining lines. In addition to the approaches used so far there is another which ought to be introduced. If the result is expanded in powers of Ω^2 and the series truncated after the first non-trivial term, we find that, as in our treatment of the Λ-effect, only expressions for non-rapid rotation, i.e. for "slow" as well as "modest" rotation, occur. Nevertheless, the resulting tensor is so complex that only some of its constituents can be given, namely

$$\begin{aligned} a &= {}^{(3)}\nu_T - \frac{8\Omega^2}{105}\iint \frac{\nu k^2(14\nu^4k^8 - 30\,\omega^2\nu^2k^4 + 4\omega^4)}{(\omega^2 + \nu^2k^4)^4}\, {}^{(3)}\hat{Q}_{ll}\, d\boldsymbol{k}\, d\omega , \\ b &= \frac{96}{105}\iint \frac{\nu k^2\omega^2(5\nu^2k^4 - 3\omega^2)}{(\omega^2 + \nu^2k^4)^4}\, {}^{(3)}\hat{Q}_{ll}\, d\boldsymbol{k}\, d\omega . \end{aligned} \tag{32}$$

As is required for very slow rotation the expression (32) for 'a' reduces to the familiar value ${}^{(3)}\nu_T$.

The rotationally induced anisotropy, which is reflected by (31), produces a striking feature. At the equator we have

$$\nu_{vv} = a , \qquad \nu_{hh} = a + b\Omega^2 ,$$

and vice versa at the poles, i.e.

$$\nu_{vv} = a + b\Omega^2, \qquad \nu_{hh} = a.$$

Hence we prefer to express the basic quantity $a + b\Omega^2$ in the explicit form ${}^{(3)}\nu_T - \check{\nu}\Omega^2$, where

$$\check{\nu} = \frac{8}{105} \iint \frac{\nu k^2(14\nu^4 k^8 - 90\nu^2 k^4 \omega^2 + 40\omega^4)}{(\omega^2 + \nu^2 k^4)^4} \, {}^{(3)}\hat{Q}_{ll} \, dk \, d\omega . \qquad (33)$$

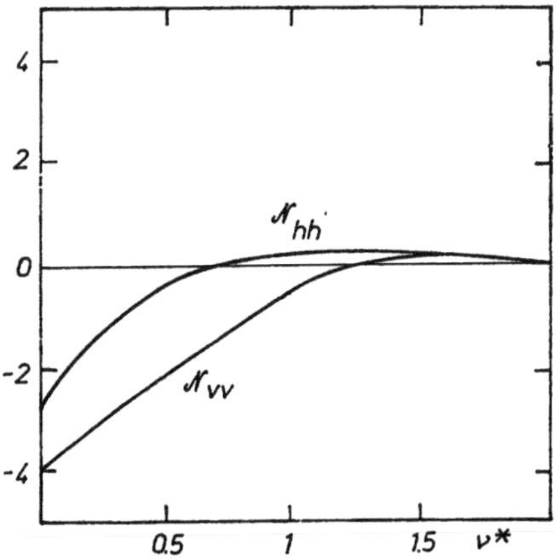

Fig. 16.3. The dependence of the integral kernels $\mathcal{N}_{vv}$ and $\mathcal{N}_{hh}$ on the normalized viscosity coefficient in the representation (34). Both kernels become negative for small ν^*; the eddy viscosity is therefore reduced by the basic rotation. For the original turbulence (which is present also when there is no rotation), we have used three-dimensional isotropic field.

The kernels of the viscosity integrals are shown in Fig. 16.3 for the equatorial region, this being the region accessible to observation. The kernels of the integrals represent the viscosities themselves only if the one-mode approximation for turbulence spectra is adopted. We write

$$\nu_{vv} = {}^{(3)}\nu_T + \nu \iint \mathcal{N}_{vv}(\nu^*) \, k^2 \omega^{-2} \, {}^{(3)}\hat{Q}_{ll} \, dk \, d\omega \, \Omega^2 , \qquad (34)$$

where the viscosity parameter is

$$\nu^* = \nu k^2 / \omega .$$

The expression for ν_{hh} is very similar to (34). Values for ν^* less than unity correspond to random waves in low viscosity media; values greater than unity are appropriate when the viscosity is high. For the solar case, the correct value appears to be slightly *less* than unity. For small values of ν^*, Fig. 16.3 shows that the kernels are negative for both ν_{vv} and ν_{hh}. The overall viscosity is therefore reduced because of the basic rotation. This effect also seems to exist for ν_{hh}. If this is true, and proves to be a general property of these viscosity coefficients, we must check whether the observed negative $Q_{\theta\phi}$-correlation in the northern hemisphere can be explained even without recourse to the Λ_H-effect. In our standard model we have always assumed that the effect of rotation on the diffusive parts of the TMT is negligible so that invariably the eddy viscosity is positive. Fig. 16.3

reminds us that even this apparently simple idea should not be accepted too uncritically.

When large values of ν^* are relevant, rotation leaves the viscosity terms practically unaffected. This statement will also hold true for the other forms of turbulence which we shall discuss in the remainder of this Chapter.

(b) *Two-dimensional Isotropic Turbulence*

This "axis" of the two-dimensional turbulence pattern (to which all motion is perpendicular) is taken to be the vertical direction parallel to gravity. As can be seen in (13.23), the order to which the anisotropy vector $\mathring{g}$ occurs in our analysis is always even. Despite the simplification due to the delta function characteristic for two-dimensional fields, we must again restrict ourselves to expressions such as

$$\begin{aligned}
\nu_{vv} &= 2 \iint \nu k_\perp^4(\nu^2 k_\perp^4 - 2\omega^2)\,(\omega^2 + \nu^2 k_\perp^4)^{-3}\, q_\perp\, \mathrm{d}\boldsymbol{k}_\perp\, \mathrm{d}\omega\;\; \Omega^2 \sin^2\theta\,,\\
\nu_{vh} &= 0\,,\\
\nu_{hv} &= 3 \iint \nu k_\perp^4(3\omega^2 - \nu^2 k_\perp^4)\,(\omega^2 + \nu^2 k_\perp^4)^{-3}\, q_\perp\, \mathrm{d}\boldsymbol{k}_\perp\, \mathrm{d}\omega\; \Omega^2 \sin^2\theta\\
&= -1.5 \iint \nu k_\perp^5(\omega^2 + \nu^2 k_\perp^4)^{-2}\,(\partial q_\perp/\partial k_\perp)\, \mathrm{d}\boldsymbol{k}_\perp\, \mathrm{d}\omega\; \Omega^2 \sin^2\theta\,,\\
\nu_{hh} &= {}^{(2)}\nu_T - \check{\nu}\Omega^2 \sin^2\theta\,,
\end{aligned} \tag{35}$$

where

$$\begin{aligned}
\check{\nu} &= \iint \frac{\nu k_\perp^4(9\omega^4 - 32\omega^2\nu^2 k_\perp^4 + 7\nu^4 k_\perp^8)}{(\omega^2 + \nu^2 k_\perp^4)^4}\, q_\perp\, \mathrm{d}\boldsymbol{k}_\perp\, \mathrm{d}\omega\\
&= \iint \left[-\frac{7\omega\nu^3 k_\perp^8}{(\omega^2 + \nu^2 k_\perp^4)^3}\frac{\partial q_\perp}{\partial\omega} + \frac{3\nu k_\perp^4 \omega^2(3\omega^2 + \nu^2 k_\perp^4)}{(\omega^2 + \nu^2 k_\perp^4)^4} q_\perp \right] \mathrm{d}\boldsymbol{k}_\perp\, \mathrm{d}\omega\,.
\end{aligned} \tag{36}$$

Note that in the one-mode approximation ${}^{(2)}\nu_T$ is negative for small viscosities ($\nu^* \ll 1$), ν_{hh} is negative, and even ν_{vv} is negative too. On the other hand, ν_{vv} becomes positive for the different type of spectrum for which $q \approx \delta(\omega)$. The values of $\check{\nu}$ always remain positive however. This statement can easily be confirmed for turbulence-type spectra. If the spectral function decreases with increasing frequency ($\omega\, \partial q/\partial\omega < 0$), the coefficients ν_{hv} become negative-definite and $\check{\nu}$, ν_T and ν_{vv} become positive-definite. It seems to be a striking fact that these strong statements can be made for wide classes of turbulence spectra. Table XVI.i. summarizes the results we obtained for the rotational influence of two-dimensional turbulence on some of the eddy viscosities.

(c) *Horizontal Turbulence (very slow rotation)*

The above results for two- and three-dimensional turbulences are the only ones that we can derive from (28) without a totally unacceptable amount of effort. Any other models would have been too complicated for complete

Table XVI.i. Signs of the viscosity terms for various spectra.

viscosity-coefficient	"turbulence"	$\tau_c = \infty$	$\nu^* > 1$	$q =$ constant	$\nu^* < 1$
ν_{vv}	+	+	+	+	−
ν_{vh}	0	0	0	0	0
ν_{hv}	−	−	−	0	+
ν_{hh}	−	−	−	−	−

analysis. For horizontal turbulence, the computations were carried out only in the limit of very slow rotation. Thus the reader should not be surprised that in what follows no Ω-terms occur: we present only the zeroth-order term for a turbulence that is intermediate between the two- and three-dimensional isotropic models. It resembles the former in that the directions of motion are everywhere perpendicular to the preferred direction $\mathring{\boldsymbol{g}}$; it resembles the three-dimensional models by having finite correlation lengths in all directions, even including $\mathring{\boldsymbol{g}}$. Is it then correct to expect an eddy viscosity whose value lies between those for the two- and three-dimensional fields?

For $\Omega \to 0$ the viscosity tensor $\mathcal{N}_{ijkl}$ is reduced to (see Krause and Rüdiger, 1974a)

$$\mathcal{N}_{ijkl} = \iint \left[\left(\delta_{jk} - \frac{2k_j k_k}{k^2} \right) \hat{Q}_{il}^{(0)} + \left(\delta_{ik} - \frac{2k_i k_k}{k^2} \right) \hat{Q}_{jl}^{(0)} - \frac{4\nu k_l k_k \hat{Q}_{ij}^{(0)}}{-i\omega + \nu k^2} \right] \times \frac{\mathrm{d}\boldsymbol{k}\, \mathrm{d}\omega}{-i\omega + \nu k^2}, \tag{37}$$

into which the spectral tensor (14.37) must be substituted. We find immediately that

$$\nu_{vv} = 0,$$

$$\nu_{vh} = \frac{2}{\Omega^3 \sin\theta \cos\theta} \iint (\mathring{\boldsymbol{k}} \cdot \boldsymbol{\Omega}_\perp)(\mathring{\boldsymbol{g}} \cdot \mathring{\boldsymbol{k}}) \times [(\boldsymbol{k}_\perp \cdot \boldsymbol{\Omega}_\perp)^2 - k_\perp{}^2 \Omega_\perp{}^2] \frac{{}^{(2)}q}{-i\omega + \nu k^2} \,\mathrm{d}\boldsymbol{k}\, \mathrm{d}\omega. \tag{38}$$

The second of these expressions vanishes only if spectral functions even in $\mathring{\boldsymbol{g}} \cdot \boldsymbol{k}$ are considered. Then ν_{hv} also vanishes but for ν_{hh} we obtain

$$\nu_{hh} = \frac{\nu}{2} \iint \left[\frac{k_\perp{}^2(\nu^2 k^4 - \omega^2)}{\omega^2 + \nu^2 k^4} + (\mathring{\boldsymbol{g}} \cdot \boldsymbol{k})^2 \right] \frac{{}^{(2)}q k_\perp{}^2}{\omega^2 + \nu^2 k^4} \,\mathrm{d}\boldsymbol{k}\, \mathrm{d}\omega, \tag{39}$$

which does indeed lie somewhere between (7) and (8). When $q \approx \delta(\mathring{\boldsymbol{g}} \cdot \boldsymbol{k})$, the two-dimensional expression (8) results. All non-delta-function-like contributions do, however, add positive values. We should stress, therefore, that the more realistic spectra tend to impart positivity to the eddy viscosity.

For example, the simple spectrum (13.65), which lacks a delta function, gives

$$\nu_{hh} = \frac{\nu}{(2\varkappa + 1)(2\varkappa + 3)(2\varkappa + 5)} \iint \frac{(2\varkappa + 5)\,\nu^2 k^4 + (2\varkappa - 3)\,\omega^2}{(\omega^2 + \nu^2 k^4)^2} \times k^4 q(k, \omega)\, d\mathbf{k}\, d\omega, \tag{40}$$

which is less negative than the two-dimensional expression (8). We conclude, therefore, that the possible negativity of the eddy viscosity ν_{hh} arises from extreme types of correlations characteristic of two-dimensional fields. The direction of motion, i.e. the fact that $u_r = 0$ for two-dimensional fields, seems to play only a minor part.

Chapter 17. The Theory of Eddy Conductivity

In our study of the mean-field dynamics of non-uniformly rotating turbulent media, we have seen that, in addition to the Λ-effect, the eddy conductivity tensor also has a role to play. The importance of this role increases with increasing Prandtl number, and only becomes small if the Prandtl number is reduced. The Prandtl number is the ratio of the scalar eddy viscosity and the scalar eddy conductivity, quantities that are readily defined for the small-scale (isotropic) components of the turbulence. Because of their short correlation times, small-scale turbulence may reasonably be assumed to be unaffected by rotational forces, which would necessarily produce anisotropies.

In Chapter 8, we presented heuristic arguments for the structure of the rotationally modified eddy conductivity tensor, χ_{ij}, which we defined implicitly by

$$\overline{u_i'T'} = -\chi_{ij}(\nabla_j T - C_p^{-1}g_j). \tag{1}$$

This tensor connects temperature-velocity correlations to the mean-fields and is, to this extent, limited by the two-scale assumption. Definition (1) reduces to the diffusion ansatz $-\chi_T \nabla \overline{T}$ if no density stratification exists, as is the case in laboratory physics. Furthermore, there is no obvious reason why only first derivatives should occur, so it may be better to postulate

$$\overline{u_i'T'} = -\chi_{ij}\overline{T}_{,j} - \chi_{ijk}\overline{T}_{,jk} - \cdots. \tag{2}$$

This form is like the one we usually adopt for eddy diffusivities. It is also closer to the non-local connection between $\overline{u_i'T'}$ and $\overline{T}$ that we believe more faithfully represents physical reality. If the turbulence is highly isotropic, the leading terms of the eddy diffusivity tensors are simply proportional to the Kronecker delta so that (cf. Krause, 1972; Parker, 1974)

$$\overline{\boldsymbol{u}'T'} = -\chi_T^{(1)}\nabla\overline{T} - \chi_T^{(3)}\nabla\Delta\overline{T} - \cdots. \tag{3}$$

The crux lies in the determination of the velocity-temperature correlations $\overline{u'T'}$ from the energy equation. Again the simplest closure technique is SOCA, and in the present Chapter we use this to calculate the tensor χ_{ij} and to solve related problems.

17.1. The Eddy Conductivity Tensor

We shall adopt as our temperature equation the following basic form (see Eq. (8.38))

$$C_v\varrho(\partial T/\partial t + \boldsymbol{u}\cdot\nabla T) - \nabla\cdot(\varrho C_v\chi\nabla T) + p\nabla\cdot\boldsymbol{u} = 0. \tag{4}$$

Suppose that the mean velocity of the medium is zero, and that the average temperature is independent of time. Then, by (4), the temperature fluctuation, T', obeys the (linearized) equation

$$C_v\bar{\varrho}\,\partial T'/\partial t = C_v\chi\nabla\cdot(\bar{\varrho}\nabla T') = -\bar{p}\nabla\cdot\boldsymbol{u}' - C_v\bar{\varrho}\boldsymbol{u}'\cdot\nabla\overline{T}, \tag{5}$$

where we have assumed that the molecular conductivity, χ, and the specific heat, C_v, are constant. Assuming the ideal gas law and incorporating conservation of mass,

$$p = (C_p - C_v)\,\varrho T, \qquad \nabla\cdot(\bar{\varrho}\boldsymbol{u}') = 0 \tag{6}$$

(acoustic waves are filtered out), we obtain

$$\partial T'/\partial t - \bar{\varrho}^{-1}\chi\nabla\cdot(\bar{\varrho}\nabla T') = \gamma[(\bar{\varrho}C_p)^{-1}\nabla\bar{p} - \nabla\overline{T}]\,\boldsymbol{u}'. \tag{7}$$

The presence of the factor γ in relation (7) is unusual, and yet it is a direct consequence of the assumption $\partial\varrho'/\partial t = 0$. Writing

$$\begin{aligned}-\bar{p}\nabla\cdot\boldsymbol{u}' &= \boldsymbol{u}'\cdot\nabla\bar{p} - \nabla\cdot(\bar{p}\boldsymbol{u}')\\ &= \boldsymbol{u}'\cdot\nabla\bar{p} + (C_v - C_p)\,\varrho\boldsymbol{u}'\cdot\nabla\overline{T} - (C_v - C_p)\,\overline{T}\,\partial\varrho'/\partial t,\end{aligned} \tag{8}$$

and ignoring the last term, we obtain (7) from (5). If, however, there is a reason why the relation $\varrho'/\varrho = -T'/T$ is valid then, as Spiegel and Veronis (1960) showed [see also Veronis (1962)], (7) becomes

$$\partial T'/\partial t - (\gamma\bar{\varrho})^{-1}\,\chi\nabla\cdot(\bar{\varrho}\nabla T') = [(\bar{\varrho}C_p)^{-1}\,\nabla\bar{p} - \nabla\overline{T}]\,\boldsymbol{u}'. \tag{9}$$

The factor γ has now disappeared from the right-hand side of the equation. Indeed, it disappears completely when the media are incompressible.

Another formulation of (7) is

$$\partial T'/\partial t - \chi(\Delta T' - \boldsymbol{G}\cdot\nabla T') = \gamma\boldsymbol{\beta}\cdot\boldsymbol{u}', \tag{10}$$

where the abbreviations

$$\boldsymbol{G} = -\nabla(\ln\varrho), \qquad \boldsymbol{\beta} = (\varrho C_p)^{-1}\,\nabla p - \nabla T, \tag{11}$$

have been introduced. Let us assume that these vectors are constant. Then the Fourier representations (14.8) lead to

$$[-i\omega + \chi k^2 + i\chi(\boldsymbol{G}\cdot\boldsymbol{k})]\,\hat{T} = \gamma\boldsymbol{\beta}\cdot\hat{\boldsymbol{u}}, \tag{12}$$

and we simply have

$$\hat{T} = \gamma\boldsymbol{\beta}\cdot\hat{\boldsymbol{u}}/\{-i[\omega - \chi(\boldsymbol{G}\cdot\boldsymbol{k})] + \chi k^2\}. \tag{13}$$

It may be observed that the definition of $\boldsymbol{G}$ differs slightly from that given in previous chapters. The required temperature-velocity correlation proves to be

$$\overline{u_i'T'} = \chi_{ij}\beta_j, \tag{14}$$

where the eddy conductivity tensor is given by

$$\chi_{ij} = \gamma\int\int\hat{Q}_{ij}\{-i[\omega - \chi(\boldsymbol{G}\cdot\boldsymbol{k})] + \chi k^2\}^{-1}\,\mathrm{d}\boldsymbol{k}\,\mathrm{d}\omega. \tag{15}$$

By merely re-expressing (15) in the form

$$\chi_{ij} = \int\int\frac{\gamma\chi k^2\hat{Q}_{ij}}{(\omega - \chi\boldsymbol{G}\cdot\boldsymbol{k})^2 + \chi^2k^4}\,\mathrm{d}\boldsymbol{k}\,\mathrm{d}\omega + i\gamma\int\int\frac{(\omega - \chi\boldsymbol{G}\cdot\boldsymbol{k})\,\hat{Q}_{ij}}{(\omega - \chi\boldsymbol{G}\cdot\boldsymbol{k})^2 + \chi^2k^4}\,\mathrm{d}\boldsymbol{k}\,\mathrm{d}\omega, \tag{16}$$

the symmetric and antisymmetric parts of the χ-tensor are easily extracted. As the symmetric part of Q_{ij} is always real and the antisymmetric part imaginary, the formulation (16) as shown actually provides both the symmetric and the antisymmetric part of the conductivity tensor. Surprisingly enough, it is the antisymmetric part that is associated with the helicity of the turbulence. The simplest helical turbulence is

$$\hat{Q}_{ij} = qK_{ij} + ip\varepsilon_{ijk}k_k \tag{17}$$

(see (13.21) for K_{ij}) and for this (14) gives

$$\chi_{ij} = \int\int\frac{\chi\gamma k^2qK_{ij}}{(\omega - \chi\boldsymbol{G}\cdot\boldsymbol{k})^2 + \chi^2k^4}\,\mathrm{d}\boldsymbol{k}\,\mathrm{d}\omega + \varepsilon_{ijl}\int\int\frac{\chi\gamma\boldsymbol{G}\cdot\boldsymbol{k}k_lp}{(\omega - \chi\boldsymbol{G}\cdot\boldsymbol{k})^2 + \chi^2k^4}\,\mathrm{d}\boldsymbol{k}\,\mathrm{d}\omega. \tag{18}$$

This shows clearly that symmetry with respect to the indices i and j is destroyed by helicity. Obviously, radially stratified media with helicity possess asymmetric conductivity tensors for which, for example, $\chi_{\theta\phi} \neq \chi_{\phi\theta}$. As yet, the implications of this antisymmetry are unknown, and we shall not study them here. It is conceivable, however, that such asymmetries might provide an independent proof of the existence of helical turbulence beneath the solar surface, whose presence is inferred from the dynamo theory.

We shall focus all our attention on the symmetric part of the conductivity tensor. It degenerates into a simple scalar when every conceivable anisotropy is absent.

17.2. The Scalar Eddy Conductivity

(a) *Two- and three-dimensional turbulence*

The simplest form of the eddy conductivity tensor (15), arises when we assume that $\boldsymbol{G} = 0$; it is

$$\chi_{ij} = \iint \hat{Q}_{ij}(-i\omega + \chi k^2)^{-1}\, \mathrm{d}\boldsymbol{k}\, \mathrm{d}\omega. \tag{19}$$

Here, as in the original paper (Eschrich and Rüdiger, 1983), we have ignored the factor γ in (15), the reason being that only spectral tensors for incompressible media are available in applications, and in this case γ disappears anyway.

To apply (19) we must prescribe the spectral tensor of the turbulence. Two quite different turbulent regimes will be studied here, namely three-dimensional isotropic turbulence where (13.20) applies, and two-dimensional turbulence where (13.25) applies, the unit vector $\hat{\boldsymbol{g}}$ being perpendicular to the plane in which the turbulent motions takes place. The two-dimensional case is extremely anisotropic and also possesses an infinite correlation length in the $\hat{\boldsymbol{g}}$-direction. We will need to distinguish precisely between these two characteristics.

For the three-dimensional turbulence, (19) becomes, after some simple algebraic reductions,

$${}^{(3)}\chi_{ij} = \frac{1}{3} \iint \chi k^{2\,(3)}\hat{Q}_{ll}(\omega^2 + \chi^2 k^4)^{-1}\, \mathrm{d}\boldsymbol{k}\, \mathrm{d}\omega\, \delta_{ij}, \tag{20}$$

whereas for the two-dimensional turbulence (see (13.26))

$${}^{(2)}\chi_{ij} = \frac{1}{2} \iint \chi k_\perp^{2\,(2D)}\hat{Q}_{ll}(\omega^2 + \chi^2 k_\perp^4)^{-1}\, \mathrm{d}\boldsymbol{k}_\perp\, \mathrm{d}\omega\, G_{ij}. \tag{21}$$

Surprisingly, there is no essential difference between these results, a fact that is by no means trivial when we recall the corresponding behaviors of the eddy viscosities deduced in Chapter 16. The analogy between the results for two- and three-dimensional fields may be taken further. For vanishing conductivity ($\chi \to 0$), we find

$${}^{(2)}\chi_{22} = \frac{1}{2}\pi \int {}^{(2D)}\hat{Q}_{ll}(\boldsymbol{k}_\perp, 0)\, \mathrm{d}\boldsymbol{k}_\perp, \qquad {}^{(3)}\chi_{22} = \frac{1}{3}\pi \int {}^{(3)}\hat{Q}_{ll}(\boldsymbol{k}, 0)\, \mathrm{d}\boldsymbol{k}, \tag{22}$$

and, consequently,

$$^{(2)}\chi_{22} = {}^{(3)}\chi_{22} \simeq \tau_c \overline{u'^2}, \tag{23}$$

where τ_c is the eddy decay time. We indeed find that the structure assumed for the turbulence model plays an extremely minor role. The required diffusivity always has the anticipated simple form (23), which has frequently and successfully been used in making estimates. The insensitivity of the eddy conductivity which we have demonstrated suggests that the value of the turbulent Prandtl number is more germane to the eddy viscosity than to the eddy conductivity (Launder, 1976).

(b) Dependence of χ_T on χ

Let us now analyse in more detail the eddy conductivity for three-dimensional isotropic turbulence

$$\chi_T = \frac{2}{3} \iint \chi k^{4\,(3)}q(\omega^2 + \chi^2 k^4)^{-1}\, \mathrm{d}\boldsymbol{k}\, \mathrm{d}\omega. \tag{24}$$

This is always positive because of the positivity of the spectral function $^{(3)}q(k, \omega)$ (Krause, 1972; Eschrich, 1978). Although we shall not specify that spectrum, we shall make some general statements about the dependence of χ_T on χ. As was mentioned earlier, the limit of $\chi \to 0$ gives

$$\chi_T^{(0)} = \frac{2}{3} \pi \int k^{2\,(3)}q(k, 0)\, \mathrm{d}\boldsymbol{k} = \frac{1}{6} \int_{-\infty}^{\infty} \overline{\boldsymbol{u}'(\boldsymbol{x}, t) \cdot \boldsymbol{u}'(\boldsymbol{x}, t + \tau)}\, \mathrm{d}\tau, \tag{25}$$

so that the value of χ_T at $\chi = 0$ depends essentially on the value of $^{(3)}q$ at $\omega = 0$. Spectra that are nonzero at $\omega = 0$ produce a non-vanishing eddy conductivity at $\chi = 0$. Spectra of "turbulence" type decrease with increasing frequency: $\omega\, \partial q/\partial\omega \leq 0$. In this case the function $\chi_T(\chi)$ also proves to be monotonically decreasing since (24) yields

$$\partial\chi_T/\partial\chi = \frac{2}{3} \iint \omega k^4 (\omega^2 + \chi^2 k^4)^{-1}\, \partial^{(3)}q/\partial\omega\, \mathrm{d}\boldsymbol{k}\, \mathrm{d}\omega. \tag{26}$$

Thus the general shapes of the two functions $^{(3)}q(k, \omega)$ (as a function of ω) and $\chi_T(\chi)$ are rather similar.

This statement is reinforced by their behavior at infinity, where they both vanish:

$$\lim_{\omega\to\infty} {}^{(3)}q = 0, \qquad \lim_{\chi\to\infty} \chi_T = 0. \tag{27}$$

The second limit follows directly from (24) after the substitution $\chi^{-1} = \varkappa$ has been made, which leads to

$$\chi_T = \xrightarrow{\varkappa\to 0} \frac{2}{3\chi} \iint {}^{(3)}q(k, \omega)\, \mathrm{d}\boldsymbol{k}\, \mathrm{d}\omega = O(\chi^{-1}). \tag{28}$$

We generally expect turbulence-type spectra to be the rule rather than the exception. The frequently used "random wave" representation, however, implies vanishing $q(k, 0)$; see Chapter 13. In fact, it seems to be rather important to know whether the spectrum contains energy at "zero frequency" or not. Only spectra having such "zero frequency" components can provide eddy quantities (such as conductivity, viscosity, α-effect, ...), for non-conducting or inviscid media. "Random wave" type turbulences do not produce such eddy quantities.

A further result can be established concerning the first derivative of the profile $\chi_T(\chi)$, which is

$$\partial \chi_T / \partial \chi = \frac{2}{3} \iint (\omega^2 - \chi^2 k^4)\,(\omega^2 + \chi^2 k^4)^{-2}\; {}^{(3)}q k^4 \,\mathrm{d}\boldsymbol{k}\,\mathrm{d}\omega\,. \tag{29}$$

For a spectrum without "zero frequency" components, e.g.

$${}^{(3)}q(k, \omega) = O(\omega^2)\,, \tag{30}$$

expression (29) proves to be positive at $\chi = 0$. "Random wave" type spectra with the properties

$$\lim_{\omega \to 0} {}^{(3)}q(k, \omega) = 0\,, \qquad \lim_{\omega \to 0} \partial^{(3)}q / \partial \omega^2 > 0\,, \tag{31}$$

thus create a χ_T-profile with almost identical features, namely

$$\lim_{\chi \to 0} \chi_T(\chi) = 0\,, \qquad \lim_{\chi \to 0} \partial \chi_T / \partial \chi > 0\,. \tag{32}$$

Bearing in mind these findings, we may conclude that both the profiles $\chi_T(\chi)$ and $q(k, \omega)$, the latter considered as a function of ω, have nearly the same shape.

(c) *Examples*

One of the simplest spectra of "turbulence" type is (13.46), a particular example of which is

$${}^{(3)}q = q_k W^2 / \pi(\omega^2 + W^2)\,, \qquad W = \text{constant} > 0\,, \tag{33}$$

which corresponds to an exponential decay with increasing time. Substituting this into (24) we obtain

$$\chi_T = \frac{2}{3}\, W \int q_k k^2 (W + \chi k^2)^{-1}\,\mathrm{d}\boldsymbol{k}\,, \tag{34}$$

which decreases monotonically with increasing χ. Its value at $\chi = 0$ is the largest attainable. On the other hand, the simplest form of "random wave" type spectrum,

$${}^{(3)}q = \frac{1}{2}\, W q_k(k)\, k^{-2} [\delta(\omega - W) + \delta(\omega + W)]\,, \tag{35}$$

contains Dirac delta functions and gives

$$\chi_T = \frac{2}{3} W \int \chi k^2 q_k(k) \, (W^2 + \chi^2 k^4)^{-1} \, \mathrm{d}\boldsymbol{k}. \tag{36}$$

Clearly, this vanishes at $\chi = 0$, and its first derivative is

$$\partial \chi_T / \partial \chi = \frac{2}{3} W \int (W^2 - \chi^2 k^4) \, (W^2 + \chi^2 k^4)^{-2} \, k^2 q_k \, \mathrm{d}\boldsymbol{k}. \tag{37}$$

Expression (37) is positive for small χ and negative for large χ. A maximum of χ_T therefore exists at an intermediate value of χ.

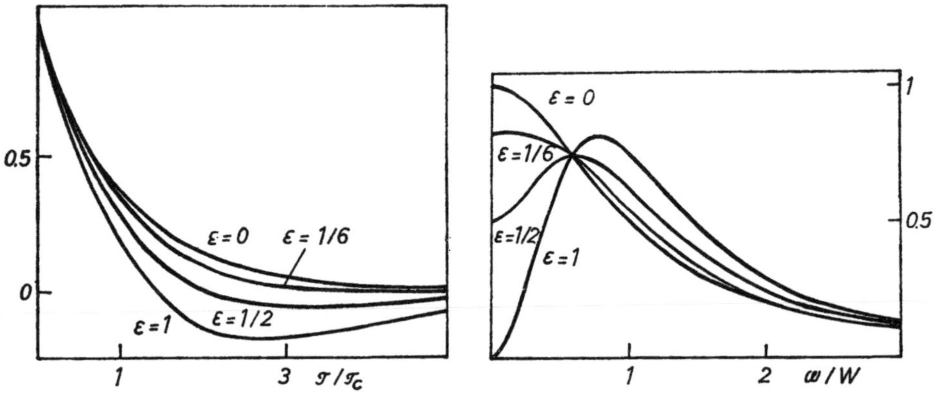

Fig. 17.1. Left-hand figures: Temporal correlations for various values of ε according to (38). Whereas $\varepsilon = 0$ describes positive decreasing correlations, $\varepsilon > 0$ also allows negative correlations. Right-hand figure: The corresponding frequency spectra. The monotonic profile ($\varepsilon = 0$) represents "turbulence" type spectra and the strongly peaked spectrum ($\varepsilon = 1$) relates to "random wave" turbulence.

Finally, we shall consider a more complex profile which even allows the temporal correlations to become negative beyond a certain time interval, e.g.

$$\overline{\boldsymbol{u}'(\boldsymbol{x}, t) \, \boldsymbol{u}'(\boldsymbol{x}, t + \tau)} \sim \left(1 - \frac{1}{2} \varepsilon \tau^2 / \tau_c^2\right) \exp\left(-|\tau| / \tau_c\right). \tag{38}$$

Its spectrum is

$$^{(3)}q = W^2 (W^2 + \omega^2)^{-1} \left[1 + \varepsilon W^2 (3\omega^2 - W^2) \, (\omega^2 + W^2)^{-2}\right] q_k / \pi. \tag{39}$$

Because this expression is necessarily positive, the parameter ε must be less than unity: $\varepsilon \leqq 1$. Fig. 17.1 illustrates both the correlation profile and the frequency spectrum for various values of ε. We observe that a peaked spectrum begins to develop for $\varepsilon \geqq 1/6$. The "turbulence" type spectra are represented by $\varepsilon \leqq 1/6$ whereas the "random wave" type belong to $\varepsilon \approx 1$.

A short calculation shows that the corresponding eddy conductivity is

$$\chi_T = \frac{2}{3} W \int [(W + \chi k^2)^{-1} - \varepsilon W^2 (W + \chi k^2)^{-3}] \, k^2 q_k \, d\boldsymbol{k}, \qquad (40)$$

which is markedly reduced as compared with (34). The value of expression (40) at $\chi = 0$ is simply

$$\chi_T(0) = \frac{2}{3} (1 - \varepsilon) \int k^2 q_k \, d\boldsymbol{k} = \frac{1}{3} (1 - \varepsilon) \, \overline{\boldsymbol{u}'^2} \tau_c, \qquad (41)$$

which again vanishes for $\varepsilon \to 1$. It is also worthwhile examining its first derivative with respect to χ:

$$\partial \chi_T / \partial \chi = \frac{2}{3} W \int [-(W + \chi k^2)^{-2} + 3\varepsilon W^2 (W + \chi k^2)^{-4}] \, k^4 q_k \, d\boldsymbol{k}. \qquad (42)$$

This is negative for large χ, and at $\chi = 0$ takes the value

$$\partial \chi_T / \partial \chi |_{\chi=0} = \frac{2}{3} \tau_c (3\varepsilon - 1) \int k^4 q_k \, d\boldsymbol{k}. \qquad (43)$$

As anticipated, sufficiently small values of $\varepsilon (\varepsilon < 1/3)$ give rise to simple, monotonically decreasing $\chi_T(\chi)$-profiles, whereas large ε yield positive derivatives for $\chi_T(\chi)$, i.e. peaked profiles. In the latter case, at a certain finite value of the conductivity χ, the eddy conductivity possesses a maximum. Since all our statements are practically independent of the actual shape of the function q_k, they should be relatively general.

(d) Application of special wave-number spectra

Our discussion so far has clarified the general nature of the eddy conductivity. It may nevertheless be also worthwhile to introduce a wave-number spectrum to facilitate further numerical discussions. For this purpose, we shall consider the monotonically decreasing spectrum

$$q_k = 2\pi^{-2} K^3 \overline{u'^2} / k^2 W (K^2 + k^2)^3, \qquad (44)$$

where K represents a characteristic wave-number whose significance is apparent from the correlations

$$\overline{\boldsymbol{u}'(\boldsymbol{x}, t) \, \boldsymbol{u}'(\boldsymbol{x} + \boldsymbol{\xi}, t)} \sim (1 + K\xi) \, e^{-K\xi}. \qquad (45)$$

On adopting the spectrum (44), the general expression (40) yields

$$\chi_T = \frac{1}{3} \overline{\boldsymbol{u}'^2} \tau_c (1 + \chi^{*1/2})^{-3} \{1 + 3\chi^{*1/2} - \varepsilon (1 + 5\chi^{*1/2} + \chi^*) (1 + \chi^{*1/2})^{-2}\}, \qquad (46)$$

where a dimensionless conductivity, normalized with respect to the spectrum characteristics,

$$\chi^* = K^2\chi/W \tag{47}$$

has been introduced. In complete accordance with (41) we find

$$\chi_T(0) = \frac{1}{3}\overline{u'^2}(1-\varepsilon)\,\tau_c. \tag{48}$$

Some characteristic examples are shown in Fig. 17.2 where the dependence of the normalized eddy conductivity $\chi_T^* = \chi_T/(\overline{u'^2}\tau_c)$ on the normalized conductivity χ^* is shown. Note the similar patterns of behavior of the

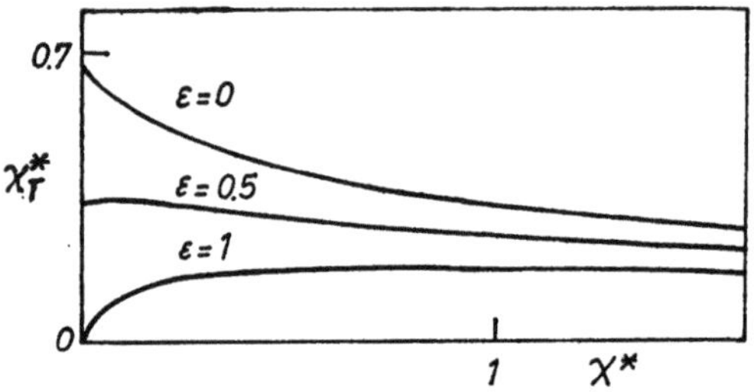

Fig. 17.2. The normalized turbulent conductivity χ^* versus the (normalized) molecular conductivity. The dependence on ε is marked only for small χ. The monotonically decreasing "turbulence" type spectrum generates monotonically decreasing χ_T-profiles. Note the contrast with the "random wave" case ($\varepsilon = 1$).

frequency spectrum and the χ_T-profile. Frequency spectra with distinct peaks (as happens for $\varepsilon > 0.33$) produce χ_T-profiles that have peaks also. On the other hand, monotonically decreasing spectra (i.e. those for $\varepsilon < 1/6$) produce monotonically decreasing χ_T-profiles. The larger the microscopic conductivity, however, the smaller are the differences between the eddy conductivities for different spectra.

17.3. The Turbulent Prandtl Number

A technique similar to the one described above can also be applied to the eddy viscosity, as we have already described in the calculations of Chapter 16. A three-dimensional turbulence implies a scalar eddy viscosity of the form (16.7) corresponding precisely to expression (20). For small microscopic viscosities we have found the expressions (16.9). A comparison of the latter with (25) immediately shows that the turbulent Prandtl number for very small diffusivities should be

$$Pr_T \equiv \nu_T/\chi_T = 2/5, \tag{49}$$

which agrees with the result of Nakano *et al.* (1979). Using the frequency profile (39), we find that (16.7) becomes

$$\nu_T = \frac{4W}{15} \int \left\{ \frac{W + 2\nu k^2}{(W + \nu k^2)^2} - \varepsilon W^2 \frac{W + 4\nu k^2}{(W + \nu k^2)^4} \right\} q_k k^2 \, dk, \tag{50}$$

from which conclusions can be drawn similar to those following from (40).

Adopting our spectral function (44), we obtain the eddy viscosity

$$\nu_T = \frac{2\overline{u'^2}}{15} \cdot \frac{\tau_c}{(1 + \nu^{*1/2})^4} \left\{ (1 + 4\nu^{*1/2} + 6\nu^*) - \frac{\varepsilon}{(1 + \nu^{*1/2})^2} \times [1 + 6\nu^{*1/2} + 15\nu^* + (5/2)\,\nu^{*3/2}] \right\} \tag{51}$$

($\nu^* = K^2\nu/W$), some characteristic profiles of which are illustrated in Fig. 17.3, in which $\nu_T^* = \nu_T/(\overline{u'^2}\tau_c)$. Again a single maximum occurs, this time for $\varepsilon > 8/35$.

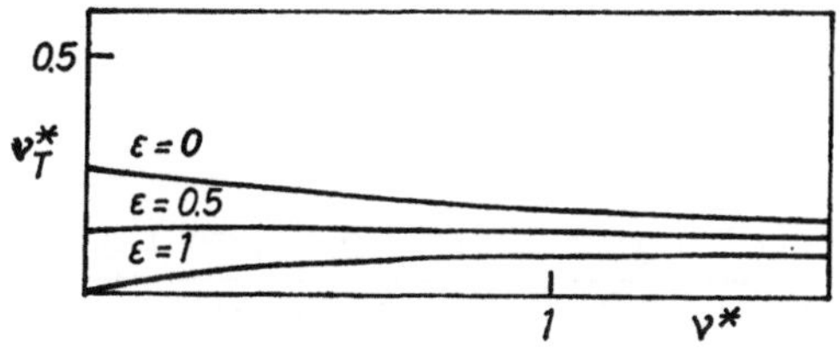

Fig. 17.3. The same as Fig. 17.2 but for the eddy viscosity.

Expressions (46) and (51) can now be used to determine the turbulent Prandtl number $Pr_T = \nu_T/\chi_T$. It has often been presumed that the best estimate for Pr_T is "1", simply because the same eddy simultaneously transports both momentum and energy. Laboratory measurements, however, lead to a slightly higher effective heat transport, that is $Pr \approx 1/2$ in flat channels and $Pr \approx 1/\sqrt{2}$ for axial flows (Szablewski, 1979). Launder (1976) reported Prandtl numbers of 0.7 for a variety of free shear flows. Near walls, tubulence seems to produce somewhat larger values (≈ 0.9), and there is even "a tendency for the turbulent Prandtl number to exceed unity as the wall is approached". Yet this near-wall peaking does not exist if the test plate used is rough.

Rotta (1972) reports values of 0.5—0.9. Earlier measurements of the distribution of heat and matter in pipe flows seemed to indicate an average turbulent Prandtl number of 0.65—0.7. The effect of the molecular Prandtl number on the turbulent Prandtl number appears to be small, as was demonstrated by some experiments in which the molecular Pr was varied from 1 to 1000. There is good experimental evidence that turbulent Prandtl numbers are less than unity. Indeed, Prandtl himself warned against setting it to unity as "the mechanism of diffusion of an element, feature or some material component is not completely identical to transport of momentum". According to Launder (1976) "the turbulent Prandtl number might be better regarded as an indicator of the state of momentum transport rather than of heat transport processes". Without proof, Szablewski (1979) mentioned the possibility that the difference between ν and χ might be due to the momentum lost through pressure fluctuations.

Theoretical works so far published also lead to turbulent Prandtl numbers of less than unity. The results depend on the technique chosen to close the equations governing the fluctuations. The so-called test-field model (Kraichnan, 1971), and the related eddy damped Markovian model (Larcheveque *et al.*, 1980), both predict a turbulent Prandtl number of 0.6, which is in good agreement with the measurements. Within the framework of the Boussinesq approximation, a spectral mode theory of eddy viscosity and turbulent conductivity has also been developed by Nakano *et al.* (1979). In their theoretical calculations, Belvedere *et al.* (1980), as well as Schmidt (1982), adopted the value 0.1. Massaguer and Zahn (1980), in their calculations of cellular convection in stratified atmospheres, favored the value 0.01.

From (46) and (51) one immediately finds

$$Pr_T = \frac{2}{5}\,\frac{(1+\chi^{*1/2})^5\,[(1-\varepsilon)\,(1+6\nu^{*1/2}+15\nu^*)+(16-2.5\varepsilon)\,\nu^{*3/2}+6\nu^{*2}]}{(1+\nu^{*1/2})^6\,[(1-\varepsilon)\,(1+5\chi^{*1/2})+(7-\varepsilon)\,\chi^*+3\chi^{*3/2}]}, \tag{52}$$

which varies from 0.4 for vanishing ν and χ to $0.8/Pr$ for very large diffusivities. In the latter case we see a distinct decrease in Pr_T for increasing Pr. Only for relatively small diffusivities does the turbulent Prandtl number prove to be independent of the microscopic Prandtl number. For both very large and very small diffusivities, the numerical value of Pr_T seems to be less than one and to be independent of ε. This, however, cannot be the whole truth for, if one of the two diffusivities is large and the other is small, one of the following two cases applies:

$$\text{i)}\ \nu^* < 1 < \chi^*, \qquad Pr_T = 2(1-\varepsilon)\,\chi^*/15, \tag{53}$$

a quantity which, of course, can be large, and

$$\text{ii)}\ \chi^* < 1 < \nu^*, \qquad Pr_T = 12/5(1-\varepsilon)\,\nu^*, \tag{54}$$

a quantity which, by contrast, is rather small. The case $\varepsilon = 1$ must, of course, be considered separately.

The general result of an analysis of (52) is that, in our theory, the turbulent Prandtl number is small for large Pr, but is large for small Pr. The latter case, for which the conditions for the validity of SOCA are comfortably satisfied, shows that large turbulent Prandtl numbers do occur in our theory. This is in stark contrast to the general belief that the turbulent transport of momentum and heat are equally potent since each is due to the same eddy. For the exceptional case when both ν^* and χ^* are small, the turbulent Prandtl number generally takes the value 0.4. From Fig. 17.4 we learn, however, that the shape of the frequency spectrum, i.e. the parameter ε, plays an indispensable role in those segments of the curves which control cases (i) and (ii) above.

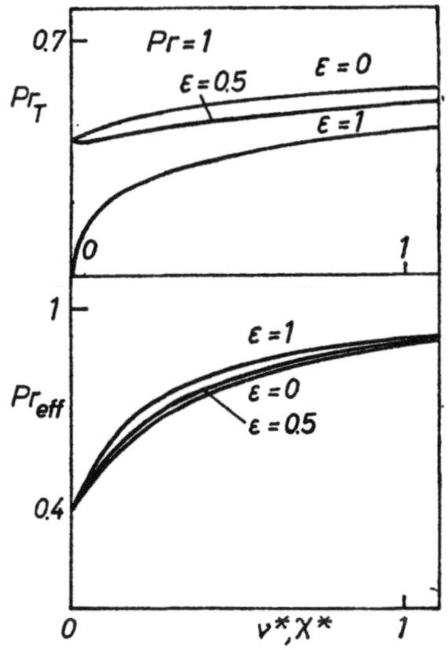

Fig. 17.4. The turbulent Prandtl number (above) and effective Prandtl number (below) for a molecular Prandtl number of unity. The turbulent Prandtl number varies between 0.4 and 0.8. The behavior of "random waves" ($\varepsilon = 1$) is exceptional because their eddy viscosity vanishes more rapidly than their turbulent conductivity. Such internal differences are greatly reduced if the "effective Prandtl number" (60) is considered.

The molecular Prandtl number of the Sun is very small. Nakano *et al.* (1979) used

$$\nu = 10^3 \text{ cm}^2/\text{s}, \qquad \chi = 10^{12} \text{ cm}^2/\text{s} \tag{55}$$

which, taken in conjunction with the scales of the granules viz.,

$$K = 3 \cdot 10^{-8} \text{ cm}^{-1}, \qquad W = 5 \cdot 10^{-3} \text{ s}^{-1}, \tag{56}$$

gave values of

$$\nu^* \simeq 2 \cdot 10^{-10}, \qquad \chi^* \simeq 0.2. \tag{57}$$

The molecular diffusivities (55) were calculated by Edmonds (1957). We see from (57) that $\nu^* \approx 0$ and $\chi^* \approx 1$ are fairly good approximations, which from (46) and (51) imply

$$\nu_T \simeq 2(1 - \varepsilon)\, \overline{u'^2}\tau_c/15 \lesssim \overline{u'^2}\tau_c/7, \tag{58}$$

and

$$\chi_T \simeq (1 - 7\varepsilon/16)\, \overline{u'^2}\tau_c/6 \lesssim \overline{u'^2}\tau_c/6. \tag{59}$$

The turbulent Prandtl number, therefore, must be less than 0.8. This value is appropriate for $\varepsilon = 0$, much smaller values would follow for $\varepsilon \approx 1$. But even for $\varepsilon = 0$, which should be the normal "turbulence" case, the value 0.8 is not the last word because, in mean-field hydrodynamics, the "effective" Prandtl number

$$Pr_{\text{eff}} = (\nu + \nu_T)/(\chi + \chi_T) \tag{60}$$

is relevant which, when $\overline{u'^2} \approx W^2/K^2$ (as is usually the case), can be rewritten as

$$Pr_{\text{eff}} = (\nu^* + \nu_T^*)/(\chi^* + \chi_T^*). \tag{61}$$

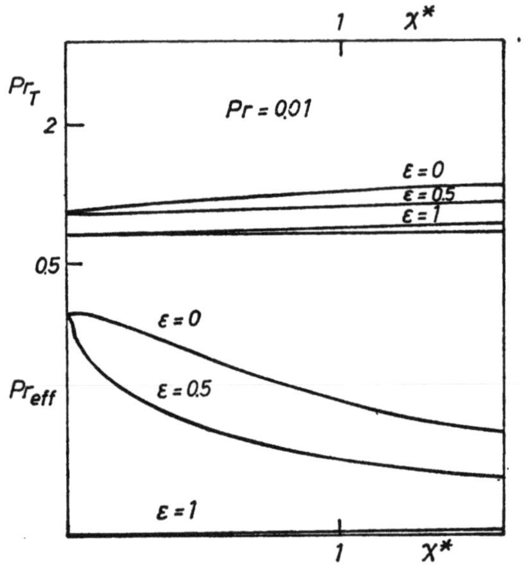

Fig. 17.5. The same as Fig. 17.4 but for small molecular Prandtl number. Note that the turbulent Prandtl number may exceed unity. "Random wave" turbulence ($\varepsilon = 1$) produces quite exotic profiles for the effective Prandtl number, while for „turbulence" type spectra it may be approximated by $Pr + 8/(15\chi^*)$.

For the Sun, we find

$$Pr_{\text{eff}} \approx 64(1 - \varepsilon)(16 - 3\varepsilon)/(5(16 - 7\varepsilon)(40 - 3\varepsilon)) < 8/25, \tag{62}$$

where the final number is the value of expression (62) for $\varepsilon = 0$. Hence, for solar mean-field hydrodynamics, a Prandtl number less than 0.3 should be an appropriate choice (see Fig. 17.5). More precise statements do not seem possible at present.

The state of the Earth's outer core undoubtedly requires that small diffusivities be adopted. The viscosity of molten iron at ordinary pressures is only 10^{-2} cm^2/s, which should be compared with the scales of random motion. Krause (1982), on the basis of dynamo theory, derived the estimates

$$L \gtrsim 10^6\text{cm}, \qquad u' > 10^{-4}\text{cm/s}, \tag{63}$$

which are consistent with the 1979 results of Gubbins *et al.* ($L \geqq 0.5 \cdot 10^6$ cm, $u' \approx 10^{-2}$ cm/s) for which $u'L \geqq 100$ cm²/s. If the Strouhal number is of order unity, we have

$$\nu^* < 10^{-4}, \tag{64}$$

which is indeed much smaller than 1. The normalized molecular conductivity proves to be only one order of magnitude larger than (64). According to our theory, we should assume a Prandtl number of 0.4 in the Earth's outer core.

17.4. The Simplest Approximation for Rotating Turbulence

Here we shall adopt the most drastic approximation possible for the first kernel in (16), setting it equal a constant correlation time τ_c, i.e.

$$\gamma\chi k^2/[(\omega - \chi \boldsymbol{G} \cdot \boldsymbol{k})^2 + \chi^2 k^4] \to \tau_c. \tag{65}$$

Then, and only then, does the simple relation

$$\chi_{ij} = \tau_c Q_{ij}, \tag{66}$$

follow, which expresses the eddy conductivity tensor by the one-point correlation tensor (cf. Daly and Harlow, 1970). This procedure reflects the true qualitative behavior of the tensors. In fact, we know the structure of the correlation tensor for rotating turbulence, and (66) seems to suggests that the same structure determines the turbulent-heat transport. For example, in a non-rotating star only the radial direction $\mathring{g}$ is preferred, and the correlation tensor is given by

$$Q_{ij} = q_{01}\delta_{ij} + q_{02}\mathring{g}_i\mathring{g}_j. \tag{67}$$

According to (66), its conductivity tensor is

$$\chi_{ij} = \tau_c q_{01}\delta_{ij} + \tau_c q_{02}\mathring{g}_i\mathring{g}_j. \tag{68}$$

If the star rotates, additional tensor combinations are available. We demonstrate this by including all the terms up to third order in Ω, that can contribute to the correlation tensor [cf. (8.66)]:

$$\begin{aligned}\tau_c^{-1}\chi_{ij} = {} & q_{00}\delta_{ij} + q_{01}\mathring{g}_i\mathring{g}_j + q_{10}(\varepsilon_{ipl}\mathring{g}_j + \varepsilon_{jpl}\mathring{g}_i)\,\mathring{g}_p\Omega_l + q_{20}\Omega^2\delta_{ij} \\ & + q_{22}(\mathring{\boldsymbol{g}} \cdot \boldsymbol{\Omega})^2\,\delta_{ij} + q_{21}\Omega^2\mathring{g}_i\mathring{g}_j + q_{25}(\mathring{\boldsymbol{g}} \cdot \boldsymbol{\Omega})\,(\mathring{g}_i\Omega_j + \mathring{g}_j\Omega_i) \\ & + q_{24}\Omega_i\Omega_j + q_{23}(\mathring{\boldsymbol{g}} \cdot \boldsymbol{\Omega})^2\,\mathring{g}_i\mathring{g}_j + q_{31}(\mathring{\boldsymbol{g}} \cdot \boldsymbol{\Omega})\,(\varepsilon_{ipl}\Omega_j + \varepsilon_{jpl}\Omega_i)\,\mathring{g}_p\Omega_l \\ & + q_{32}(\mathring{\boldsymbol{g}} \cdot \boldsymbol{\Omega})^2\,(\varepsilon_{ipl}\mathring{g}_j + \varepsilon_{jpl}\mathring{g}_i)\,\mathring{g}_p\Omega_l.\end{aligned} \tag{69}$$

Expression (69) is restricted to terms which are symmetric in i and j; all the q's represent turbulent intensities.

Some of the components of the eddy conductivity tensor are not needed if only axisymmetric ($\partial/\partial\phi = 0$) problems are to be solved. From the expressions of Chapter 8 we find that only three correlations are required: the auto-correlations Q_{rr} and $Q_{\theta\theta}$, and the cross-correlation $Q_{\theta r}$. It should be noted that, in the rotation problem, only the cross-correlations $Q_{r\phi}$ and $Q_{\theta\phi}$ appear. This is a striking, formal difference between energy transport and angular momentum transport. Obviously, the six components of the one-point correlation tensor fall into two distinct groups. The first consists of the three components that have no ϕ index. They change sign when Ω is reversed, and they control the transport of angular momentum. The remaining component, $Q_{\phi\phi}$, never appears in problems having symmetry with respect to the axis of rotation. It is also a striking fact that the group involved in the heat transport contains the auto-correlations. They, of course, exist even in the absence of rotation, this being in sharp contrast to the cross-correlations that transport angular momentum.

Let us turn to the components of the correlation tensor that we need. As usual we write

$$\chi_{ij} = \chi_{ij}^{(0)} + \tilde{\chi}_{ij}, \tag{70}$$

where the effects of rotation are confined to the part having a tilde. We want to express this as a series expansion, i.e. as a disturbance. We shall therefore adopt the representation (69), which leads [as in (8.67)] to

$$\begin{aligned} \chi_{rr}^{(0)} &= \tau_c(q_{00} + q_{01}), \qquad \chi_{\theta\theta}^{(0)} = \tau_c q_{00}, \\ \tilde{\chi}_{rr} &= \tau_c \Omega^2 [q_{20} + q_{21} + q_{22} + q_{23} + q_{24} + 2q_{25} \\ &\quad - (q_{22} + q_{23} + q_{24} + 2q_{25}) \sin^2\theta], \\ \tilde{\chi}_{\theta\theta} &= \tau_c \Omega^2 [q_{20} + q_{24} + (q_{22} - q_{24}) \cos^2\theta], \\ \tilde{\chi}_{r\theta} &= -\tau_c \Omega^2 (q_{25} + q_{24}) \cos\theta \sin\theta. \end{aligned} \tag{71}$$

Note that the (odd) coefficients q_{11}, q_{31} and q_{32}, do not appear. The even coefficients $q_{20} - q_{25}$ form the meridional components of the conductivity tensor which we introduced in the chapters of this book that dealt with the mean-field model. A comparison of (8.71) with (71) yields the simple relations

$$\begin{aligned} VV^{(1)} &= \Omega^2 \tau_c \chi_T^{-1} (q_{22} + q_{23} + q_{24} + 2q_{25}), \\ HV^{(1)} &= VH^{(1)} = \Omega^2 \tau_c \chi_T^{-1} (q_{24} + q_{25}), \quad HH^{(1)} = \Omega^2 \tau_c \chi_T^{-1} (q_{22} - q_{24}). \end{aligned} \tag{72}$$

The coefficient q_{21} does not contribute. In order to determine the quantities $VV^{(1)} - HH^{(1)}$, one must determine the q coefficients in the representation (69) of the correlation tensor for rotating turbulence. We shall do this for several types of turbulence:

(a) Three-dimensional isotropic turbulence

If no preferred direction exists, as is the case for three-dimensional turbulences, we have contributions only from q_{24}, and hence

$$VV^{(1)} = HV^{(1)} = VH^{(1)} = -HH^{(1)} = \Omega^2 \tau_c \chi_T^{-1} q_{24}. \tag{73}$$

This is quite a general property for all isotropic turbulences and, as we have seen, its consequences are far from trivial. Let us consider this topic in more detail. The problem is greatly smiplified if we confine ourselves to the simple spectral tensor (13.20) for three-dimensional turbulence.

The coefficient of $\Omega_i \Omega_j$ in the correlation tensor is q_{24}. As a series expansion in powers of Ω we derive from (14.35)

$$\hat{Q}_{ij} = [1 + (3\omega^2 - \nu^2 k^4)(\omega^2 + \nu^2 k^4)^{-2} (2\hat{\boldsymbol{k}} \cdot \boldsymbol{\Omega})^2]\, {}^{(3)}q k^2 K_{ij}. \tag{74}$$

Integration over Fourier space yields

$$Q_{ij} = q_{01} \delta_{ij} + (\Omega_i \Omega_j - 2\Omega^2 \delta_{ij})\, q_{24}, \tag{75}$$

where

$$q_{24} = (8/15) \iint (\nu^2 k^4 - 3\omega^2)(\omega^2 + \nu^2 k^4)^{-2} k^2\, {}^{(3)}q \,\mathrm{d}\boldsymbol{k}\, \mathrm{d}\omega, \tag{76}$$

which is not sign-definite. It is negative for constant spectra (flat profiles) and positive for delta functions $\delta(\omega)$, i.e. for steep profiles. A partial integration leads to

$$VV^{(1)} = -\frac{8}{15} \frac{\Omega^2 \tau_c}{\chi_T} \iint \frac{\omega k^2}{(\omega^2 + \nu^2 k^4)^2} \frac{\partial}{\partial \omega} [(\omega^2 + \nu^2 k^4)\, {}^{(3)}q]\, \mathrm{d}\boldsymbol{k}\, \mathrm{d}\omega, \tag{77}$$

from which we again see that flat frequency profiles produce negative $VV^{(1)}$, while very steep profiles do the opposite. Our standard profile (33) yields

$$q_{24} = (32\pi/15\nu\tau_c) \int_0^\infty (\nu k^2 - W)(\nu k^2 + W)^{-2} k^2 q_k \,\mathrm{d}k. \tag{78}$$

The k-spectrum therefore determines the sign. For small k the kernel is negative; for large k it is positive. If large k components are absent from the spectrum, as is the case for decreasing spectra of turbulence-type, then obviously q_{24} is negative. This is also transparent if (78) is rewritten as

$$q_{24} = (32\pi/15\nu\tau_c) \int_0^\infty k(W + \nu k^2)^{-1}\, \partial(k^2 q_k)/\partial k \,\mathrm{d}k, \tag{79}$$

which is clearly negative for decreasing $k^2 q_k$.

To sum up, the correlation quantity q_{24} will often be negative, as will be the tensor components $VV^{(1)}$, $HV^{(1)}$ and $VH^{(1)}$. This is in sharp contrast to the positivity of $VV^{(1)}$ and $HV^{(1)}$ indicated by the mean-field model of

Chapters 10 and 11. As shown below, this positivity can be achieved only through the anisotropies of the turbulence. The order of magnitude of q_{24} should be $\tau_c^2\overline{u'^2}/\nu^*$, so that the order of magnitude of $VV^{(1)}$, $HV^{(1)}$ and $VH^{(1)}$ should be $\Omega^2\tau_c^2/(\nu^*\chi_T^*)$. The Strouhal number may be assumed to be of order unity. In arriving at this estimate, the effects of the shape of the wave-number spectrum has not been taken into account. For example, for the spectrum

$$q_k \sim k^{-2}, \tag{80}$$

(79) gives a very small value of q_{24}. Quite apart from this, the order of magnitude of the computed q_{24} is small only if the correlation time is short. For the solar isotropic mode (i.e. the granulation), this is in fact the case.

(b) *Horizontal turbulence*

The situation is quite different for turbulence having (13.23) as spectral tensor, for which motion takes place only in planes perpendicular to the preferred direction $\mathring{g}$. All the terms in (69) are then present, and must be combined as indicated by (72). The effect of rotation on the turbulence defined by the spectral tensor (13.23) is given in (14.37). A series expansion of the even part in powers of Ω is

$$\begin{aligned}\hat{Q}_{ij}^{\text{even}} &= {}^{(2)}\hat{Q}_{ij} + (2\mathring{k}\cdot\boldsymbol{\Omega})^2\,\frac{{}^{(2)}q(k,\omega)}{\omega^2+\nu^2k^4}\Big\{k^2\mathring{g}_i\mathring{g}_j + (\mathring{g}\cdot\boldsymbol{k})^2\,k_ik_j \\ &\quad - (\mathring{g}\cdot\boldsymbol{k})\,(\mathring{g}_ik_j+\mathring{g}_jk_i) + 2\,\frac{\omega^2-\nu^2k^4}{\omega^2+\nu^2k^4} \\ &\quad \times\,[k^2\delta_{ij} - k_ik_j - (\mathring{g}\cdot\boldsymbol{k})^2\delta_{ij} - k^2\mathring{g}_i\mathring{g}_j + (\mathring{g}\cdot\boldsymbol{k})\,(\mathring{g}_ik_j+\mathring{g}_jk_i)]\Big\}.\end{aligned} \tag{81}$$

An extended calculation and final transformation to real space yields, in the language of (69),

$$q_{2i} = (4/105)\iint {}^{(2)}qk^2\mathcal{K}_i(\omega^2+\nu^2k^4)^{-2}\,\mathrm{d}\boldsymbol{k}\,\mathrm{d}\omega, \qquad i = 0, 1, \ldots, 5 \tag{82}$$

where

$$\begin{aligned}&\mathcal{K}_0 = 43\omega^2 - 41\nu^2k^4, \qquad \mathcal{K}_3 = 0,\\ &\mathcal{K}_1 = -19\omega^2 + 65\nu^2k^4, \qquad \mathcal{K}_5 = 18\omega^2 - 38\nu^2k^4,\\ &\mathcal{K}_2 = \mathcal{K}_4 = -26\omega^2 + 30\nu^2k^4.\end{aligned} \tag{83}$$

If these coefficients are manipulated as indicated by (72), the final results are strikingly simple:

$$VV^{(1)} = HV^{(1)} = -(64/105)\,\Omega^2\tau_c\chi_T^{-1}\iint {}^{(2)}qk^2(\omega^2+\nu^2k^4)^{-1}\,\mathrm{d}\boldsymbol{k}\,\mathrm{d}\omega. \tag{84}$$

As can be seen, the heat flux modes $VV^{(1)}$ and $HV^{(1)}$ become negative for ${}^{(2)}q > 0$ and positive for ${}^{(2)}q < 0$. The latter case corresponds to vertical-type turbulence; the former to horizontal-type. This conclusion is general. Obviously, the simplified theory of eddy conductivity, which is based on the approximation (65), yields positive flux modes $VV^{(1)}$ and $HV^{(1)}$ only if the turbulence has negative components in its spectral function ${}^{(2)}q$, i.e. if it is of vertical-type.

(c) *Axisymmetric turbulence*

The foregoing results can also provide the turbulent energy transport for the more general models of § 13.3b. There a combination of isotropic and anisotropic fields was used to model more general anisotropic, but axially symmetric, turbulence fields. Taking (13.65) and assuming $\varkappa = 0$, we find that (76) and (84) give

$$VV^{(1)} = HV^{(1)} = \frac{8\Omega^2\tau_c}{105\chi_T} \iint \frac{(8\alpha + 7)\,\nu^2k^4 + (8\alpha - 21)\,\omega^2}{(\omega^2 + \nu^2k^4)^2}\, qk^2\,\mathrm{d}k\,\mathrm{d}\omega\,. \quad (85)$$

Remembering that q is positive semi-definite, it is clear that negative α gives negative VV and HV. According to Chapter 11, positive VV and HV can be produced only by positive α. There is practically no way in which positive fluxes VV and HV can be created by turbulence with negative α, i. e. by horizontal-type turbulence. The turbulence must be of the vertical type, i.e. its vertical intensity must exceed its horizontal intensity. To produce positive VV and HV, the frequency spectrum cannot be arbitrary. It need not be as steep, however, as was required in (76). Our standard example (13.46) yields

$$VV^{(1)} = HV^{(1)} = \frac{8\Omega^2}{105\nu\chi_T} \int \frac{(8\alpha + 7)\,\nu k^2 + (8\alpha - 7)\,W}{(W + \nu k^2)^2}\, q_k\,\mathrm{d}k\,, \quad (86)$$

which is certainly positive for $\alpha > 7/8$. The same condition arises if we set $q =$ constant, i.e. if we assume an extremely flat spectrum. A very steep spectrum of turbulence type follows from the choice $q \sim \delta(\omega)$. Using (85), this provides values of $VV^{(1)}$ and $HV^{(1)}$ that are positive for $\alpha > -7/8$. For very steep frequency spectra, even weakly horizontal turbulence generates a positive $VV^{(1)}$, but dominating vertical intensities tend to be preferred.

We see that the presence of anisotropy in the turbulence field, although it is responsible for the Λ-effect, changes the sign of VV and HV from negative to positive. Since here the three-dimensional isotropic part of the spectral tensor (which plays no role in the theory of the Λ-effect) makes a negative contribution, we may expect that VV and VH, even if positive, will be small. For example, with the same value of α (namely $\alpha = 1$) that we used in § 15.4 [see (15.47)] to explain the Λ-modes we required, we can again produce the small positive values of $VV^{(1)}$ and $HV^{(1)}$ that were required in the mean-field model of differential rotation developed in Chapter 11.

If our basic equations and our analysis of them are free from serious error, we may now dismiss horizontal-type turbulence for the Sun. By (84), the positivity of $VV^{(1)}$ and $HV^{(1)}$ that we require can only be achieved using turbulences in which the vertical motion exceeds the horizontal one. This conclusion should, however, only be regarded as tentative. Clearly the mean-field model of Chapters 10 and 11 is still not sophisticated enough for us to be dogmatic in our claims. Much work still remains to be done to improve our approach, so that a detailed analysis of the "microscopic" background makes more sense. So far we have demonstrated only that one consistent model does exist: *turbulence having dominating vertical motion and non-flat cells provides the required turbulent transport of both momentum and energy.*

Epilogue

We only need to look at the
planets and suns to see there
are still plenty of nuts to crack.
Goethe, 1827.

It is clear that the existing analytical theory of non-uniform rotation rests on two pillars. Observations form the first of these. As far as the Sun is concerned, these are now very accurate, and the temporal variations, in particular, are becoming increasingly well resolved, so that new insights can be expected in the near future. Also, the indirect determination of the rotation beneath the surface of the Sun now leads to consistent results. We can therefore be optimistic that the main task of the theory, namely to determine the rotation inside the Sun, will be achieved in the not too distant future. We shall then finally know not only the character of the turbulence in the SCZ but also the mechanism of magnetic field generation. We shall also be able to assess the accuracy of those bulwarks of present day studies: mixing length theory and dynamo theory.

There has been little progress in measuring the horizontal cross-correlation $\overline{u_\theta' u_\phi'}$, which is hard to observe, but which is of much greater importance than is often realized. The analysis of non-magnetic tracers provided us with no information about this quantity. Its values, as a function of position on the solar surface would be of as great a significance as the observations of the rotation law and the meridional circulation have been in the past. It is our hope that future satellite experiments will determine the flow patterns in the granulation so that averaged quantities such as Ω, $\overline{u_\theta}$ and $\overline{u_\theta' u_\phi'}$ can all be determined.

The Sun is but one star of many that can be studied. Apart from these, only planetary atmospheres might be considered by similar methods, but (needless to say) our theory is inadequate for these. In Part II, we could specify solar values for only five of the turbulence parameters, namely $V^{(0)}$, $V^{(1)}, \ldots, H V^{(1)}$, and it was possible to compare only these (in Part III) with simple turbulence formulae. It would be an almost trivial task to use precisely the same values for these parameters when modeling stars along the entire main sequence or even red giants, and this has been done by some authors. Whether such investigations make sense depends on the validity of the implied assumption that developed turbulence does not depend on the spectral class of a star, and that global quantities, such as the depth of the convection zone, angular velocity etc., determine the mean-field flow pattern. We do not know whether this is justified.

In common with heuristic theories that involve several free parameters, it is hard to make predictions for other stars by using our methods. We are more competent to analyse the principal factors and to separate the competing mechanisms. In this way, we have isolated no less than six effects that can generate differential rotation. Two of them [viz: $V^{(1)}$, $H^{(1)}$] do this directly, and the remaining four achieve it through the meridional circulation that they produce. We have also been able to clarify analytically the role played by the thickness of the convection zone. Although numerical methods seldom provide physical insight, their predictions (when successful) are more precise. An important relationship subsists between analytical and numerical theories (cf. also Gilman, 1979; Belvedere *et al.*, 1980). Who would have thought a few years ago that such numerical modeling would have provided more and more stringent tests of the theory?

Estimates of *stellar* differential rotation have been published based on both photometric and spectroscopic observations. Gray (1977), as well as Bruning (1981), developed Fourier techniques for the rotational broadening of line contours; these techniques should be successful principally for early-type stars. Both the amplitude and the position of the zeros of the sidelobes of the Fourier transform carry information about the value, $\omega_2(1)$ of the differential rotation at the surface. Because of the disturbingly large number of lines in the spectra of late-type stars, the method has so far been confined to the A and F stars. Neither Gray (1977, 1982) nor Wöhl (1983) found any evidence that such stars have a steeper differential rotation than the Sun.

Differential rotation in binary systems of the type RS CVn can even be deduced by a normal light curve analysis. The cooler of the two components is of spectral class G or K and probably possesses starspots that create a modulation of the normal eclipsing light curve. To interpret this migrating wave-like distortion a possible activity cycle, the drift of the activity zone, as well as the differential rotation, must be taken into account (Rodonò, 1981; Dorren and Guinan, 1983). The resulting rotation laws are still, however, somewhat discordant, and we have thought it better not to enter into details.

There are prospects of considerable successes from CaII photometry. The emission in the Calcium lines H and K, which indicates the chromospheric activity of stars (see Eberhard and Schwarzschild, 1913; Wilson, 1978), is rotationally modulated. Its study has already led to important insights into stellar rotation (Vaughan *et al.*, 1981). LaBonte (1982) used the Sun to calibrate the method. In the disk-integrated light of H and K, the rotation period determined proved to belong to the latitude having maximal magnetic flux density. As is well known, this migrates slowly towards the equator during the cycle, and the latitude dependence of the angular velocity should therefore be directly observable. This method should give stellar differential rotations equally easily. Strangely enough, LaBonte was unable to vindicate these ideas for the Sun. Both the amplitude and sign of the periodic variation were incorrect. Nevertheless, Horne *et al.* (1983) provided observational evidence of differential rotation for two stars through H and K photometry, finding variations over latitude of 5% and 11%.

Certainly there are still discrepancies in the data for stellar differential rotation and magnetism, for which a wide range of observations already exists, but we are optimistic that in the not too distant future the situation will improve. Theories are in prospect that may really explain the stellar data.

Our second pillar, turbulence theory, is also under construction. As we have shown, we know the global flow pattern of the solar convection zone when the magnitudes of the five parameters $V^{(0)}, \ldots, HV^{(1)}$ have been determined. The values of these constants are absolutely crucial for the models, and their determination requires a theory of turbulence. The SOCA used in Part III is, of course, somewhat simplistic. The additional assumption of two-scaledness of flow produced local relations that further simplified the true situation. In recent years there has, however, been some rethinking in the theory of turbulence and related phenomena such as convection, dynamos, etc... New, numerically-discovered insights into the complexity of the solutions of non-linear equations have given rise to new ideas on the nature of random phenomena. For example, the Landau-Hopf picture of turbulence, which had previously been widely accepted, was not confirmed by the new scenarios. After only a few steps of the bifurcation sequence visualized by Landau and Hopf, transition to chaos occurs. This is in dramatic contrast to the earlier idea that, as the energy density of turbulence increased, an endless succession of splitting and separation of modes took place.

Yet we remain far from a general *modus operandi*, and even some qualitative aspects are incompletely understood. It is fortunate for us that few of the difficulties of turbulence theory impinge on the theory of differential rotation. Often we need only the structure of the correlation tensor of the "original" turbulence (where by original we mean "uninfluenced by rotation"), and this structure can usually be derived heuristically by general considerations, independently of any particular theory. Occasionally, it is sufficient to make sure that the sign of the spectral function complies with Bochner's theorem, and worthwhile results follow. Needless to say though, by following this path many important questions remain unanswered, such as, for example, the interconnection between the Λ-effect for rotation and the α-effect for MHD. The latter arises from those terms of the correlation tensor which are odd in $\boldsymbol{g}$, while the former stems from those that are even.

No wonder, then, that a unified theory encompassing both the differential rotation of the Sun *and* the solar dynamo has not yet been developed!

Looked at from this new standpoint, differential rotation is a dissipative structure generated spontaneously in a rotating system having energy input and heat output. Given the global structures "temperature gradient" and "basic rotation", and with a convective instability as the catalyst, a new large-scale structure is created: a combination of *meridional circulation and differential rotation:*

Appendix

A. Representations in spherical coordinates (r, θ, ϕ)

If $\partial/\partial\phi = 0$, then for a vector $\boldsymbol{A} = (A_r, A_\theta, A_\phi)$ and a scalar Ψ:

$$\nabla\Psi = (\partial\Psi/\partial r, r^{-1}\,\partial\Psi/\partial\theta, 0) \tag{A.1}$$

$$\nabla\cdot\boldsymbol{A} = r^{-2}\,\partial(r^2A_r)/\partial r + (r\sin\theta)^{-1}\,\partial(A_\theta\sin\theta)/\partial\theta \tag{A.2}$$

$$\left.\begin{aligned}(\nabla\times\boldsymbol{A})_r &= (r\sin\theta)^{-1}\,\partial(A_\phi\sin\theta)/\partial\theta\\ (\nabla\times\boldsymbol{A})_\theta &= -r^{-1}\,\partial(rA_\phi)/\partial r\\ (\nabla\times\boldsymbol{A})_\phi &= r^{-1}[\partial(rA_\theta)/\partial r - \partial A_r/\partial\theta]\end{aligned}\right\} \tag{A.3}$$

$$\left.\begin{aligned}&\mathring{\boldsymbol{\Omega}} = \begin{pmatrix}\cos\theta\\ -\sin\theta\\ 0\end{pmatrix}, \quad \mathring{\boldsymbol{g}} = \begin{pmatrix}1\\ 0\\ 0\end{pmatrix}, \quad \mathbf{s} = r\sin\theta\begin{pmatrix}\sin\theta\\ \cos\theta\\ 0\end{pmatrix}\\ &\boldsymbol{\Omega}_\perp = \begin{pmatrix}0\\ -\sin\theta\\ 0\end{pmatrix}\Omega, \quad \boldsymbol{k}_\perp = \begin{pmatrix}0\\ k_\theta\\ k_\phi\end{pmatrix}\end{aligned}\right\} \tag{A.4}$$

B. Legendre Functions

Differential relations: with $\mu = \cos\theta$

$$\frac{1}{\sin\theta}\frac{\mathrm{d}}{\mathrm{d}\theta}\left[\frac{\mathrm{d}P_n(\mu)}{\mathrm{d}\theta}\sin\theta\right] + n(n+1)\,P_n(\mu) = 0 \tag{A.5}$$

$$\frac{1}{\sin\theta}\frac{\mathrm{d}}{\mathrm{d}\theta}\left[\frac{\mathrm{d}P_n^m(\mu)}{\mathrm{d}\theta}\sin\theta\right] + \left[-\frac{m^2}{\sin^2\theta} + n(n+1)\right]P_n^m(\mu) = 0 \tag{A.6}$$

$$P_n^m(\mu) = \sin^m\theta \, \mathrm{d}^m P_n(\mu)/\mathrm{d}\mu^m \tag{A.7}$$

$$P_n^1 = -\mathrm{d}P_n/\mathrm{d}\theta \tag{A.8}$$

$$\frac{1}{\sin\theta}\frac{\mathrm{d}}{\mathrm{d}\theta}\,[P_n^1(\mu)\sin\theta] - n(n+1)\,P_n(\mu) = 0 \tag{A.9}$$

$$\frac{\mathrm{d}}{\mathrm{d}\theta}\left\{\frac{1}{\sin\theta}\frac{\mathrm{d}}{\mathrm{d}\theta}\,[P_n^1(\mu)\sin\theta]\right\} + n(n+1)\,P_n^1 = 0 \tag{A.10}$$

$$\frac{\mathrm{d}}{\mathrm{d}\theta}\left[\sin^3\theta\,\frac{\mathrm{d}}{\mathrm{d}\theta}\left(\frac{P_n^1}{\sin\theta}\right)\right] = -(n-1)\,(n+2)\,P_n^1\sin^2\theta \tag{A.11}$$

Recursions:

$$(2n+1)\sin\theta\,P_n = -P_{n-1}^1 + P_{n+1}^1 \tag{A.12}$$

$$(2n+1)\cos\theta\,P_n^1 = (n+1)\,P_{n-1}^1 + nP_{n+1}^1 \tag{A.13}$$

$$(2n+1)\sin\theta\,P_n^1 = n(n+1)\,(P_{n-1} - P_{n+1}) \tag{A.14}$$

$$\cos^2\theta\,P_n^1 = \frac{n(n+1)}{(2n-1)\,(2n+1)}\,P_{n-2}^1 + \frac{2n^2+2n-3}{(2n-1)\,(2n+3)}\,P_n^1 + \frac{n(n+1)}{(2n+1)\,(2n+3)}\,P_{n+2}^1 \tag{A.15}$$

$$(2n+1)\sin\theta\,\mathrm{d}P_n^1/\mathrm{d}\theta = -(n+1)^2\,P_{n-1}^1 + n^2P_{n+1}^1 \tag{A.16}$$

$$(2n+1)\,\mathrm{d}(P_n^1\sin\theta)/\mathrm{d}\theta = n(n+1)\,(-P_{n-1}^1 + P_{n+1}^1) \tag{A.17}$$

$$(2n+1)\sin^2\theta\,\mathrm{d}[(\sin\theta)^{-1}\,P_n^1]/\mathrm{d}\theta = -(n+1)\,(n+2)\,P_{n-1}^1 + n(n-1)\,P_{n+1}^1 \tag{A.18}$$

Legendre polynomials: with $\mu = \cos\theta$

$$P_0(\mu) = 1, \quad P_1(\mu) = \mu, \quad P_2(\mu) = \frac{1}{2}\,(3\mu^2 - 1) \tag{A.19}$$

$$P_3(\mu) = \frac{1}{2}\,(5\mu^3 - 3\mu), \qquad P_4(\mu) = \frac{1}{8}\,(35\mu^4 - 30\mu^2 + 3) \tag{A.20}$$

$$P_5(\mu) = \frac{1}{8}\,(63\mu^5 - 70\mu^3 + 15\mu) \tag{A.21}$$

$$P_6(\mu) = \frac{1}{16}\,(231\,\mu^6 - 315\mu^4 + 105\,\mu^2 - 5) \tag{A.22}$$

C. Fourier integrals

$$\int F(k)\,\mathrm{d}\boldsymbol{k} = 4\pi \int_{-\infty}^{\infty} k^2 F(k)\,\mathrm{d}k \tag{A.23}$$

$$\int k_i k_j F(k)\,\mathrm{d}\boldsymbol{k} = \frac{1}{3}\,\delta_{ij} \int k^2 F(k)\,\mathrm{d}\boldsymbol{k} \tag{A.24}$$

$$\int k_i k_j k_k k_e F(k)\,\mathrm{d}\boldsymbol{k} = (1/15)\,(\delta_{ij}\delta_{ke} + \delta_{ik}\delta_{je} + \delta_{ie}\delta_{jk}) \int k^4 F(k)\,\mathrm{d}\boldsymbol{k} \tag{A.25}$$

$$\int (\boldsymbol{k}\cdot\mathring{\boldsymbol{g}})^{2m}\,F(k)\,\mathrm{d}\boldsymbol{k} = (2m+1)^{-1} \int k^{2m} F(k)\,\mathrm{d}\boldsymbol{k} \tag{A.26}$$

$$\begin{aligned}
&\int (\boldsymbol{k}\cdot\mathring{\boldsymbol{g}})^{2m}\,k_i k_j k_k k_e F(k)\,\mathrm{d}\boldsymbol{k}\\
&= [\delta_{ik}\delta_{ej} + \delta_{ie}\delta_{kj} + \delta_{ij}\delta_{ke} + 4m(m-1)\,\mathring{g}_i\mathring{g}_j\mathring{g}_k\mathring{g}_e\\
&\quad + 2m(\mathring{g}_i\mathring{g}_j\delta_{ek} + \mathring{g}_i\mathring{g}_e\delta_{jk} + \mathring{g}_i\mathring{g}_k\delta_{je} + \mathring{g}_j\mathring{g}_e\delta_{ik} + \mathring{g}_j\mathring{g}_k\delta_{ie} + \mathring{g}_e\mathring{g}_k\delta_{ij})]\\
&\quad \int k^{2m+4} F(k)\,\mathrm{d}\boldsymbol{k}/(2m+1)\,(2m+3)\,(2m+5)
\end{aligned} \tag{A.27}$$

D. Frequency integrals

$$\int_{-\infty}^{\infty} (\omega^2 + w^2)^{-1}\,\mathrm{d}\omega = \pi/w \tag{A.28}$$

$$\int_{-\infty}^{\infty} [(\omega^2 + w^2)\,(\omega^2 + v^2)]^{-1}\,\mathrm{d}\omega = \pi/wv(w + v) \tag{A.29}$$

$$\int_{-\infty}^{\infty} \omega^2[(\omega^2 + w^2)\,(\omega^2 + v^2)]^{-1}\,\mathrm{d}\omega = \pi/(w + v) \tag{A.30}$$

$$\int_{-\infty}^{\infty} \frac{\mathrm{d}\omega}{(\omega^2 + v^2)^2\,(\omega^2 + w^2)} = \frac{\pi}{2}\,\frac{2v + w}{v^3 w (w + v)^2} \tag{A.31}$$

$$\int_{-\infty}^{\infty} \frac{\omega^2\,\mathrm{d}\omega}{(\omega^2 + v^2)^2\,(\omega^2 + w^2)} = \frac{\pi}{2}\,\frac{1}{v(w + v)^2}\,. \tag{A.32}$$

$$\int_{-\infty}^{\infty} \frac{\omega^4\,\mathrm{d}\omega}{(\omega^2 + v^2)^2\,(\omega^2 + w^2)} = \frac{\pi}{2}\,\frac{v + 2w}{(w + v)^2} \tag{A.33}$$

References

Altrock, R. C. and Canfield, R. C., Observations of photospheric pole-equator temperature differences. *Solar Phys.* **23**, 257 (1972).

Antonucci, E., Coronal and chromospheric rotation. In: *Workshop on Solar Rotation,* p. 204, Catania (1978).

Appenzeller, I. and Schröter, E. H., Center-to-limb variations of the intensity and the wavelength of several Fraunhofer lines along the Sun's polar and equatorial diameter. *Astrophys. J.* **147**, 1100 (1967).

Arlt, K. and Rüdiger, G., A numerical study on turbulent heat transfer in convective envelopes. *Astron. Nachr.* **305**, 229 (1984).

Baker, N. and Temesváry, S., *Tables of convective stellar envelope models.* New York (1966).

Balthasar, H., Schüssler, M. and Wöhl, H., On changes of the rotation velocities of stable, recurrent sunpots and their interpretation with a flux tube model. *Solar Phys.* **76**, 21 (1982).

Balthasar, H. and Wöhl, H., Differential rotation and meridional motions of sunspots in the years 1940—1968. *Astron. Astrophys.* **92**, 111, (1980).

Becker, U., Die Eigenbewegung der Sonnenflecken in Breite. *Z. Astrophysik* **34**, 129 (1954).

Beckers, J. M., Material motions in sunspot umbrae. *Astrophys. J.* **213**, 900 (1977).

Beckers, J. M., Observations of solar rotation and meridional flows at the Sacramento Peak Observatory. In: *Workshop on Solar Rotation*, p. 166, Catania (1978).

Beckers, J. M., Dynamics of the solar photosphere. In: *The Sun as a Star* (ed. S. D. Jordan), p. 11, (1981).

Belopolsky, A., Über die Bewegungen auf der Sonnenoberfläche. *Astron. Nachr.* **124**, 17 (1890).

Belvedere, G., Godoli, G., Motta, S., Paternò, L. and Zappala, R., Surface Reynolds stress determined from the analysis of facular motions and the maintenance of the sun's differential rotation. *Solar Phys.* **46**, 23 (1976).

Belvedere, G. and Paternò, L., Convection in a rotating deep compressible spherical shell: application to the sun. *Solar Phys.* **54**, 289 (1977).

Belvedere, G., Paternò, L. and Stix, M., Differential rotation set up by latitude dependent heat transport. *Geophys. Astrophys. Fluid Dyn.* **14**, 209 (1980).

Belvedere, G., Zappala, R. A., D'Arrigo, C., Motta, S., Pirronella, V., Godoli, G. and Paternò, L., Differential rotation of the sun and surface angular momentum

transport deduced from chromospheric tracers during the period 1967–77. In: *Workshop on Solar Rotation*, p. 189, Catania (1978).
Berthold, G., *Der Magister Johann Fabricius und die Sonnenflecken*. Verlag von Veit et Comp, Leipzig (1894).
Biermann, L., Untersuchungen über den inneren Aufbau der Sterne. *Z. Astrophysik* **5**, 117 (1932).
Biermann, L., Zur Theorie der Granulation und der Wasserstoffkonvektionszone der Sonne. *Astron. Nachr.* **264**, 395 (1938).
Biermann, L., Bemerkungen über das Rotationsgesetz in irdischen und stellaren Instabilitätszonen. *Z. Astrophysik* **28**, 304 (1951).
Bisshopp, F. E., On the thermal instability of a rotating fluid sphere. *Philos. Mag.* **3**, 1342 (1958).
Bochner, S., Monotone Funktionen, Stieltjessche Integrale und harmonische Analyse. *Math. Ann.* **108**, 378 (1933).
Bogart, R. S. and Gierasch, P. J., Meridional circulation and the redistribution of heat in the solar convective envelope. *Astrophys. J.* **228**, 624 (1979).
Boussinesq, M. J., *Théorie de l'écoulement tourbillonnant et tumultueux des liquides dans les lits rectilignes a grande section*, Comptes Rendus de l'Acad. des Sciences, Gauthier-Villars, Paris (1897).
Braginsky, S. I., On the nearly axially-symmetrical model of the hydrodynamic dynamo of the earth. *Phys. Earth Planet. Interiors* **11**, 191 (1976).
Bruning, D. H., Fourier analysis of differential rotation in late-type stars. *Astrophys. J.* **248**, 274 (1981).
Bumba, V., Observations of solar magnetic and velocity fields. In: *International School of Physics "Enrico Fermi" XXXIX Plasma Astrophysics* p. 77 (1967).
Busse, F. H., Differential rotation in stellar convection zones. *Astrophys. J.* **159**, 629 (1970).
Busse, F. H., Differential rotation in stellar convection zones II. *Astron. Astrophys.* **28**, 27 (1973).
Busse, F. H., Generation of mean flows in a rotating convection layer. *Z. Naturforsch.* **37a**, 752 (1982).
Busse, F. H. and Cuong, P. G., Convection in rapidly rotating spherical fluid shells. *Geophys. Astrophys. Fluid Dyn.* **8**, 17 (1977).
Caccin, B., Falciani, R., Moschi, G. and Rigutti, M., Temperature differences between the equator and the poles of the sun. *Solar Phys.* **13**, 33 (1970).
Carrington, R. C., *Observations of the spots on the sun*. London (1863).
Chan, K. L., Sofia, S. and Mayr, H. G., Mixing length, shears, and differential rotation. In: *The internal solar angular velocity* (ed. B. R. Durney and S. Sofia), p. 347, Reidel, Dordrecht (1987).
Chandrasekhar, S., *Hydrodynamic and hydromagnetic stability*. Clarendon Press, Oxford (1961).
Christensen-Dalsgaard, J., Dziembowski, W. and Gough, D., How deep is the solar convection zone? *Lecture Notes in Physics* **125**, 313 (1979).
Claverie, A., Isaak, G. R., Mc Leod, C. P., van der Raay, H. B. and Roca Cortes, T., Rapid rotation of the solar interior. *Nature* **293**, 443 (1981).
Cocke, W. J., On the solar differential rotation and meridional currents. *Astrophys. J.* **150**, 1044 (1967).

Coffey, H. E. and Gilman, P. A., Sunspot motion statistics for 1965—67. *Solar Phys.* **9**, 423 (1969).

Cowling, T. G., The condition for turbulence in rotating stars. *Astrophys. J.* **114**, 272 (1951).

Cox, J. P., *Theory of stellar pulsations*. Princeton University Press, Princeton—New Jersey (1980).

Cram, L. E., Durney, B. R. and Guenther, D. B., Preliminary observations of velocity fields of the solar poles. *Astrophys. J.* **262**, 442 (1983).

Csada, I. K., The differential rotation and the large-scale meridional motion of the stars. *Contributions Konkoly Observatory* **22**, 215 (1949).

Daly, B. J. and Harlow, F. H., Transport equations in turbulence. *Physics of Fluids* **13**, 2634 (1970).

Deinzer, W. and Stix, M., On the eigenvalues of Krause-Steenbeck's solar dynamo. *Astron. Astrophys.* **12**, 111 (1971).

Deubner, F.-L., Observations of low wavenumber non-radial eigenmodes of the sun. *Astron. Astrophys.* **44**, 371 (1975).

Deubner, F.-L., Ulrich, R. K. and Rhodes Jr., E. J., Solar p-mode oscillations as a tracer of radial differential rotation. *Astron. Astrophys.* **72**, 177 (1979).

Dorren, J. D. and Guinan, E. F., Starspots, differential rotation and a possible six-year spot cycle on lambda andromedae. *Lecture Notes in Physics* **193**, 259 (1983).

Drazin, P. and Reid, W., *Hydrodynamic stability*. Cambridge University Press, Cambridge (1981).

Drobyshevski, E. M., Outward transport of angular momentum by gas convection and the equatorial acceleration. *Solar Phys.* **51**, 473 (1977).

Dupree, A. K. and Henze, Jr. W., Solar rotation as determined from OSO-EUV spectroheliograms. *Solar Phys.* **27**, 271 (1972).

Durney, B., Nonaxisymmetric convection in a rotating spherical shell. *Astrophys. J.* **161**, 1115 (1970).

Durney, B., Differential rotation, meridional velocities, and pole-equator difference in temperature of a rotating convective spherical shell. *Astrophys. J.* **163**, 353 (1971).

Durney, B., On a model of a slowly rotating solar convection zone. *Astrophys. J.* **244** 678 (1981).

Durney, B., On the large-scale dynamics of rapidly rotating convection zones. *Astrophys. J.* **269**, 671 (1983).

Durney, B. and Latour, J., On the angular momentum loss of late-type stars. *Geophys. Astrophys. Fluid Dyn.* **9**, 241 (1978).

Durney, B., Mihalas, D. and Robinson, R. D., A preliminary interpretation of stellar chromospheric CaII emission variations within the framework of stellar dynamo theory. *Publ. Astron. Soc. Pac.* **555**, 537 (1981).

Durney, B. and Roxburgh, I. W., Inhomogeneous convection and the equatorial acceleration of the sun. *Solar Phys.* **16**, 3 (1971).

Durney, B. and Spruit, H. C., On the dynamics of stellar convection zones: The effect of rotation on the turbulent viscosity and conductivity. *Astrophys. J.* **234**, 1067 (1979).

Duvall, T. L., Large-scale solar velocity fields. *Solar Phys.* **63**, 3 (1979).

Duvall, T. L., The equatorial rotation rate of the supergranulation cells. *Solar Phys.* **66**, 213 (1980).

Duvall, T. L., Dziembowski, W. A., Goode, P. R., Gough, D. O., Harvey, J. W. and Leibacher, J. W., Internal rotation of the sun. *Nature* **310**, 22 (1984).

Eberhard, G. and Schwarzschild, K., Über Umkehr der Kalziumlinien H und K in Sternspektren. *Sitzungsbericht Preuss. Akademie Wiss.* 1913, 308 (1913).

Eddy, J. A., The Maunder minimum. *Science* **192**, 1189 (1976).

Edmonds, F. N., The coefficients of viscosity and thermal conductivity in the hydrogen convection zone. *Astrophys. J.* **125**, 535 (1957).

Elsässer, K., Anisotrope Reibung. *Z. Astrophys.* **63**, 65 (1966).

Emden, R., *Gaskugeln*. Verlag B. G. Teubner, Leipzig u. Berlin (1907).

Emden, R., Zum Rotationsgesetz der photosphärischen Schichten. *Z. Astrophysik* **12**, 233 (1936).

Eschrich, K.-O., Wärmetransport und Wärmeerzeugung in einem turbulent bewegten Gas. *Astron. Nachr.* **299**, 137 (1978).

Eschrich, K.-O. and Rüdiger, G., A second-order correlation approximation for thermal conductivity and Prandtl number of free turbulence. *Astron. Nachr.* **304**, 174 (1983).

Fabricius, J., *De maculis in Sole observatis Narratio*, Witebergae, Anno MDCXI.

Falciani, R., Rigutti, M. and Roberti, G., Temperature variations in the solar photosphere. II. A search for equator-to-pole differences in photospheric temperature. *Solar Phys.* **35**, 277 (1974).

Faye, M., Sur la constitution physique du soleil. *Comptes Rendus de l'Acad. des Sciences*, Paris **60**, 138 (1865).

Fenstermacher, P. R., Swinney, H. L. and Gollub, J. P., Dynamical instabilities and the transition to chaotic vortex flow. *J. Fluid Mech.* **94**, 103 (1979).

Foukal, P., Magnetic coupling of the active chromosphere to the solar interior. *Astrophys. J.* **173**, 439 (1972).

Foukal, P., Spectroscopic evidence for a higher rotation rate of magnetized plasma at the solar photosphere. *Astrophys. J.* **203**, L 145 (1976).

Gailitis, A. and Rüdiger, G., On the radial rotation law in the solar supergranulation layer. *Astrophys. Lett.* **22**, 89 (1982).

Galilei, G., *Istoria e demonstraziani interna alle machi solari e loro accidenti comprese in tre lettere*, Roma (1613).

Gião, A. and Wehrlé, P., Sur le rotations de astres fluides. *Beitr. Physik freien Atmosphäre* **19**, 237 (1932).

Gierasch, P. J., The differential rotation of the solar surface. *Astrophys J.* **190**, 199 (1974).

Gilman, P. A., Nonlinear Boussinesq convection model for large scale circulations. *Solar Phys.* **27**, 3 (1972).

Gilman, P. A., Solar Rotation. *Ann. Rev. Astron. Astrophys.* **12**, 47 (1974).

Gilman, P. A., Nonlinear dynamics of Boussinesq convection in a deep rotating spherical shell I. *Geophys. Astrophys. Fluid Dyn.* 8, 93 (1977).

Gilman, P. A., Nonlinear dynamics of Boussinesq convection in a deep rotating spherical shell II: Effects of temperature boundary conditions. *Geophys. Astrophys. Fluid Dyn.* **11**, 157 (1978).

Gilman, P. A., Global circulation of the sun: Where are we and where are we going? *Proc. XVII General Assembly IAU*, Montreal (1979a).

Gilman, P. A., Differential rotation in stars with convection zones. *Lecture Notes in Physics* **114**, 19 (1979b).

Gilman, P. A., Convective dynamos for rotating stars. In: *Second Cambridge Workshop on cool stars, stellar systems, and the sun* (ed. M. S. Giampapa and L. Golub), p. 165 (1981).

Gilman, P. A., What would a dynamo theorist like to know about the dynamics of the solar convection zone? *Snowmass conference on Solar Seismology from Space*, Boulder (1983).

Gilman, P. A., The solar dynamo: Observations and theories of solar convection, global circulation, and magnetic fields. In: *Physics of the sun* (ed. P. A. Sturrock), p. 95 (1986).

Gilman, P. A. and Foukal, P. V., Angular velocity gradients in the solar convection zone. *Astrophys. J.* **229**, 1179 (1979).

Gilman, P. A. and Glatzmaier, G. A., Compressible convection in a rotating spherical shell. I. Anelastic equations. *Astrophys. J. Suppl.* **45**, 335 (1981).

Gilman, P. A. and Howard, R., Variation in solar rotation with the sunspot cycle. *Astrophys. J.* **283**, 385 (1984a).

Gilman, P. A. and Howard, R., On the correlation of longitudinal and latitudinal motions of sunspots. *Solar Phys.* **93**, 171 (1984b).

Glatzmaier, G. A., Numerical simulations of stellar convective dynamos I. The model and method. *J. Comp. Phys.* **55**, 461 (1984).

Glatzmaier, G. A., Numerical simulation of stellar convective dynamos III. At the base of the convection zone. *Geophys. Astrophys. Fluid Dyn.* **31**, 137 (1985).

Godoli, G. and Mazzucconi, F., Age, life-time and differential rotation of sunspots. In: *Workshop of Solar Rotation*, p. 135, Catania (1978).

Godoli, G. and Mazzucconi, F., Angular velocity of sunspots along the butterfly diagram. *Astron. Astrophys.* **116**, 188 (1982).

Gollub, J. P. and Benson, S. V., Many routes to turbulent convection. *J. Fluid Mech.* **100**, 449 (1980).

Golub, L. and Vaiana, G. S., Differential rotation rates for short-lived regions of emerging magnetic flux. *Astrophys. J.* **219**, L 55 (1978).

Gough, D. O., The current state of mixing-length theory. *Lecture Notes in Physics* **71**, 15 (1976).

Gough, D. O., The estimation of fluxes due to small scale turbulences. In: *Workshop on Solar Rotation*, p. 337, Catania (1978).

Gray, D. F., A quest for differential stellar rotation in A stars. *Astrophys. J.* **211**, 198 (1977).

Gray, D. F., Observational evidence against differential rotation in F stars. *Astrophys. J.* **258**, 201 (1982).

Greenspan, H. P., *The theory of rotating fluids.* Cambridge University press (1969).

deGroot, S. R., *Thermodynamics of irreversible processes.* North Holland Publ. Comp. Amsterdam (1952).

Gubbins, D., Masters, D. J. and Jacobs, J. A., Thermal evolution of the earth's core. *Geophys. J. Roy astron. Soc.* **59**, 57 (1979).

Hansky, A., Photographics de la granulation. *Pulkowo Mitt.* **1**, 81 (1905).

Harzer, P., Über die Rotationsbewegungen der Sonne. *Astron. Nachr.* **127**, 18 (1891).

Hathaway, D. H., Nonlinear simulations of solar rotation effects in supergranules. *Solar Phys.* **77**, 341 (1982).

Hathaway, D. H., A convective model for turbulent mixing in rotating convection zones. *Astrophys. J.* **276**, 316 (1984).

Hathaway, D. H., Gilman, P. A. and Toomre, J., Convective instability when the temperature gradient and rotation vector are oblique to gravity. I. Fluids without diffusion. *Geophys. Astrophys. Fluid Dyn.* **13**, 289 (1979).

Hathaway, D. H. and Somerville, R. C. J., Three-dimensional simulation of convection in layers with tilted rotation vectors. *J. Fluid Mech.* **126**, 75 (1983).

Hathaway, D. H., Toomre, J. and Gilman, P. A., Convection instability when the temperature gradient and rotation vector are oblique to gravity. II. Real fluids with effects of diffusion. *Geophys. Astrophys. Fluid Dyn.* **15**, 7 (1980).

Hence, W. and Dupree, A. K., Solar rotation in the chromosphere and corona. *Solar Phys.* **33**, 425 (1973).

Herschel, W., Observations tending to investigate the nature of the Sun. *Phil. Trans. Roy. Soc.* **265** (1801), *Collected Scientific Papers*, Vol II, 147, *Roy. Soc. and Roy. Astr. Soc.* (1912).

Hide, R., Rotation of the Earth and planets. *Phil. Trans. Roy. Soc.* A **313**, 107 (1984).

Hill, H. A., Yakowitz, D. S., Rosenwald, R. D. and Campbell, W., The internal rotation of the sun inferred from the rotational splitting of acoustic and gravity mode multiplets. *Proc. of the 4th EPS Meeting on Solar Physics "Hydromagnetics of the Sun"*, p. 187 (1984).

Horne, J. H., Balinuas, S. L., Noyes, R. W., Duncan, D. K. and Vaughan, A. H., Time series measurements of chromospheric emission and possible evidence for differential rotation. *Lecture Notes in Physics* **193**, 143 (1983).

Howard, R., A possible variation of the solar rotation with the activity cycle. *Astrophys. J.* **210**, L 159 (1976).

Howard, R., The rotation of the Sun. *Rev. Geophys. Space Phys.* **16**, 721 (1978).

Howard, R., Evidence for large-scale velocity features on the Sun. *Astrophys. J.* **228**, L 45 (1979).

Howard, R., Solar Rotation. *Ann. Rev. Astron. Astrophys.* **22**, 131 (1984).

Howard, R., Adkins, J. M., Boyden, F., Cragg, T. A., Gregory, T. S., LaBonte, B. J. and Webster, L., Solar rotation results at Mount Wilson. IV. Results. *Solar Phys.* **83**, 321 (1983).

Howard, R., Gilman, P. A. and Gilman, P. I., Rotation of the sun measured from Mt. Wilson white-light images. *Astrophys. J.* **283**, 373 (1984).

Howard, R. and Harvey, J., Spectroscopic determinations of solar rotation. *Solar Phys.* **12**, 23 (1970).

Howard, R. and LaBonte, B. J., The sun is observed to be a torsional oscillator with a period of 11 years. *Astrophys. J.* **239**, L 33 (1980).

v. Humboldt, A., *Kosmos. Entwurf einer physischen Weltbeschreibung*. I. G. Cottascher Verlag, Stuttgart und Tübingen (1850).

Iroshnikov, R. S., The origin of the solar rotation. *Astron. J. (SSSR)* **46**, 97 (1969).

Janssen, J., Étude de la surface solaire par la photographie. *Annales de l'Observatoire d'Astronomie Physique de Paris*, Tome Premier, 103 (1896).

Jeans, J. H., *Astronomy and Cosmogony*. University Press, Cambridge (1929).

Kichatinov, L. L., A mechanism for differential rotation based on angular momentum transport by compressible convection. *Geophys. Astrophys. Fluid Dyn.* **38**, 273 (1987).

Kippenhahn, R., Differential rotation in stars with convective envelopes. *Astrophys. J.* **137**, 664 (1963).

Köhler, H., Differential rotation caused by meridional circulation. *Mitt. der Astron. Gesellschaft* **21**, 91 (1966).

Köhler, H., Differentielle Rotation als Folge anisotroper turbulenter Viskosität. *Thesis*, Göttingen (1969).

Kraichnan, R. H., An almost-Markovian Galileian-invariant turbulence model. *J. Fluid Mech.* **47**, 513 (1971).

Kraichnan, R. H., Eddy viscosity in two and three dimensions. *J. Atmos. Sci.* **33**, 1521 (1976).

Krause, F., *Eine Lösung des Dynamoproblems auf der Grundlage einer linearen Theorie der magnetohydrodynamischen Turbulenz.* Habilitationsschrift (1967).

Krause, F., Zur Hydrodynamik des mittleren Geschwindigkeitsfeldes in einem turbulenten Medium. *Mber. Dtsch. Akad. Wiss. Berlin* **11**, 188 (1969).

Krause, F., Heat flow and magnetic field diffusion in turbulent fluids. *Astron. Nachr.* **294**, 83 (1972).

Krause, F., Überlegungen zur Berechnung von Sternmodellen mit Magnetfeldern. *Astron. Nachr.* **297**, 241 (1976).

Krause, F., Progress in dynamo theory: Kinematic and dynamic models. *Phil. Trans. Roy. Soc.* A **306**, 255 (1982).

Krause, F. and Rädler, K.-H., *Mean-field magnetohydrodynamics and dynamo theory.* Akademie-Verlag, Berlin and Pergamon Press, Oxford (1980).

Krause, F. and Roberts, P. H., Some problems of mean-field electrodynamics. *Astrophys. J.* **191**, 977 (1973).

Krause, F. and Rüdiger, G., On the Reynolds stresses in mean-field hydrodynamics, I. Incompressible homogeneous isotropic turbulence. *Astron. Nachr.* **295**, 93 (1974a).

Krause, F. and Rüdiger, G., On the Reynolds stresses in mean-field hydrodynamics. II. Two-dimensional turbulence and the problem of negative viscosity. *Astron. Nachr.* **295**, 185 (1974b).

Kröner, E., *Statistical continuum mechanics.* Springer-Verlag, Udino (1971).

Kubicela, A. and Karabin, M., Some results of the photospheric large-scale velocity research from Belgrade observations. *Solar Phys.* **52**, 199 (1977).

Kuhn, J. R., Observations of global scale photospheric Fraunhofer line shifts. *Astrophys. J.* **264**, 689 (1983).

Küveler, G. and Wöhl, H., Daily variations of the photospheric equatorial rotation velocity of the sun and its absolute values in 1981 and 1982 as determined from measurements using a two-dimensional photodiode array. *Astron. Astrophys.* **123**, 29 (1983).

LaBonte, B. J., Solar calibration of stellar rotation tracers. *Astrophys. J.* **260**, 647 (1982).

LaBonte, B. J., Howard, R. and Gilman, P. A., An improved search for large-scale convection cells in the solar atmosphere. *Astrophys. J.* **250**, 796 (1981).

LaBonte, B. J. and Howard, R., Torsional waves on the sun and the activity cycle. *Solar Phys.* **75**, 161 (1982a).

LaBonte, B. J. and Howard, R., Solar rotation measurements at Mount Wilson III. Meridional flow and limbshift. *Solar Phys.* **80**, 361 (1982b).

Larcheveque, M., Chollet, J. P., Herring, J. R., Lesieur, M., Newman, G. R. and Schertzer, D., Two-point closure applied to a passive scalar in decaying isotropic turbulence. In: *Turbulent Shear Flows II* (ed. L. Bradbury), p. 50, (1980).

Latour, J., Spiegel, E. A., Toomre, J. and Zahn, J. P., Stellar convection theory I. The anelastic model equations. *Astrophys. J.* **207**, 233 (1976).

Launder, B. E., Heat and mass transport. *Topics in Applied Physics* (ed. P. Bradshaw) **12**, 232 (1976).

Lebedinski, A. I., Rotation of the sun. *Astron. J. (SSSR)* 18, 10 (1941).

Livingston, W. C., On the differential rotation with height in the solar atmosphere. *Solar Phys.* **9**, 448 (1969).

Livingston, W. and Duvall, T. L., Solar rotation, 1966—1978. *Solar Phys.* **61**, 219 (1979).

Lustig, G., Solar rotation 1947—1981 determined from sunspot data. *Astron. Astrophys.* **125**, 355 (1983).

Maeder, A., Stellar evolution III.: the overshooting from convection cores. *Astron. Astrophys.* **40**, 303 (1975).

Marcus, Ph. S., Stellar convection. III. Convection at large Rayleigh numbers. *Astrophys. J.* **240**, 203 (1980).

Massaguer, J. M. and Zahn, J.-P., Cellular convection in a stratified atmosphere. *Astron. Astrophys.* **87**, 315 (1980).

Mestel, L., Notes concerning the final informal discussion. In: *Workshop on Solar Rotation*, p. 364, Catania (1978).

Monin, A. S., The global hydrodynamics of the sun. *Usp. Fiz. (SSSR)* **132**, 123 (Sov. Phys. Usp. **23**, 594) (1980).

Monin, A. S. and Simuni, L. M., Statistical hydromechanics of the solar differential rotation. *Proc. Nat. Acad. Sci. USA* **79**, 3903 (1982).

Moss, D. and Vilhu, O., Models of stellar differential rotation on the lower main sequence. *Astron. Astrophys.* **119**, 47 (1983).

Nakano, T., Fukushima, F., Unno, W. and Kondo, M., Viscosity, conductivity and power spectra of the turbulent convection in Boussinesq fluids. *Publ. Astron. Soc. Japan* **31**, 713 (1979).

Newton, H. W. and Nunn, M. L., The sun's rotation derived from sunspots 1934 to 1944 and additional results. *Mon. Not. Roy. astron. Soc.* **111**, 413 (1951).

Nickel, G. H., The maintenance of solar differential rotation by two-dimensional turbulence. *Solar Phys.* **10**, 472 (1969).

Noyes, R. W., Ayres, T. R. and Hall, D. N. S., Equator-pole temperature difference and the solar oblateness. *Solar Phys.* **28**, 343 (1973).

Noyes, R. W., Weiss, N. O. and Vaughan, A. H., The relation between stellar rotation rate and activity cycle periods. *Astrophys. J.* **287**, 769 (1984).

Oberbeck, A., Über die Wärmeleitung der Flüssigkeiten bei Berücksichtigung der Strömungen infolge von Temperaturdifferenzen. *Ann. Phys. Chem.* **7**, 271 (1879).

Orszag, S. A., Analytical theories of turbulence. *J. Fluid Mech.* **41**, 363 (1970).

Parker, E. N., Hydrodynamic dynamo models. *Astrophys. J.* **122**, 283 (1955).

Parker, E. N., The nature of sunspot phenomenon. I: Solutions of the heat transport equation. *Solar Phys.* **36**, 249 (1974).

Paternò, L., Survey of solar roatation. In: *Workshop on Solar Rotation*, p. 11, Catania (1978).

Pérez Garde, M., Vàzquez, M., Schwan, H. and Wöhl, H., Large scale motions as determined by Doppler shift measurements using a linear photodiode array. *Astron. Astrophys.* **93**, 67 (1981).

Plaskett, H. H., Limb darkening and solar rotation II. *Mon. Not. Roy. astron. Soc.* **148**, 149 (1970).

Prandtl, L., Bericht über Untersuchungen zur ausgebildeten Turbulenz. *ZAMM* **5**, 136 (1925).

Prandtl, L., Meteorologische Anwendung der Strömungslehre. *Beitr. Physik d. freien Atmosphäre* **19**, 188 (1932).

Proudman, J., On the motion of solids in liquids possessing vorticity. *Proc. Roy. Soc.* A **92**, 408 (1916).

Rädler, K.-H., Remarks on the influence of a non-uniform rotation on axisymmetric and non-axisymmetric magnetic fields. *International Workshop, Alsovice*, p. 46, (1979).

Rädler, K.-H., Investigations of spherical mean-field dynamo models. *Astron. Nachr.* **307**, 89 (1986).

Reynolds, O., On the dynamical theory of incompressible viscous fluids and the determination of the criterion. *Scientific Papers* II, 535 (1895).

Ribes, E., The large-scale solar variability through the cycle. *Adv. Space Res.* **6**, 221 (1986).

Richardson, R. S. and Schwarzschild, M., On the possibility of a 22-year oscillation of the sun. *Accademia Nazionale de Lincei, Atti de Convegni* 11, 228 (1953).

Roberts, P. H., On the thermal instability of a rotating fluid sphere containing heat sources. *Phil. Trans. Roy. Soc.* **263**, 93 (1968).

Roberts, P. H., Kinematic dynamo models. *Phil. Trans. Roy. Soc.* A **272**, 663 (1972).

Roberts, P. H. and Soward, A. M., On first order smoothing theory. *J. Math. Physics* **16**, 609 (1974).

Roberts, P. H. and Soward, A. M., A unified approach to mean field electrodynamics. *Astron. Nachr.* **296**, 49 (1975).

Rodonò, M., Progress and problems in RS CVn star research. In: *Photometric and Spectroscopic Binary Systems* (ed. E. B. Carling and Z. Kopal), p. 285 (1981).

Rotta, J., *Turbulente Strömungen.* Teubner Verlag, Stuttgart (1972).

Roxburgh, I. W., Convection and stellar structure. *Astron. Astrophys.* **65**, 281 (1978).

Rüdiger, G., On the Reynolds stresses in mean-field hydrodynamics III. Two-dimensional turbulence and the problem of differential rotation. *Astron. Nachr.* **295** (1974).

Rüdiger, G., Superrotation und zweidimensionale Turbulenz. *Gerlands Beitr. Geophysik* **84**, 92 (1972).

Rüdiger, G., Negative viscosity in the Gulf stream? *Tellus* **28**, 183 (1976).

Rüdiger, G., Two-dimensional stochastic motions and the problem of differential rotation. *Solar Phys.* **51**, 257 (1977a).

Rüdiger, G., A theory of differential rotation based on the discussion of turbulent transport of angular momentum. *Astron. Nachr.* **298**, 245 (1977b).

Rüdiger, G., Autokorrelation in homogener rotierender axialsymmetrischer Turbulenz. Tagungsbeitrag Eisenach (1978a).

Rüdiger, G., One-position cross-correlations in a SOCA-theory of rotating anisotropic turbulence. In: *Workshop on Solar Rotation*, p. 269, Catania (1978b).

Rüdiger, G., Reynolds stresses and differential rotation I. On recent calculations of zonal fluxes in slowly rotating stars. *Geophys. Astrophys. Fluid Dyn.* **16**, 239 (1980).

Rüdiger, G., On turbulent heat transport in rotating convection zones. *Astron. Nachr.* **303**, 293 (1982a).

Rüdiger, G., Reynolds stresses and differential rotation II. Mean-field models. *Geophys. Astrophys. Fluid Dyn.* **21**, 1 (1982b).

Rüdiger, G., On Reynolds stress theories of solar differential rotation. In: *Stellar and Planetary Magnetism* (ed. A. M. Soward), p. 133, Gordon and Breach, New York (1983a).

Rüdiger, G., Reynolds stresses and differential rotation III. Zonal momentum fluxes in rapidly rotating stars. *Geophys. Astrophys. Fluid Dyn.* **25**, 213 (1983b).

Rüdiger, G., Tuominen, I., Krause, F. and Virtanen, H., Dynamo-generated flows in the sun. *Astron. Astrophys.* **166**, 306 (1986).

Rutten, R. J., The solar temperature distribution with latitude. *Solar Phys.* **28**, 347 (1973).

Ruzmaikin, A. A. and Vainshtein, S. I., The nature of one mechanism of solar differential rotation. *Astrophys. and Space Sci.* **59**, 215 (1978).

Sakurai, T., Hydrodynamical investigation of the solar differential rotation. *Publ. Astron. Soc. Japan* **18**, 174 (1966).

Sampson, R. A., On the rotation and mechanical state of the sun. *Memoirs of the Royal Astron. Soc.* **LI**, 123 (1895).

Scherrer, P. H., Wilcox, J. M. and Svalgaard, L., The rotation of the sun: Observations at Stanford. *Astrophys. J.* **241**, 811 (1980).

Schmidt, W., Der Massenaustausch bei der ungeordneten Strömung in freier Luft und seine Folgen. *Sitzungsber. Kais. Akad. Wiss. Wien*, Abt. IIa **126**, 757 (1917).

Schmidt, W., *Der Massenaustausch in freier Luft und verwandte Erscheinungen.* Verlag von H. Grand, Hamburg (1925).

Schmidt, W., Models of solar differential rotation. *Geophys. Astrophys. Fluid Dyn.* **21**, 27 (1982).

Schmidt, W. and Stix, M., Two comments of the sun's differential rotation. *Astron. Astrophys.* **118**, 1 (1983).

Schröter, E. H., The solar differential rotation — Present status of observations. *Mitt. Kiepenheuer-Inst.* Nr. 251 (1985).

Schröter, E. H. and Wöhl, H., Differential rotation, meridional and random motions of the solar Ca^{+}-network. *Solar Phys.* **42**, 3 (1975).

Schröter, E. H. and Wöhl, H., Differential rotation and giant cell circulation of solar Ca^{+}-network. *Solar Phys.* **49**, 19 (1976).

Schröter, E. H. and Wöhl, H., A critical review of observational data for the rotation of the solar photosphere and chromosphere obtained by tracer techniques. In: *Workshop on Solar Rotation*, p. 35, Catania (1978).

Schröter, E. H., Wöhl, H., Soltau, D. and Vàzquez, M., An attempt to compare the differential rotation of the Ca^{+}-network with that of the photospheric plasma. *Solar Phys.* **60**, 181 (1978).

Schüssler, M., The solar torsional oscillation and dynamo models of the solar cycle. *Astron. Astrophys.* **94**, L 17 (1981).

Schwabe, S. H., Sonnen-Beobachtungen im Jahre 1843. *Astron. Nachr.* **21**, 233 (1844).

Schwarzschild, K., Über das Gleichgewicht der Sonnenatmosphäre. *Nachr. Königl. Ges. d. Wiss. Göttingen* **1**, 41 (1906).

Siedentopf, H., Konvektion in Sternenatmosphären III. *Astron. Nachr.* **255**, 157 (1933).

Simon, G. W. and Weiss, N. O., Supergranules and the hydrogen convection zone. *Z. Astrophysik* **69**, 435 (1968).

Snodgrass, H. B. and Howard, R., Torsional oscillations of low mode. *Solar Phys.* **95**, 221 (1985).
Spiegel, E. A., A generalization of the mixing-length theory of turbulent convection. *Astrophys. J.* **138**, 216 (1963).
Spiegel, E. A. and Veronis, G., On the Boussinesq approximation for a compressible fluid. *Astrophys. J.* **131**, 442 (1960).
Spörer, G., Über die Periodizität der Sonnenflecken seit dem Jahre 1618, vornehmlich... *Vierteljahresschrift der Astron. Ges.* **22**, 323 (1887).
Spruit, H. C., A model of the solar convection zone. *Solar Phys.* **34**, 277 (1974).
Spruit, H. C., Appearance at the solar surface of disturbances in the heat flow associated with differential rotation. *Astron. Astrophys.* **55**, 151 (1977).
Starr, V. P., *Physics of Negative Viscosity Phenomena.* Mc Graw Hill, New York (1968).
Staude, J., Models of the convective zone. *Bull. Astron. Inst. Czechosl.* **27**, 365 (1977).
Steenbeck, M. and Krause, F., Meridionale Strömungen im Innern der Sonne als Ursache der differentiellen Rotation. *Mber. Dtsch. Akad. Wiss. Berlin* **7**, 335 (1965).
Steenbeck, M. and Krause, F., Zur Dynamotheorie stellarer und planetarer Magnetfelder I. Berechnung sonnenähnlicher Wechselfeldgeneratoren. *Astron. Nachr.* **291**, 49 (1969).
Stix, M., Comments on the solar dynamo. *Astron. Astrophys.* **37**, 121 (1974).
Stix, M., Dynamo theory and the solar cycle. In: *Proc. IAU Symp. No 71, Basic Mechanism of Solar Activity* (eds. V. Bumba and J. Kleczek), p. 367, D. Reidel, Dordrecht (1976a).
Stix, M., Differential rotation and the solar dynamo. *Astron. Astrophys.* **47**, 243 (1976b).
Stix, M., Coronal holes and the large scale solar magnetic field. *Astron. Astrophys.* **59**, 73 (1977).
Stix, M., Screening effects in the solar convection zone. *Astron. Astrophys.* **93**, 339 (1981).
Stix, M., Die differentielle Rotation und der magnetische Zyklus der Sonne. *Mitt. der Astron. Gesellschaft* **60**, 95 (1983).
Szablewski, W., *Turbulente Scherströmungen. Freie Turbulenz.* Akademie-Verlag, Berlin (1979).
Tassoul, J.-L., *Theory of rotating stars.* University Press, Princeton (1978).
Tayler, R. J., Convection in rotating stars. *Mon. Not. Roy. astron. Soc.* **165**, 39 (1973).
Taylor, G. I., Eddy motion in the atmosphere. *Phil. Trans. Roy. Soc.* A **215**, 1 (1915).
Taylor, G. I., Experiments with rotating fluids. *Proc. Roy. Soc.* A **100**, 114 (1921).
Timothy, A. F., Krieger, A. J. and Vaiana, G. S., The structure and evolution of coronal holes. *Solar Phys.* **42**, 135 (1975).
Tritton, D. J., Turbulence in rotating fluids. In: *Rotating Fluids in Geophysics* (ed. P. H. Roberts and A. M. Soward), Academic Press, New York (1978).
Tuominen, I., Eleven-year cycle in solar rotation, mean meridional motions and angular momentum transport, presented in: *IAU Symp.* No 106 (unpublished) (1982).
Tuominen, I., Krause, F., Rüdiger, G., Virtanen, H., Cyclic flows in the solar convection zone due to the dynamo reaction. *Proc. of the 4th EPS Meeting on Solar Physics "Hydromagnetics of the Sun"*, p. 225, 1984.

Tuominen, I. and Virtanen, H., Oscillatory motions in the sun corresponding to the solar cycle. *Astron. Nachr.* **305**, 225 (1984).

Tuominen, J., Die systematische Strombewegung der Sonnenflecke in heliographischer Breite. *Z. Astrophysik* **21**, 96 (1941).

Tuominen, J., On the latitude drift of sunspot groups. *Z. Astrophysik* **51**, 91 (1961).

Tuominen, J. and Kyröläinen, J., On the latitude drift of sunspot groups and solar rotation. *Solar Phys.* **79**, 161 (1982).

Tuominen, J., Tuominen, I. and Kyröläinen, J., Eleven-year cycle in solar rotation and meridional motions as derived from the positions of sunspot groups. *Mon. Not. Roy. astron. Soc.* **205**, 691 (1983).

Ulrich, R. H., A non-local mixing-length theory of convection for use in numerical calculations. *Astrophys. J.* **207**, 564 (1976).

Unno, W., Osaki, Y., Ando, H. and Shibahashi, H., *Nonradial Oscillations in Stars.* Tokyo University Press, Tokyo (1979).

Unsöld, A., Konvektion in der Sonnenatmosphäre. *Z. Astrophysik* **1**, 138 (1930).

Vainshtein, S. I., Functional approach to the turbulent dynamo. *Soviet Phys.-JETP* **35**, 725 (1972).

Vainshtein, S. I. and Kichatinov, L. L., The macroscopic magnetohydrodynamics of inhomogeneously turbulent cosmic plasma. *Geophys. Astrophys. Fluid Dyn.* **24**, 273 (1983).

Vandakurov, Yu. V., Equilibrium problems in a rotating convection zone. *Solar Phys.* **45**, 501 (1975).

Vauclair, S., Vauclair, G., Schatzmann, E., and Michaud, G., Hydrodynamical instabilities in the envelopes of main sequence stars: Constraints by the Lithium, Beryllium and Boron observations. *Astrophys. J.* **223**, 567 (1978).

Vaughan, A. M., Baliunas, S. L., Middelkoop, F., Hartmann, I. W., Mihalas, D., Noyes, R. W. and Preston, G. W., Stellar rotation in lower main-sequence stars measured from time variations in H and K emission-line fluxes I. Initial results. *Astrophys. J.* **250**, 276 (1981).

Vaughan, A. M. and Preston, G. W., A survey of chromospheric Ca II H and K emission in field stars in the solar neighborhood. *Publ. Astron. Soc. Pacific* **92**, 385 (1980).

Veronis, G., The magnitude of dissipation terms in the Boussinesq approximation. *Astrophys. J.* **135**, 655 (1962).

Wagner, W. J., Solar rotation as marked by extreme ultraviolet coronal holes. *Astrophys. J.* **198**, L 141 (1975).

Ward, F., The general circulation of the solar atmosphere and the maintenance of the equatorial acceleration. *Astrophys. J.* **141**, 534 (1965).

Ward, F., Determination of the solar rotation rate from the motion of identifiable features. *Astrophys. J.* **145**, 416 (1966).

Ward, F., The latitudinal motion of sunspots and meridional circulations. *Solar Phys.* **30**, 527 (1973).

Wasiutyński, J., Studies in hydrodynamics and structure of stars and planets. *Astrophys. Norveg.* **4** (1946).

Weiss, N. O., Convection and the differential rotation of the sun. *Observatory* **85**, 37 (1965).

Wilcox, J. M. and Howard, R., Differential rotation of the photospheric magnetic field. *Solar Phys.* **13**, 251 (1970).

Wilcox, J. M., Schatten, V. H., Tananbaum, A. S. and Howard, R., Photospheric magnetic field rotation: rigid and differential. *Solar Phys.* **19**, 255 (1970).

Wilsing, J., Über das Rotationsgesetz der Sonne und über die Periodizität der Sonnenflecke. *Astron. Nachr.* **127**, 236 (1891).

Wilson, O. C., Chromospheric variations in main-sequence stars. *Astrophys. J.* **226**, 379 (1978).

Wittmann, A., Observation of solar granulation — a review. *Mitt. Kiepenheuer-Institut Freiburg* **179**, 29 (1979).

Wöhl, H., Possibilities and attempts to determine the differential rotation on F-type main-sequence stars. In: *Solar and Stellar Magnetic Fields: Origins and Coronal Effects* (ed. J. O. Stenflo), p. 155 (1983).

Wolf, R., Beobachtungen der Sonnenflecken im Jahre 1873... *Astron. Mitt.* XXXVI, Zürich (1872).

Wolf, R., *Geschichte der Astronomie.* Verlag Oldenburg, München (1877).

Yoshimura, H., Solar cycle Lorentz force waves and the torsional oscillations of the sun. *Astrophys. J.* **247**, 1102 (1981).

Yoshimura, H. and Kato, S., Differential rotation of convective spherical shells. *Publ. Astron. Soc. Japan* **23**, 57 (1971).

Zahn, J. P., Stellar convection theory. *Lecture Notes in Physics* **114**, 1 (1979).

Zöllner, F., Über das Rotationsgesetz der Sonne und der großen Planeten. In: *Wissenschaftliche Abhandlungen*, Bd. 4, Commissionsverlag von L. Staackmann, Leipzig (1881).

Index

Bei Fragen zur Produktsicherheit wenden Sie sich bitte an:
If you have any questions regarding product safety,
please contact:

Walter de Gruyter GmbH
Genthiner Straße 13
10785 Berlin
productsafety@degruyterbrill.com